COMBAT
Stress Injury

ROUTLEDGE PSYCHOSOCIAL STRESS SERIES
Charles R. Figley, Ph.D., Series Editor

1. *Stress Disorders among Vietnam Veterans*, Edited by Charles R. Figley, Ph.D.
2. *Stress and the Family Vol. 1: Coping with Normative Transitions*, Edited by Hamilton I. McCubbin, Ph.D., and Charles R. Figley, Ph.D.
3. *Stress and the Family Vol. 2: Coping with Catastrophe*, Edited by Charles R. Figley, Ph.D., and Hamilton I. McCubbin, Ph.D.
4. *Trauma and Its Wake: The Study and Treatment of Post-Traumatic Stress Disorder*, Edited by Charles R. Figley, Ph.D.
5. *Post-Traumatic Stress Disorder and the War Veteran Patient*, Edited by William E. Kelly, M.D.
6. *The Crime Victim's Book, Second Edition*, by Morton Bard, Ph.D., and Dawn Sangrey.
7. *Stress and Coping in Time of War: Generalizations from the Israeli Experience*, Edited by Norman A. Milgram, Ph.D.
8. *Trauma and Its Wake Vol. 2: Traumatic Stress Theory, Research, and Intervention*, Edited by Charles R. Figley, Ph.D.
9. *Stress and Addiction*, Edited by Edward Gottheil, M.D., Ph.D., Keith A. Druley, Ph.D., Steven Pashko, Ph.D., and Stephen P. Weinsteinn, Ph.D.
10. *Vietnam: A Casebook*, by Jacob D. Lindy, M.D., in collaboration with Bonnie L. Green, Ph.D., Mary C. Grace, M.Ed., M.S., John A. MacLeod, M.D., and Louis Spitz, M.D.
11. *Post-Traumatic Therapy and Victims of Violence*, Edited by Frank M. Ochberg, M.D.
12. *Mental Health Response to Mass Emergencies: Theory and Practice*, Edited by Mary Lystad, Ph.D.
13. *Treating Stress in Families*, Edited by Charles R. Figley, Ph.D.
14. *Trauma, Transformation, and Healing: An Integrative Approach to Theory, Research, and Post-Traumatic Therapy*, by John P. Wilson, Ph.D.
15. *Systemic Treatment of Incest: A Therapeutic Handbook*, by Terry Trepper, Ph.D., and Mary Jo Barrett, M.S.W.
16. *The Crisis of Competence: Transitional Stress and the Displaced Worker*, Edited by Carl A. Maida, Ph.D., Norma S. Gordon, M.A., and Norman L. Farberow, Ph.D.
17. *Stress Management: An Integrated Approach to Therapy*, by Dorothy H. G. Cotton, Ph.D.
18. *Trauma and the Vietnam War Generation: Report of the Findings from the National Vietnam Veterans Readjustment Study*, by Richard A. Kulka, Ph.D., William E. Schlenger, Ph.D., John A. Fairbank, Ph.D., Richard L. Hough, Ph.D., Kathleen Jordan, Ph.D., Charles R. Marmar, M.D., Daniel S. Weiss, Ph.D., and David A. Grady, Psy.D.
19. *Strangers at Home: Vietnam Veterans Since the War*, Edited by Charles R. Figley, Ph.D., and Seymour Leventman, Ph.D.
20. *The National Vietnam Veterans Readjustment Study: Tables of Findings and Technical Appendices*, by Richard A. Kulka, Ph.D., Kathleen Jordan, Ph.D., Charles R. Marmar, M.D., and Daniel S. Weiss, Ph.D.
21. *Psychological Trauma and the Adult Survivor: Theory, Therapy, and Transformation*, by I. Lisa McCann, Ph.D., and Laurie Anne Pearlman, Ph.D.
22. *Coping with Infant or Fetal Loss: The Couple's Healing Process*, by Kathleen R. Gilbert, Ph.D., and Laura S. Smart, Ph.D.
23. *Compassion Fatigue: Coping with Secondary Traumatic Stress Disorder in Those Who Treat the Traumatized*, Edited by Charles R. Figley, Ph.D.
24. *Treating Compassion Fatigue*, Edited by Charles R. Figley, Ph.D.
25. *Handbook of Stress, Trauma and the Family*, Edited by Don R. Catherall, Ph.D.
26. *The Pain of Helping: Psychological Injury of Helping Professionals*, by Patrick J. Morrissette, Ph.D., RMFT, NCC, CCC.
27. *Disaster Mental Health Services: A Primer for Practitioners*, by Diane Myers, R.N., M.S.N., and David Wee, M.S.S.W.
28. *Empathy in the Treatment of Trauma and PTSD*, by John P. Wilson, Ph.D., and Rhiannon B. Thomas, Ph.D.
29. *Family Stressors: Interventions for Stress and Trauma*, Edited by Don. R. Catherall, Ph.D.
30. *Handbook of Women, Stress and Trauma*, Edited by Kathleen Kendall-Tackett, Ph.D.
31. *Mapping Trauma and Its Wake*, Edited by Charles R. Figley, Ph.D.
32. *The Posttraumatic Self: Restoring Meaning and Wholeness to Personality*, Edited by John P. Wilson, Ph.D.
33. *Violent Death: Resilience and Intervention Beyond the Crisis*, Edited by Edward K. Rynearson, M.D.

COMBAT Stress Injury

Theory, Research, and Management

Edited by Charles R. Figley and William P. Nash

Foreword by Jonathan Shay

Routledge
Taylor & Francis Group
New York London

Routledge is an imprint of the
Taylor & Francis Group, an informa business

The opinions and assertions contained herein are the private views of the editors and contributing authors and are not to be construed as official or as reflecting the views of any military or government branches.

Routledge
Taylor & Francis Group
270 Madison Avenue
New York, NY 10016

Routledge
Taylor & Francis Group
2 Park Square
Milton Park, Abingdon
Oxon OX14 4RN

Routledge is an imprint of Taylor & Francis Group, an Informa business

Printed in the United States of America on acid-free paper
10 9 8 7 6 5 4 3 2

International Standard Book Number-10: 0-415-95433-9 (Hardcover)
International Standard Book Number-13: 978-0-415-95433-4 (Hardcover)

Library of Congress Cataloging-in-Publication Data

Combat stress injury : theory, research, and management / Charles R. Figley, William P. Nash, editors.
p. ; cm.
Includes bibliographical references.
ISBN 0-415-95433-9 (hc : alk. paper)
1. War neuroses. 2. Psychology, Military. 3. Post-traumatic stress disorder. 4. Veterans--Psychology. I. Figley, Charles R., 1944- II. Nash, William P., 1952-
[DNLM: 1. Combat Disorders. 2. Stress Disorders, Post-Traumatic. 3. Veterans--psychology. WM 184 C729 2007]

RC550.C66 2007
616.85'212--dc22 2006016582

Visit the Taylor & Francis Web site at
http://www.taylorandfrancis.com

and the Routledge Web site at
http://www.routledgementalhealth.com

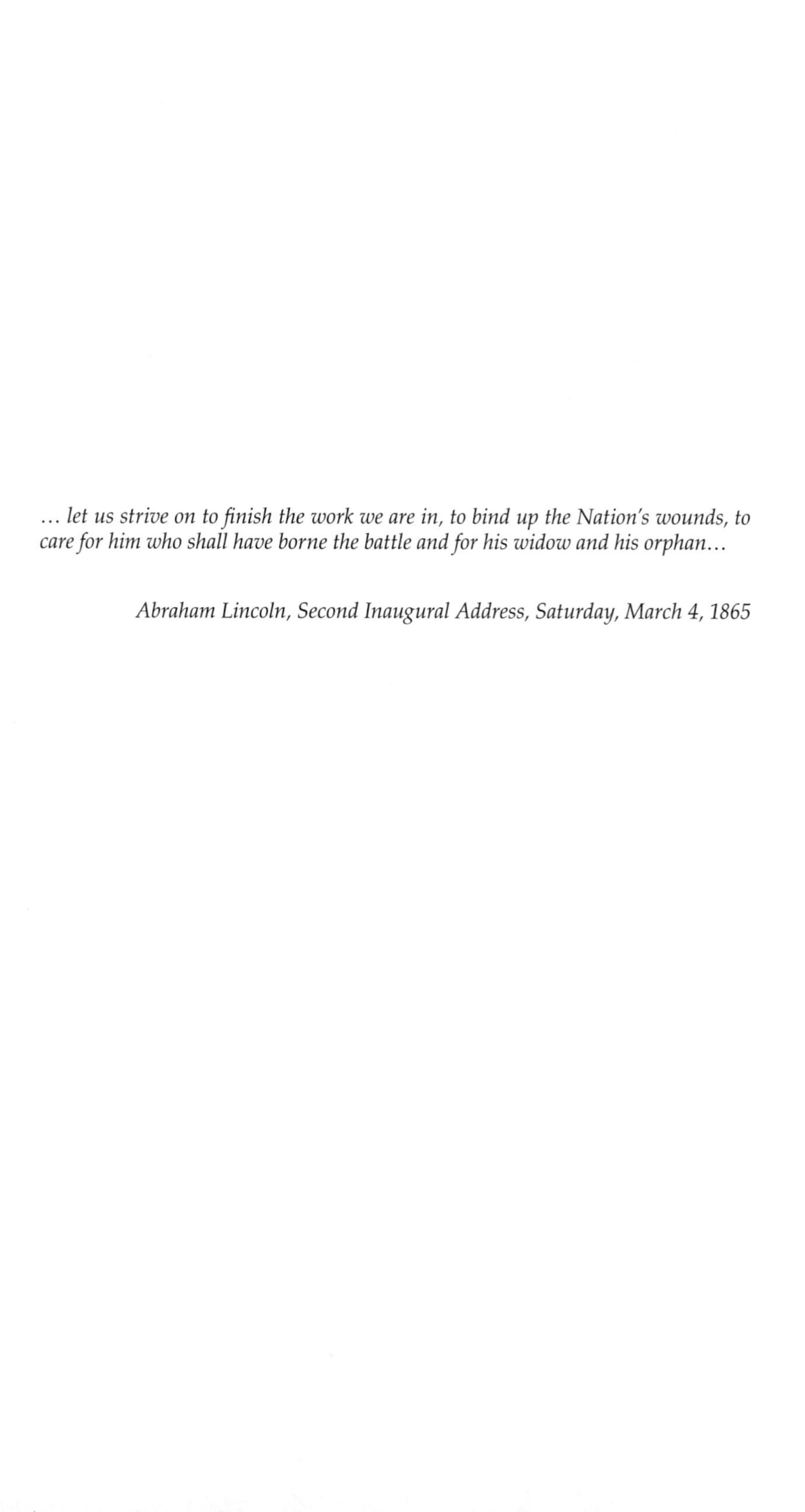

… let us strive on to finish the work we are in, to bind up the Nation's wounds, to care for him who shall have borne the battle and for his widow and his orphan…

Abraham Lincoln, Second Inaugural Address, Saturday, March 4, 1865

Contents

About the Editors

Charles R. Figley, Ph.D., is a Fulbright Fellow; former U.S. Marine sergeant and Vietnam War veteran; founding editor of the *Journal of Traumatic Stress*; founding president of the International Society for Traumatic Stress Studies; and currently a College of Social Work professor, and director of the Florida State University Traumatology Institute, Tallahassee, Florida.

William P. Nash, M.D., is a senior U.S. Navy psychiatrist who served with the First Marine Expeditionary Force in the Al Anbar Province of Iraq during 2004–2005, and is combat/operational stress control coordinator, Headquarters, Marine Corps (Manpower & Reserve Affairs), Quantico, Virginia.

Contributors

Dewleen Baker, M.D., is director of the PTSD and Stress Disorders Program at the San Diego, California, Veterans Administration Medical Center.

Joseph A. Boscarino, Ph.D., M.P.H., is a former senior scientist at the New York Academy of Medicine and is currently the Senior Investigator-II at the Center for Health Research & Rural Advocacy, Geisinger Health System in Danville, Pennsylvania.

Nancy Clayton, M.D., who served on active duty as a naval psychiatrist from 1998 to 2006, is in private practice in Northern Virginia.

Kathy Darte, M.N., is the co-manager of the Operational Stress Injury Social Support (OSISS) Program, National Defense Headquarters, Ottawa, Canada.

Rachel Dekel, Ph.D., is an associate professor in the School of Social Work, Bar-Illan University, Ramat-Gan, Israel.

Kent D. Drescher, M.Div., Ph.D., is the coordinator of assessment and data management, National Center for PTSD Inpatient Programs, Menlo Park, California.

David W. Foy, Ph.D., is a professor of psychology at the Graduate School of Education & Psychology, Pepperdine University, Encino, California.

Ken Graap, M.Ed., is president and CEO of Virtually Better, Inc., Decatur, Georgia.

Neil Greenberg, BM, BSc, MMedSc, ILTM, MRCPsych, is Surgeon Commander, Royal Navy and is currently the King's Centre for Military Health Research Military Liaison Officer, Academic Department of Psychological Medicine, GKT School of Medicine & Institute of Psychiatry, Weston Education Centre, London, United Kingdom.

Stephane Grenier, Lt. Colonel, M.S.C., who was awarded in 2006 the Meritorious Service Cross, is founder and project manager of the Operational Stress Injury Social Support (OSISS) program, and project manager, Defence Communication Services Agency (DCSA), National Defence Headquarters, Ottawa, Ontario.

Alexandra Heber, M.D., FRCPC, is a consulting psychiatrist for the Canadian Forces Health Services Centre, Operational Trauma and Stress Support Centre, Ottawa, Ontario.

Deborah Hemel, B.A., is a 2005–2006 Fulbright Fellow in the Department of Overseas Studies, University of Haifa, Israel.

Yair Hilel is a former member of the Israeli Army who was injured from a terrorist attack and currently is a Ph.D. candidate in the Psychology Department of the University of Haifa, Israel, focusing his dissertation on metamemory for traumatic events as a risk factor for PTSD.

Noa Idar, M.A., is a Ph.D. candidate in the Psychology Department of the University of Haifa, Israel. Her M.A. thesis was on quality of life among injured soldiers with and without PTSD.

Ehud M. Klein, M.D., is the chair of the Department of Psychiatry, Rambam Medical Center, Haifa, Israel, and professor in the Department of Psychiatry, Faculty of Medicine, Technion–Israel Institute of Technology, Haifa, Israel.

Danny Koren, Ph.D., is the chair of the Clinical Program in the Psychology Department, University of Haifa, Israel.

Judith A. Lyons, Ph.D., is a clinical psychologist with the G.V. ("Sonny") Montgomery VA Medical Center, the VA South Central Mental Illness Research, Education, and Clinical Center (MIRECC), and the University of Mississippi Medical Center, Jackson, Mississippi.

Cameron March, Major (rtd), is acting major, Royal Marines, Ministry of Defence, Fareham, Hampshire, United Kingdom.

Bret A. Moore, Psy.D., is a U.S. Army captain who served in Iraq in 2005–2006. He was the Officer in Charge of a combat stress control preventive team with the 85th Medical Detachment. Currently he is a clinical psychologist, 85th Medical Detachment, Combat Stress Control, Fort Hood, Texas.

Jeffrey M. Pyne, Ph.D., is with the Center for Mental Health and Outcomes Research, Central Arkansas Veterans Healthcare System, Little Rock, Arkansas.

Greg M. Reger, Ph.D., is a U.S. Army captain who served in Iraq 2005–2006. Currently he is a clinical psychologist, 98th Medical Detachment, Combat Stress Control, Fort Lewis, Washington.

Don Richardson, M.D., FRCPC, is a consultant psychiatrist at Operational Stress Injury Clinic, Parkwood Hospital, St. Joseph's Health Care, London, Ontario; adjunct professor in the Department of Psychiatry, University of Western Ontario; and a consultant psychiatrist at Veterans Affairs Canada, Hamilton, Ontario.

Albert "Skip" Rizzo, Ph.D., is research scientist and research assistant professor, University of Southern California Institute for Creative Technologies and School of Gerontology, Marina del Rey, California.

Barbara O. Rothbaum, Ph.D., is the former president of the International Society for Traumatic Stress Studies; and currently a professor in psychiatry, and director, Trauma and Anxiety Recovery Program, Emory University School of Medicine, Atlanta, Georgia.

Mark W. Smith, CDR, CHC, USN, is an instructor at National Security Affairs; he was the chaplain who worked with the returning First Marine Expeditionary Force Command Element from Iraq in the spring of 2005.

Zahava Solomon, Ph.D., is a former major, Israeli Defense Force, and pioneer in the study of war veterans and their wives. She was the first to document secondary traumatic stress reactions in the former population. Currently she is professor, School of Social Work, Tel Aviv University, Israel.

James L. Spira, Ph.D., M.P.H., ABPP, is an associate clinical professor, Department of Psychiatry, University of California, San Diego, and the Mental Health Services, Naval Medical Center, San Diego, California.

Brenda K. Wiederhold won the 2005 Satava Award for "demonstrating unique vision and commitment to the improvement of medicine through advanced technology"; and is currently CEO of the Interactive Media Institute in San Diego, California.

Series Editor's Foreword

As editor of the Routledge Psychosocial Stress Book Series, I welcome and celebrate the latest addition to the series with the inclusion of *Combat Stress Injury: Theory, Research, and Management* that I edited with William Nash, M.D. This book is about war. As noted in the book's introductory chapter, it is a book that is dedicated to the young men and women who fought, are fighting, and will fight in Afghanistan and Iraq. It is a thank-you note to them.

The Prussian general and war theorist Carl Gottfried von Clausewitz used the term "friction" to describe the physical, mental, and emotional stressors of combat. Clausewitz first entered combat as a cadet at the age of 13, and rose to the rank of major-general at 38. His book, *On War* (Clausewitz, 1976/1832), published a year after his death due to cholera, is one of the most influential works of military philosophy in the Western world. In Chapter VII, "The Friction in War," he offers a rather simple lesson, but one that is often not carried forward to subsequent generations, on the concept of war friction. He notes that no amount of training or preparation prepares combatants for the friction or the unexpected and distressing experiences of combat—not even those veterans of other battles. The friction of war occupies the mind and distracts the warfighter from the true mission. Combat friction, or simply combat stress, can easily lead to stress injuries unless the warfighter's leaders are fully aware that these are the consequences of war, having little to do with courage, fear, allegiance to duty, or competence. And it is equally important to understand that attending to the stress injury does not imply a lack of courage, weakness, or competence, and furthermore such attention will decrease the likelihood that the injury will become a stress disorder during and following military service.

Consistent with Clausewitz's warnings about the frictions of combat, the first book in the series, *Stress Disorders Among Vietnam Veterans: Theory, Research, and Treatment* (Figley, 1978), pointed out what this current book emphasizes again: Combat stress is like no other as are the memories it creates, and the subsequent consequences of these memories. The *Stress Disorders* book has been credited with ushering in the modern era of the field of traumatology, the study and treatment of trauma. The book was the first to discuss combat-related traumatic stress and outline what became the diagnosis of PTSD. Twenty-four scholar-practitioners from psychiatry, psychology, and social work collaborated to publish 14 chapters and a comprehensive introduction on the psychosocial consequences of fighting in the Vietnam War. Inconsistent with the prevailing view of postwar readjustment that discounted the trauma of war for the theory of predisposition, the book pointed to combat stress as the primary causative factor.

Today, this combat stress paradigm is the prevailing view both inside and outside the military because the evidence is overwhelming and the paradigm-informed approaches to prevention and treatment are promising. The shift in thinking began in the 1980s with special importance from the publication of a government report that was edited and expanded for publication in this series (Kulka et al., 1990). It reported on the largest and best study of its kind on the epidemiology of combat-related stress disorders. The report was alarming in terms of the number of war veterans who were at risk of developing debilitating mental health problems that affect the veterans' ability to function at home and in the workplace. Yet, the report became a useful guide to designing programs that would both help control and manage combat stress and also mitigate the unwanted effects after military service.

This new book in the series is yet another extraordinary contribution to the field and may usher in yet another paradigm shift in the way we understand, study, and manage combat stress and its distressing memories. As noted in the first chapter by the editors, the book suggests that everyone in combat experiences combat stress and a large and unknown portion experience combat injuries. As with a physical injury, stress injuries deserve little condemnation (there is no shame in spraining an ankle, for example) and sufficient attention in order to avoid stress disorders during and following combat service.

The early genius of Clausewitz, together with several books in this series, including its latest addition on combat stress injury, enables us to understand and quantify frictions of war that become debilitating memories for the warfighters. With more time and attention, building on the lessons in this new book, perhaps we will be able to help these warfighters cope more effectively during and following war. If so, they, their families, and the nation will be grateful.

Charles R. Figley, Ph.D.
Series Editor

REFERENCES

Clausewitz, C. von. (1982). *On War* (A. Rapoport, Ed., J. J. Graham, Trans.). New York: Penguin Classics. (Original work published 1832)

Clausewitz, C. von. (1976). *On War* (M. E. Howard & P. Paret, Eds. and Trans.). Princeton: Princeton University Press. (Original work published 1832)

Figley, C. R. (Ed). (1978). *Stress Disorders Among Vietnam Veterans: Theory, Research, and Treatment*. New York: Brunner/Mazel.

FOREWORD

JONATHAN SHAY

The editors of this volume say in their introduction that it is directed to, among others, military leaders. The following anecdote is a forecast that their message of psychological force protection will be welcome. In 1996, I was invited to teach for a day at the Amphibious Warfare School at Quantico. This is the Marine Corps' officers' professional school at the captain level. The fire in my belly then, as now, was preventing psychological and moral injury in military service. But what I didn't know then, being quite new at it, was why the prevention message is so palatable to senior military leaders. During the midday luncheon I was more or less kidnapped by Lt. Gen. Paul van Riper, then the Deputy Commandant for Combat Development [the rough equivalent of the Army's Deputy Chief of Staff for Operations, G-3], and all he wanted to talk about was prevention. After answering his barrage of questions I said, "General, I'm delighted that you are so interested in prevention of psychological injury, but somewhat startled. Why so interested?" Perhaps a bit pained to have to explain something so obvious to this dumb civilian he said, "It's a simple matter of readiness. If any of my Marines are injured, whether physically or psychologically, I can't replace them so easily. So if you can tell me how to protect them, I'm listening."*

When a military service member's arm is shot off, do we say he or she suffers from Missing Arm Disorder? That would be ludicrous, and we do not. I have been agitating for some time to acknowledge that the diagnostic entity we now call Posttraumatic Stress Disorder is an *injury,* not

* With his permission, I have told this story publicly before. Lt. General Paul van Riper, personal communication with author, October 2, 1996, and e-mail permission, April 2, 1998.

a malady, disease, sickness, illness, or disorder.* This excellent volume of papers by a multinational, multidisciplinary group of mental health, military, and other professionals shows that many others have drawn this conclusion as well. Canadian Forces has adopted the term Operational Stress Injury, in the hope of reducing the barrier of stigma that service members must surmount within themselves and among their peers and seniors to ask for help with distress and disruption. Words *do* matter. To be injured in the line of duty is honorable; to return to duty after recovery from injury is the norm. To be sick in the line of duty is…what?…unlucky? Who wants to share a fighting hole with an unlucky comrade?

Few realms of human practice show as vividly as military functioning does, that we are just one creature—body, mind, society, culture—at every moment. None of these four avatars of the human being—brain, mind, society, culture—have ontological priority over the others: that is, no one of them is really real, the others epiphenomenal bubbles on the stream or Platonic shadows on the wall. What I am doing in the moment of writing this foreword, and what you are doing in the moment of reading it, is physiological, psychological, social, and cultural, all at every instant. Dr. Nash's masterful reviews in the early chapters leave no doubt of this, although, because he is a nicer person than I am, he sees our collective cup of life- and spirit-saving knowledge in this area as half full, rather than half empty, as I do. We have a long way to go.

Several chapters in this book provide specific insights into essential *social* nutrition that apparently only military peers and families provide—or don't provide—to service members, and report on potentially toxic metabolites of the service member's military experience within the family. Other excellent chapters address the neurophysiology, neurochemistry, and pharmacology of stress injuries. One chapter provides a flyover of the unmapped, mostly unresearched terrain of soldiers' and veterans' spiritual lives, a matter that expresses culture in addition to much else. And it goes without saying that there are chapters here that address the mind in classic voices of psychology. Asserting the ontologic parity of brain, mind, society, culture, is not a squishy "whatever floats your boat," bringing all disciplined research and thinking to a halt. On the contrary, it opens previously unseen paths of treatment, prevention, and research. The unity of body, mind, and spirit is nowhere more vivid, nor more painful, than in the results of the rigorous work reported here on the downstream medical morbidity and premature mortality among combat veterans with psychological injury (PTSD). The chapter, "The Mortality Impact of Combat

* The US Department of Veterans Affairs requested the National Academy of Sciences, through its Institute of Medicine, to review the scientific merit and clinical utility of the diagnostic entity PTSD. My statement before this committee, "The Pen and the Dollar Bill: Two Philosophical Stage Props," can be downloaded as a PDF file at http://www.iom.edu/?id=32930

Stress 30 Years after Exposure," is an illustration of the scientific fertility of breaking out of disciplinary stovepipes.

Military functioning, like clinical functioning, is a *practice,* which is to say: it always happens in real time, its outcome cannot be certain until it is actually over, and it is always under the star of urgency.* Military families cannot wait; active service members cannot wait; line military leaders cannot wait; injured veterans cannot wait. This book is necessarily a report of work in progress, intelligently infused with urgency to get good and actionable information and approaches out there, in preference to perching like gods in eternity waiting for the final, perfect answer to be seen in retrospect.

I want to applaud the practical chapters in this book, such as the chapter of advice for families and friends of service members returning from an active theater. Another chapter amply demonstrates that a service member's psychological injury from war stresses and sometimes damages the family. The basic point is that *valid* survival skills and adaptations in the war zone can persist into life after war and put family, neighbors, employers in a quandary. If *everyone,* including the returning service member himself or herself, has this knowledge and understanding of where these patterns of thought, feeling, and action came from and what they were once good for, many destructive misunderstandings and misattributions will be avoided all around. The returning service member's demoralizing sense of being a freak—the only person with such quirks—might also be avoided. Providing usable education is always the first and most important task of those aiming to protect service members and their families and to prevent the life-blighting complications of out-of-place adaptations to combat which have persisted into life in garrison or civilian life. Two chapters in this volume (one from the UK Royal Marines and the other from Canadian Forces) build on the wisdom that *peers* are often the *only* people from whom a psychologically injured service member or veteran can take in the desperately needed education, social support, and trustworthy referrals for professional help.

I cannot end this Foreword without a word on sleep loss. I hope readers will note Dr. Nash's charming report that when he was deployed with Marines to Iraq he stashed a jar of prazosin capsules in his back pack, so as to be able to offer safe, practical help with *nightmares.* Discovery of this use for an old and long off-patent alpha-1 adrenergic blocking antihypertensive is one of the tiny handful of true breakthroughs in the pharmacotherapy of psychological injury. Nightmares wreck restorative sleep, during the few hours a deployed service member has.

Sleep is fuel for the frontal lobes of the brain, where emotional and ethical self-restraint live and complex cognitive and social judgments are executed—neuroscience research is clear on this, as is military research

* Bourdieu, Pierre. *The Logic of Practice.* Trans. by R. Nice. Stanford University Press, 1990.

on how increasing sleep loss deteriorates discrimination between friend, foe, and protected person. The extremely sleep-deprived brain cannot tell the difference and ceases to care, while, tragically, the overtrained capacity to shoot straight hardly wavers. Sleep loss promotes disastrous *leadership* errors and malpractice, which in turn are fertile sources of psychological and moral injury to the *troops*. Once psychologically injured, sleep loss (a symptom of injury) perpetuates the injury and blocks spontaneous healing, while fatally greasing the slide into alcohol and drug abuse—which often begin as desperate efforts at sleep. If the enemy destroys sleep, well, good for him, that's his job, just like destroying our fuel, ammo, and water. But all too often American military personnel live by policies, practices, and culture that *inflict sleep loss on ourselves*.*

Charles Figley's path-breaking multiauthor volumes on combat trauma in the early '80s are now joined by this volume edited by and containing chapters by Charles Figley and William Nash. I have every expectation that this will open new paths as well.

Jonathan Shay, M.D., Ph.D. is a staff psychiatrist at the Department of Veterans Affairs Outpatient Clinic, Boston. Between September 1, 2004 and August 31, 2005, he was Chair of Ethics, Leadership, and Personnel Policy in the Office of the U.S. Army Deputy Chief of Staff for Personnel, and earlier was Visiting Scholar-at-Large at the U.S. Naval War College (2001) and performed the Commandant of the Marine Corps Trust Study (1999–2000) for the Commandant, General James Jones. He is the author of *Achilles in Vietnam: Combat Trauma and the Undoing of Character* and of *Odysseus in America: Combat Trauma and the Trials of Homecoming*. The latter has a foreword authored jointly by Senators John McCain and Max Cleland.

* Why we would continue to do this to ourselves in the face of decades of good science and wise military counsel on the effects of sleep loss is addressed in my paper, "Ethical Standing for Commander Self-Care: The Need for Sleep." *Parameters: U.S. Army War College Quarterly*, 28:93-105, Summer 1998. http://www.carlisle.army.mil/usawc/Parameters/98summer/shay.htm.

Acknowledgments

We acknowledge, first and foremost, Dr. Figley's graduate research assistant, Kristy N. Scarpa. Inspired by the military war service in Afghanistan of her boyfriend, Byron Conley, and blessed with a gift of organization and energy, she coordinated the review process for most of the chapters efficiently and effectively. Thank you, Kristy, for a job well done. We thank those colleagues who served as reviewers, in addition to those who contributed chapters as authors or co-authors; in alphabetical order, they are: Keith Armstrong, Rebekah Bradley, Brian Bride, Jeffrey B. Clark, Kent Drescher, Mathew Friedman, Christopher Frueh, Thomas Gaskin, Cheryl Goodson, Geoffrey Grammer, Fred Gusman, Col Charles Hoge, Terence Keane, Cdr John Kennedy, Capt Robert Koffman, Robert E. Lee, Steve Lewis, Brett T. Litz, Gary E. May, Shad Meshad, CA Morgan III, John Parker, Murray Raskind, Martell Teasley, and Bruce Thyer. To my (CRF) wife, Kathleen Regan Figley, and for her reviews, love, support, and encouragement. To my (CRF) daughters, Laura Figley and Jessica Chynoweth, and son-in-law, Mike Chynoweth, for their love and encouragement. To my (WPN) wife, Nancy Clayton, for her unwavering encouragement and passion for the well-being of those Marines fortunate enough to know her. To my (WPN) sons, Zachary, Daniel, and Thomas, for their generosity with my time, and to my (WPN) oldest son, Aaron, for sharing his love for the Marine Corps with me.

This book is dedicated to the memory of our mothers, Geni Figley and Helen M. Nash.

1

Introduction: For Those Who Bear the Battle

CHARLES R. FIGLEY AND WILLIAM P. NASH

The nature of war is destruction, both of property and human life. It has always been known that combat takes the lives of warfighters on the battlefield and later, through wounds that are too serious to heal. It is also common knowledge that of those who survive the dangers of battle, some carry physical wounds forward through the rest of their lives. This book draws attention to another type of wound, a mental and emotional wound, that can also generate long-term disability, and can also lead to early death (Boscarino, this volume, chapter 5; Boscarino, 2004, 2006). Our purpose in describing injuries caused by stress rather than steel is to promote a greater scientific understanding of their causes and cures, and to promote more effective methods to prevent and manage stress injuries before, during, and after operational deployments. Like physical injuries, the natural course for most injuries caused by stress is to heal over time. But also like physical injuries, stress injuries heal more quickly and completely if they are promptly recognized and afforded the proper care, if only a brief period of rest.

The chapters in *Combat Stress Injury* report on the theories, research, and intervention programs that may save the lives and the mental health of the warfighters who serve their country in battle. Thirty professionals from medicine, psychiatry, psychology, nursing, social work, human development, engineering, physiology, clergy, and military science collaborated to produce a book that is highly readable and of immediate use to those who train, lead, and care for warfighters. Of special interest to the contributors were the military commanders who bear primary responsibility for the management of all hazards faced by warfighters, both physical and mental. Military leaders have always understood their unique responsibility to balance the sometimes competing priorities of mission accomplishment and conservation of personnel. An important focus of the chapters in this book has been to place tools into the hands of military leaders to help them preserve the fighting strength of their troops while also protecting their long-term health and that of their families.

Combat Stress Injury is also written for the men and women who attend to those in harm's way: mental health professionals, the clergy, and medical professionals. These professionals want to be as prepared as possible to help the warrior recover from and even be inspired by their combat stress injuries. We hope that this book will serve as an important resource for the latest and most comprehensive information and guide to helping.

Although not written for policy and law makers, we hope that the knowledge available in *Combat Stress Injury* will lead to significant improvements in the way modern warfare is conceived and the warrior trained and managed. As a result these improvements will increase both warrior proficiency and resiliency in and after battle.

Finally, and most importantly, *Combat Stress Injury* is written for the warriors themselves and their families who wish to understand the stress-related psychosocial and medical consequences of battle. They will learn that there are positive as well as negative consequences to combat, and they have a role in managing these consequences. The various chapters, we hope, will help warriors and their families master the many challenges of deployment and homecoming. Deployment to a war zone is a transformative process for everyone connected with this enterprise; it is hoped that information contained in this volume will promote the enduringly positive life transformations that are possible while mitigating some of the negative changes that may be unavoidable. At the very least, we hope to prevent a repetition of the unfortunate experiences of previous cohorts of veterans, especially Vietnam veterans, who struggled with combat stress injuries in silence and misunderstanding (Figley & Leventman, 1980).

It is not known how many of the more than half million warfighters who have already served in Iraq or Afghanistan have experienced persistent stress injury symptoms, but veterans support organizations such as the U.S. Department of Veterans Affairs are gearing up to provide services to veterans of these conflicts. Families of veterans, also, have long been known to be greatly affected by the stress of war, both physically and emotionally (Figley, 1978). To be most effective, programs and services to serve this generation of warriors and their families must be built on a solid understanding of 21st century warfare, warrior culture, and stress injury science. We believe this book offers useful information to guide the development of such services, and to educate the current generation of service members, families, and support personnel.

The editors of this book met by happenstance. A mutual acquaintance, David Wee, introduced them after hearing a paper presented by the second co-editor, Dr. Bill Nash. From their first e-mail exchanges the editors felt a connection through their love of the Marine Corps and a devotion to serving warfighters and their families. From those initial e-mail exchanges in May 2005, the editors conceived of a book that would draw upon the best minds and work available that answered five fundamental questions: (1) What are the positive and negative short- and long-term consequences of war fighting for the warfighter? (2) What are the pre-combat factors that affect these consequences? (3) What are the factors during and following combat that affect these consequences? (4) What are the psychosocial and medical programs, treatments, and interventions that mitigate the negative consequences of combat and enhance the positive consequences? (5) What can be done to utilize the answers to these questions in order to more effectively educate, train, lead, and care for our future military combatants?

To present the answers to these fundamental questions, the editors organized the 14 other chapters into three sections. Section I: Theoretical Orientation to Combat Stress Management contains three chapters that serve as the conceptual anchor for the volume. All three of these theoretical chapters were written by the second editor, William Nash, M.D., a senior U.S. Navy psychiatrist who served with the First Marine Expeditionary Force in the Al Anbar Province of Iraq during 2004 and 2005. In chapter 2, The Stressors of War, he describes the specific physical and mental challenges of the current war in Iraq as well as the attitudes of military personal toward the concept of "combat stress" that form an important cultural context in which they experience these stressors. It is vital for the reader first to have a clear understanding of the challenges of current warfare in order to appreciate the relevance of the rest of the book.

In chapter 3, Combat/Operational Stress Adaptations and Injuries, a model is presented for differentiating stress adaptations from stress injuries and illnesses. Stress adaptations include the spectrum of so-called normal and reversible reactions to the abnormal challenges of combat and military operations, while stress injuries and illnesses are a group of very specific syndromes with highly predictable symptoms and courses over time. The concepts of stress injury and stress illness in no way imply permanence or even chronicity, but to deny that combat and operational stresses can injury the brain and mind is to invalidate the suffering of those brave warriors who have been injured by combat and operational stress.

In chapter 4, Competing and Complementary Models of Combat Stress Injury, Dr. Nash and Dewleen Baker, M.D. (director of the PTSD and Stress Disorders Program at the San Diego, California, Veterans Administration Medical Center) review the major competing and complementary models of combat and operational stress and stress injury. After traversing the landscape of enduring military leadership theories, twenty-first-century neurobiological models, and the psychological theories that underpin all modern psychotherapies for posttraumatic stress disorder (PTSD), the authors leads us to two important conclusions: (1) no single conceptual model of overwhelming stress and its management answers all our questions regarding resiliency, prevention, and treatment;

but (2) each of these models contributes uniquely to our understanding of these problems.

Section II: Research Contributions to Combat Stress Injuries and Adaptation includes the most important scientific findings relevant to answering the primary questions of this book. The section contains three chapters, starting with chapter 5, The Mortality Impact of Combat Stress 30 Years after Exposure: Implications for Prevention, Treatment, and Research, by Joseph A. Boscarino, Ph.D., M.P.H. (senior investigator at the Center for Health Research & Rural Advocacy, Geisinger Health System in Danville, Pennsylvania). Building on his 1997 reports and utilizing the same extraordinary data set of Vietnam veterans, Dr. Boscarino analyzed mortality rates of those diagnosed with PTSD. Among other findings in his chapter, Dr. Boscarino reported that Vietnam "theater" veterans (N = 7,924) with diagnosed PTSD (10.6%) were far more likely than other veterans to die. Controlling for all major demographic factors he found that adjusted post-war mortality hazards ratios [HR] were 1.7 ($p = 0.034$) for cardiovascular-related death; 1.9 ($p = 0.018$) for cancer-related death; and 2.3 ($p = 0.001$) for all types of external-caused deaths (including motor vehicle accidents, accidental poisonings, suicides, homicides, and injuries of undetermined intent). For all causes of death, the HR was 2.2 ($p < 0.001$).

Chapter 6, Combat Stress Management: The Interplay Between Combat, Physical Injury, and Psychological Loss, is written by Danny Koren, Ph.D. (professor in the Psychology Department, University of Haifa, Israel), Yair Hilel (former member of the Israeli Army and current Ph.D. candidate in the Psychology Department of the University of Haifa), Noa Idar (also a Ph.D. candidate in the Psychology Department of the University of Haifa), Deborah Hemel, B.A. (2005-2006 Fulbright Fellow in the Department of Overseas Studies, University of Haifa), and Ehud Klein, M.D. (professor in the Department of Psychiatry, Faculty of Medicine, Technion–Israel Institute of Technology, Haifa, Israel). Among other things, they present the findings of their study that isolated the unique contribution of *physical* injury to the subsequent development of PTSD. They compared 60 injured soldiers to 40 noninjured soldiers, controlling for rank, military occupational status, length of service, and combat situation using various standard mental health measures including the Structured Clinical Interview for DSM-IV (SCID). Among their findings was that the injured had PTSD by a 10 (injured):1 (noninjured) ratio, and had significantly higher scores on all clinical measures, compared to controls. Consistent with earlier findings (Figley, 1978), the presence of PTSD was not related to severity of injury. However, the lack of a relationship between PTSD and severity of the trauma was surprising. They discuss the implications of this and other findings about the impact of combat-related injury as a major risk factor—not a protective factor—for PTSD, that bodily injury contributes to the appraisal of the traumatic event as more dangerous, and that this heightened level of perceived threat is far more complicated than once thought. The authors suggest improvements in the treatment protocols to offset these predictable reactions and enable the injured to be far more resilient through effective combat and combat injury stress management.

Chapter 7, Secondary Traumatization Among Wives of War Veterans with PTSD, is written by Rachel Dekel (associate professor at the School of Social Work, Bar-Illan University, Ramat-Gan, Israel) and Zahava Solomon (professor at the School of Social Work, Tel Aviv University, Israel, and former Major, Israeli Defense Force). In their chapter they review the extensive research that demonstrates the secondary effects of combat stress in combatants on combatant wives. Among other things, they and others have found that a husband's impairment and a wife's sense of burden predicted both of the latter's emotional distress and the overall marital adjustment, compared with the general population in Israel. Since perceived caregiver burden was more closely associated with distress than the level of the combatant's impairment the authors discuss ways of lowering emotional distress by changing the perception of burden by the wives.

Section III: Combat Stress Management Programs is appropriately the largest and final section with eight chapters. Collectively these chapters demonstrate that combat stress can be managed both through prevention and training programs prior to combat, effective stress reduction methods during operations, and especially the desensitization program immediately following to long after combat exposure.

Chapter 8, Historical and Contemporary Perspectives of Combat Stress and the Army Combat Stress Control Team, was written by Bret A. Moore, Psy.D. (U.S. Army Captain, Clinical & Aeromedical Psychologist, 85th Combat Stress Control, Fort Hood, Texas) and Greg Reger, Ph.D. (U.S. Army Captain, Clinical & Aeromedical Psychologist, 98th Combat Stress Control, Fort Lewis, Washington). After differentiating between physical and psychological war casualties and the history and terms relevant to combat stress, Captains Moore and Reger discuss the history and mission of the Combat Stress Control Teams, of which they are members, and the vital roles played by the various behavioral health professionals in managing combat operational stress reactions (COSR). The latter section of the chapter presents three case studies to illustrate the way these teams provide treatment for, prevent, and consult with command/leadership about COSR. Moore and Reger were both deployed to Iraq in support of Operation Iraqi Freedom III during the construction of this chapter.

Chapter 9, Virtual Reality Applications for the Treatment of Combat-Related PTSD, is written by Skip Rizzo, Ph.D. (research scientist and research assistant professor at the University of Southern California Institute for Creative Technologies, Marina del Rey, California), Barbara Rothbaum, Ph.D. (professor of psychiatry and director of the Trauma and Anxiety Recovery Program, Emory University School of Medicine, Atlanta), and Ken Graap (president and CEO of Virtually Better, Inc., Decatur, Georgia). Their chapter focuses on the use and advantages of virtual reality (VR) methods for managing combat stress injuries and the more serious combat-related PTSD. After briefly reviewing the definitions and theoretical basis for using VR in a cognitive behavioral PTSD treatment, they review the empirical evidence in treating Vietnam warfighters using the "Virtual Vietnam" scenario. In the final section of the chapter the authors present a detailed overview of current work developing a "Virtual Iraq" scenario for the assessment and treatment of Iraq War PTSD. This approach is believed to offer certain advantages over imaginal and in vivo exposure methods for the treatment of anxiety disorders and early findings from the PTSD VR literature have been encouraging.

Chapter 10, Experiential Methods in the Treatment of Combat PTSD, was written by James L. Spira, Ph.D., M.P.H, ABPP (associate clinical professor in the Department of Psychiatry, University of California, and the Mental Health Services, Naval Medical Center, San Diego), Jeffrey M. Pyne (Center for Mental Health and Outcomes Research, Central Arkansas Veterans Healthcare System, Little Rock), and Brenda Wiederhold (Interactive Media Institute, San Diego). This chapter is a step-by-step guide to using virtual reality in the treatment of PTSD following a discussion of other more traditional treatment approaches and a description of the benefits of VR-assisted treatments. The primary focus of this chapter will be the role of experiential psychotherapy treatments that teach cognitive, affective, and behavioral control to better cope with combat-related PTSD. In particular, the authors focus on self-help skills during exposure therapy, particularly those utilizing virtual reality systems, to assist returning troops gain control over PTSD symptoms.

Chapter 11, Medication Management of Combat and Operational Stress Injuries in Active Duty Service Members, is written by Nancy M. Clayton, M.D. (U.S. Navy Lieutenant Commander, Medical Corps) and William P. Nash, M.D. This chapter concisely and clearly surveys the major classes of medication proven to be useful in the treatment and prevention of combat and operational stress injuries, and gives practical guidance for the use of such medications by active duty service members. Current clinical practice guidelines are cited for all major classes of medications. This chapter demonstrates that while pharmacotherapy cannot be considered a "cure" for operational stress injuries, the benefits of medications are often substantial, and they should always be considered in a management program.

Chapter 12, The Royal Marines' Approach to Psychological Trauma, is written by Major (rtd) Cameron March (Acting Major, Royal Marines) and Dr. Neil Greenberg (Surgeon Commander, Royal Navy). The trauma risk management (TRiM) program was developed in 1997 by the Royal Marines. It is a postincident protocol for use by hierarchical organizations, such as the Royal Marines, to systematically manage the psychological consequences of potentially traumatic events. It assumes that service personnel are, in the main, resilient individuals and the best immediate management strategies for such personnel are to effectively bolster peer support and to enable/assist those who need professional help to find it at an early stage. TRiM practitioners are nonmedically trained personnel drawn from all ranks. They receive specific training which does not intend to turn them into counselors, but instead aims to build on their innate human management skills gained by life experience and military service. Suitably trained TRiM practitioners are able to manage all psychological aspects of adverse incidents. They use established protocols and a structured risk assessment procedure to identify those at risk of developing problems and ensure such personnel are correctly managed. This is mainly done by making use of established management systems which are an inherent part of military culture. They are supported and supervised by medical and mental health services from within the Royal Navy, but are an accepted source of first-line support and are an integral part of all Royal Marines units. This chapter explains the history behind the TRiM system, how it works, and where its future lies.

Chapter 13, The Operational Stress Injury Social Support Program, is written by Lieutenant Colonel Stephane Grenier (project manager of the Operational Stress Injury Social Support [OSISS] program and the DCSA, National Defense Headquarters, Ottawa), Kathy Darte, M.N. (co-manager of the OSISS program), Alexandra Heber, M.D., FRCPC (consulting psy chiatrist for the Canadian Forces Health Services Centre, Operational Trauma and Stress Support Centre), and Don Richardson, M.D., FRCPC (consultant psychiatrist at Operational Stress Injury Clinic, Parkwood Hospital, St. Joseph's Health Care, London, Ontario; adjunct professor in the Department of Psychiatry, University of Western Ontario; and consultant psychiatrist at Veterans Affairs Canada, Hamilton, Ontario). The authors trace in this chapter the development of a highly successful and innovative social support program designed to ease the transition of Canadian Forces veterans to civilian life. The first part of the chapter discusses the creation of the Operational Stress Injury Social Support Program, which began with the postdeployment experiences and ideas of the first author. The next part of the chapter defines operational stress injury, discusses the initial beginning of the program followed by a discussion of the concept and literature review of the benefits of social support activities. The following two sections discuss the OSISS peer support model and the research that documents its successful impact, particularly the avoidance of stigma and reframing PTSD using the concept of operational stress injuries. Collectively, the chapter is a good illustration of one person's inspiration emerging from his pain and suffering resulting in a social service program that literally saved lives.

Chapter 14, Spirituality and Readjustment Following War-Zone Experiences, is written by Kent Drescher, M.Div., Ph.D. (coordinator, Assessment and Data Management, National Center for PTSD Inpatient Programs, Menlo Park, California; and an ordained Presbyterian minister), Commander Mark Smith, CHC, USN (instructor at National Security Affairs), and David Foy, Ph.D. (professor of psychology at the Graduate School of Education & Psychology, Pepperdine University, Encino, California). The authors point out that the experience of veterans of prior wars indicates that war-zone trauma frequently impacts the spirituality of survivors. Early data indicated that most veterans reported difficulty reconciling their experiences in the war zone with their religious faith, and there are reasons to expect veterans of the current wars in Iraq and Afghanistan may, as well. This chapter is designed to offer suggestions as to ways these issues of trauma and spirituality might be addressed with returnees from Iraq or Afghanistan, with a goal of making spirituality a healing resource, as opposed to an obstacle to full recovery and return to functioning.

Chapter 15, *The Returning Warrior: Advice for Families and Friends*, by Judith A. Lyons, Ph.D. (Psychology Service, V. ["Sonny"] Montgomery VA Medical Center, and the University of Mississippi Medical Center in Jackson) builds on the groundwork laid by Dekel and Solomon in chapter 7. Dr. Lyons provides a practical explanation of the research literature for family and friends of warfighters and for clinicians who are assisting them. The chapter identifies issues that frequently emerge in reestablishing relationships after a combat deployment. General information and advice are offered for relationships undergoing the normal strains of readjustment. Specific resources are identified for relationships that may need more formal intervention.

We urge the reader to carefully consider the perspectives offered here: that collectively the contributors are calling into question the current views of combat operational stress leading so seemingly automatically to the medical condition of PTSD. Rather, we hope that there is a more serious consideration for the concept of combat stress injuries; that these stress injuries provide a critical window of opportunity to prevent the unwanted and long-term negative effects that include but are not limited to mental disorders; that there will be a new outlook of optimism and respect for those who bear these injuries and that they will not be separated from those with more physical injuries. In doing so, the current fear of being diagnosed with PTSD will be greatly reduced, if not eliminated; that much can be done to prevent an injury from becoming a disorder.

REFERENCES

Boscarino, J. A. (2006). Posttraumatic stress disorder and mortality among U.S. Army veterans 30 years after military service. *Annals of Epidemiology, 16*, 248–256.

Boscarino, J. A. (2004). Posttraumatic stress disorder and physical illness: results from clinical and epidemiologic studies. *Annals of the New York Academy of Sciences, 1032*, 141–153.

Figley, C. R. (1978). Psychosocial adjustment among Vietnam veterans: An overview of the research. In C. R. Figley (Ed.), *Stress Disorders among Vietnam Veterans: Theory, Research, and Treatment* (pp. 57–70). New York: Brunner/Mazel.

Figley, C. R., & Leventman, S. (Eds.). (1980). *Strangers at Home: Vietnam Veterans Since the War.* New York: Brunner/Mazel.

Section I

Theoretical Orientation to Combat Stress Management

This first section provides an overview of the theoretical and conceptual basis for understanding and managing combat stress injuries. The field of traumatology and psychosocial stress generally lacks a clear and unified theory of the stress-injury-disorder-treatment-recovery process. These chapters not only provide such a theoretical base by explaining the relationships among these domains, they provide the building blocks for a theory that can help to predict stress injuries and disorders emerging from combat operations. More importantly, these models can lead directly to effective methods for preventing and mitigating combat-related stress disorders among warfighters and those who care for them.

2

The Stressors of War

WILLIAM P. NASH

> Everything in war is very simple, but the simplest thing is difficult. The difficulties accumulate and end by producing a kind of friction that is inconceivable unless one has experienced war.
>
> *Clausewitz (1982/1832)*

One does not need to have ever worn a military uniform to have a conception of the stressors produced by modern warfare. Popular movies and novels, which reflect our fascination with the experience of war, have taught us much about what it is like to participate in combat and the forward support of combat. So also have broadcast and print news media, which have provided us a window into the realities of war through the eyes, ears, and lenses of reporters. But the picture of war painted by artists and journalists must always be approximate and somewhat distorted. We who are duty bound to help prevent, identify, and treat combat and operational stress injuries must have a fuller, more accurate appreciation for the specific physical and mental challenges of participation in military deployment. Our theories, research, and tactics for the management of combat and operational stress reactions depend, for their validity, on the accuracy of our understanding of those stressors.

The stressors themselves tell only part of the story of the impact of war on individuals. Equally fundamental is an appreciation of the shared attitudes, beliefs, and expectations that prevail within military units as part of their shared culture and traditions. These culturally shaped attitudes and beliefs form a lens through which combat and operational stressors can be either filtered or magnified for individual warriors. Like the stressors of combat, the cultural contexts within which individuals experience war have few parallels in modern civilian life.

Observational study of the specific stressors of combat and operational military deployment is an area that has been largely neglected by empirical research. Most historical works on warfare have focused more on the experience of those who made the major decisions in war—the generals and politicians—than on the men and women on the frontlines who carried out those decisions, day to day. And most clinical works have focused, naturally, on the experience of individuals who were suffering from a negative reaction to combat and military operations, rather than on the stressors of warfighters as a population. Nevertheless, previous historical and clinical works have described a great many of the stressors experienced in previous wars by American, British, and Canadian soldiers and marines (Bourne, 1969; Copp & McAndrew, 1990; Dean, 1997; Figley, 1978; Grinker & Spiegel, 1945; Hendin & Haas, 1984; Kulka et al., 1990; Moran, 1967; Shay, 1994, 2002; Shephard, 2001). Belenky (1986) and Solomon (1993), among others, have described the experiences of Israeli troops in the 1973 Arab-Israeli War and the 1982 war in Lebanon. Schneider (1986) offered a rare glimpse into the stressors of German Wehrmacht troops in WWII, and the prevailing German national and military cultures that shaped their experience of those stressors. Hoge et al. (2004) conducted landmark research on self-reported post-deployment stress problems in ground combatants who participated in the war in Afghanistan and the 2003 invasion of Iraq, and the cultural attitudes that discouraged those same U.S. soldiers and Marines from accessing care.

More complete lists of combat and operational stressors can be found in the *U.S. Army Combat Stress Control Handbook* (Department of the Army, 2003), the U.S. Marine Corps (2000) reference publication *Combat Stress*, and chapters in *War Psychiatry* (Jones, 1995a,b), from the U.S. Army's *Textbook of Military Medicine* series. All these works were written before the current conflicts in Southwest Asia, however.

This chapter has two goals. The first is to briefly describe one narrow but important aspect of current Western military culture—its prevailing attitudes about "combat stress" and its recognition and management. The second is to describe in a systematic way the physical, cognitive, emotional, social, and spiritual challenges characteristic of current military conflicts.

MILITARY ATTITUDES TOWARD COMBAT STRESS

Effective military leaders are experts in the stress of combat and operational deployment. *Generating* combat stress in their adversaries on the battlefield and mitigating operational stress in their own troops through leadership, training, and unit cohesion are among the most basic tools of the warfighter. Because the creation and management of stress reactions are so fundamental to their craft, military leaders tend not to be neutral to the concept of "combat stress." And they tend to view quite differently stress experienced by their adversaries, their own troops, and themselves.

Four military cultural attitudes toward combat stress will be examined in the first part of this chapter: (1) combat stress seen as a weapon, (2) combat stress seen as "friction" to be overcome and banished from awareness, (3) combat stress seen as a challenge to leadership, and (4) combat stress seen as a test of personal competence.

Combat Stress as a Weapon

> "Combat stress" is what *we* inflict on the *enemy*!
>
> *Marine commander in Iraq*

Military leaders have long known that human factors are crucial determinants of victory or defeat on the battlefield. War is a clash of opposing human wills, fueled by emotion, and influenced as much by mental and moral forces as by technology and material factors. It is seldom the physical destruction of people or equipment that brings victory, but destruction of adversaries' will to go on fighting because of the bombs, bullets, and other hardships they endure. Combat stressors are weapons whose targets are the hearts and minds of individual opposing warriors.

Given the central role in warfighting played by human responses to the stressors of war, it is not surprising that the military has intentionally developed strategies and tactics specifically to increase the physical, mental, and emotional stress experienced by adversaries. Stress-inducing factors such as chaos, uncertainty, surprise, hopelessness, physical hardship, isolation, and sleep deprivation are studied by military leaders. These factors are weighed, if only unconsciously, in every military decision in combat. One fairly recent development of combat stress as a weapon is the concept of "psychological operations," or PSYOPS. Psychological operations are military operations designed to influence the emotions, motives, objective reasoning, and ultimately the behavior of adversaries and their governments, to reduce their will to fight (Chairman of the Joint Chiefs of Staff, 2003). The tools of the PSYOPS trade include leaflets, media broadcasts, and combat loudspeakers to communicate demoralizing information to opponents. These psychological operations are integrated with other combat operations to produce the maximum cognitive and emotional stress in opposing forces.

How might the conscious, intentional wielding of "combat stress" as a weapon affect how military personnel perceive their own stress? First and foremost, experienced warriors understand that "combat stress" is not a by-product or side effect of war that can be sanitized away; war *is* stress. And the greater the tempo of operations and intensity of combat, the greater must be the stress experienced by all combatants on both sides.

Furthermore, the knowledge that stress in war is a weapon may discourage sympathy for those affected by combat stress. To inflict suffering on an adversary, the warrior must avoid identifying with that adversary or feeling remorse or sympathy for the suffering imposed (Grossman, 1995). Of course, warriors view themselves and their comrades-in-arms differently than they view their adversaries, but the callousness that warriors must develop and maintain toward their adversaries' suffering can not easily be turned on and off. It may be asking too much of warriors, at times, to acknowledge their own or their comrades' vulnerability to combat stress at the same time they are exploiting their adversary's vulnerability to almost the same stressors.

Combat Stress as "Friction" to Be Overcome and Banished from Awareness

> In the air, I only see what I need to see to get the job done and get out of there. The only bullets flying through the air are *mine*, going *that* way. I don't see anything coming *this* way unless I need to do something about it.
>
> *Marine Cobra attack helicopter pilot*

Carl von Clausewitz (1982/1832), a Prussian general whose 18th- and 19th-century writings on the theory and practice of war are still revered by Western military leaders, used the term "friction" to describe the physical, mental, and emotional stress of combat. Current U.S. Marine Corps (1995, 1997) doctrinal publications use the same term in the same context to denote both that stress in war is inevitable and that it is merely an obstacle on the path to victory. The warrior ethos places success on the battlefield above all else. Victory in combat is the only way warriors can discharge their responsibilities to the nation and their own honor, while also protecting themselves and their comrades from further harm. This single-mindedness in war is lifesaving, even if it is also destructive.

Since most of the terrors, horrors, and hardships of war are unavoidable, it is imperative that warfighters learn to perform effectively despite the "friction" generated by these stressors. To be most effective, warriors must strive to become tough and resistant to the steady stream of physical and mental stressors that impact on them. Ideally, warriors must even learn to ignore combat and operational stress—to not even allow a conscious awareness of stressors and their impact. Could an athlete be victorious in a sporting contest while pondering the distance to be covered or the weight of equipment worn or the dangers of the sport? The genius of great athletes lies in their ability to perform as if there is no distance or weight or danger. The genius of great warriors is to fight as if there is no terror, horror, or hardship. In their minds, there can be none—at least, not until the fight is over.

Even when combat and operational stress cannot be ignored, some warriors—especially elite ground combatants—sometimes have little interest in finding ways to reduce their own stress. Such warriors know, if only unconsciously, that searching for ways to become more comfortable or safe in war can be not only a distraction from the real business at hand, but also a serious hazard to success and even survival. Searching for greater comfort or safety increases warriors' conscious awareness of their own discomfort and suffering, and such awareness, in itself, may erode warriors' confidence and courage, two attributes that are crucial to military success and survival.

Confidence in battle can be defined as trust in oneself and one's peers, leaders, and equipment to perform under stress (Center for Army Leadership, 2004). Courage has a broader range of possible definitions, but it was perhaps most eloquently defined by Lord Moran, Winston Churchill's personal physician during WWII, who wrote *The Anatomy of Courage* based on his experiences as a battalion surgeon during the First World War. In that monograph, Moran wrote:

> Courage is a moral quality; it is not a chance gift of nature like an aptitude for games. It is a cold choice between two alternatives, the fixed resolve not to quit; an act of renunciation which must be made not once but many times by the power of the will. Courage is will power. (Moran, 1967)

The will to go on fighting that Lord Moran equated with courage may be impossible without the simultaneous confidence that one *can* go on fighting. And confidence in oneself to be able to endure even a little bit longer may be seriously weakened by a full conscious awareness of one's own stressors and stress responses. Helping professionals such as chaplains, medical officers, and mental health workers must always consider the possibility that they may do more harm than good by asking warriors in an operational theater to become more aware of their own stressors and stress reactions.

Combat Stress as a Leadership Challenge

> A brave captain is as a root, out of which, as branches, the courage of his soldiers doth spring
>
> *Sir Philip Sidney, quoted by Grossman (1995, p. 85)*

When the concept of "stress" enters the conscious awareness of military leaders as a *problem*, rather than just as an unavoidable component of warfare or as "friction" to be overcome and ignored, it is most commonly seen by them as a challenge to their own leadership skills. Military leaders rarely view operational stress as primarily a medical or spiritual problem to be solved by doctors or chaplains. Military leaders are the owners of combat and operational stress and its management, for two important reasons.

The first reason is because troop morale, confidence, and the will to fight—long known to be essential for the *prevention* of disabling stress reactions—all depend heavily on the personal relationship between leaders and their troops. The first job of every military leader is to motivate and empower subordinates to function effectively and endure regardless of the stress experienced, for the sake of their mission as well as their own survival. The tools at the leader's disposal include concrete objects such as weapons and protective gear, processes like training and tactics, and intangible assets such as traditions and esprit de corps. But the tools of military leadership are just that—tools. They are no substitute for the *essence* of leadership, which is the personal relationship between leader and subordinates. It is the leader's moral courage, fortitude, and will power that enable members of the unit to endure and maintain their combat power despite the unrelenting stress of war. Dave Grossman, a military scientist and former U.S. Army Ranger, perhaps put it best in his popular and influential book, *On Killing*, when he described great military leaders as those who have the ability to encourage others to draw from their "well of fortitude" to replenish their own reservoirs of emotional stamina (Grossman, 1995, p. 85). Replenishment of courage and confidence in this way is a very personal transaction, occurring between trusted leader and valued subordinate. Helping professionals outside the direct military chain of command, including chaplains and mental health providers, can do little to facilitate this essential leadership function because they operate outside of it.

Second, military commanders bear full, personal responsibility for deciding when and how to spend the resources of war placed at their disposal by the nation. The most precious resources of war, of course, are the lives and well-being of the men and women who fight and support the war. Since losses cannot be avoided altogether, it is the responsibility of unit leaders to ensure that each and every casualty, including stress casualties, contributes substantially to strategic goals. There are times when entire units must be pushed beyond their breaking points just because there is no reasonable alternative. But only a unit commander who has a grasp of strategic objectives and the demands of particular tactical situations can make such a decision. No one but a unit commander is in a position to weigh operational requirements against troop welfare, moment to moment.

One consequence of the responsibility military leaders bear for the combat and operational stress of their own troops is how those leaders view the stress "symptoms" displayed by their troops, when they arise. To the extent leaders see themselves as personally responsible for the management of combat stress in their units, they may also see themselves as personally to blame for the appearance of disabling combat stress reactions in men and women under their command. And commanders to whom small unit leaders report may also see combat stress problems in a unit as evidence of a relative failure of leadership in that unit. The potential for self-blame among leaders, and possible negative career repercussions for them, place a significant negative valence on combat stress problems. They also may provide motivations to avoid recognizing stress problems as they arise. The best leaders likely have the clearest idea which combat stress reactions should be considered leadership problems rather than unavoidable stress injuries. But there exists nowhere a clear taxonomy for discriminating leadership-responsive stress adaptations from stress wounds that cannot be prevented or healed by leadership alone. Chapter 3 of this volume was written, in part, to fill this void.

Combat Stress as a Test of Personal Competence

In his interesting recent book, *War and Gender: How Gender Shapes the War System and Vice Versa*, Joshua Goldstein (2001) explores the social psychology of war as a test of "manhood" and a rite of initiation among males in many cultures. Certainly, participation in war is not the only way a man can prove to himself and others that he is strong and brave (Gilmore, 1991). And women have never been immune to the lure of combat as a proving ground. Yet, war is likely the toughest challenge a person can face, especially for the teenagers and young adults who comprise most of the fighting forces in all-volunteer military services. It is far beyond the scope of this chapter to explore the forces that compel men and women to test themselves on the fields of battle. Rather, two specific consequences of war perceived as a test of personal worth and competence will be briefly explored.

First, to the extent participation in war *is* a test, the stressors of war are absolutely essential. In fact, the more potentially overwhelming the stressors endured, the greater the value to an individual's self-concept for having mastered them. Warriors who are unconsciously testing themselves in the crucible of war might be reluctant to intentionally reduce their combat and operational stress in any way. To do so would only diminish their personal triumph. On the other hand, of course, it is also self-defeating for warriors submitting themselves to testing on the battlefield to intentionally heap additional, unnecessary stressors on themselves just to make their own challenges greater. As the Austrian psychiatrist Viktor Frankl (1984/1946) wrote after surviving years in Auschwitz and other Nazi concentration camps during World War II, suffering can ennoble an individual only if it is unavoidable.

Second, to the extent participation in war is perceived by warriors as a test of their personal strength, courage, and competence, admitting to combat stress "symptoms" may be tantamount to admitting failure. Even if some stress symptoms are understood to be due to unavoidable stress injuries, and not merely personal weakness or cowardice, developing stress symptoms can bring with it considerable shame. Warriors volunteer for, train for, and expect themselves to conquer all the stressors of war, even the worst terrors and horrors of ground combat. Therefore, it is hard for warriors to not perceive stress symptoms of any kind as evidence of personal weakness and failure. The fact that stress injuries are invisible makes it even more difficult for warriors to forgive themselves for developing symptoms of stress injuries. Why, they inevitably ask themselves, were *they* unable to "handle" what others seemingly endured without any sign of strain? It is difficult for stress-injured warriors, however heroic they may have been in other ways, to accept in themselves evidence of being damaged by combat and operational stress without feeling like they have failed the test of war.

THE SPECIFIC STRESSORS OF DEPLOYMENT AND COMBAT IN MODERN MILITARY OPERATIONS

The mental and physical challenges of participating in any war, of course, are many and varied in their intensity and duration. Some are so powerful in their impact that they would be expected to overwhelm nearly anyone almost immediately. Others are more subtle and annoying rather than overwhelming. Clearly, the stressors which are potentially the most toxic are those with the greatest force and impact—typically those involving the terror of near-death experiences or the horror of seeing others die in violent and painful ways. But it is important not to lose sight of the additive and relentless nature of operational stressors. Many of the stressors of deployment are not overwhelming in themselves, but as these stressors persist—day after day, month after month—their effect on the mental and physical functioning of troops may accrue until a breaking point is reached. Ultimately, it may be a relatively minor stressor, such as bad news from back home, a mild illness, or conflict with a peer or superior that serves as "the straw that breaks the camel's back." But it is always the sum of all stressors over time that weighs down the proverbial camel to the point of damage.

In the following sections, specific stressors of deployment in support of Operations Iraqi Freedom and Enduring Freedom will be listed and briefly described. For convenience, these stressors will be divided into five groups: physical, cognitive, emotional, social, and spiritual. These groupings are admittedly arbitrary, and few stressors fit neatly and entirely in just one of these groups. The stressors described in each group also by no means comprise complete lists. But it is hoped that this catalogue of stressors will be useful to those who deploy in support of U.S. and allied combat operations, or who wish to understand better what was endured by the veterans of recent conflicts in Southwest Asia.

Physical Stressors

Heat and Cold. Summertime temperatures in Southwest Asia often top 120°F, and lows in the winter can reach below freezing. The effects of the heat are greatly amplified by the protective gear required to be worn by all personnel outside of hardened and safe structures, including Kevlar (helmet) and flak (armored vest). The new ceramic SAPI (small arms protective insert) plates in the front and back of each flak vest have saved countless lives by stopping everything up to an AK-47 7.62-mm round, but they also have turned the armored vest into a small oven, effectively raising ambient temperature an additional 10 to 20 degrees. Very few vehicles are air conditioned; most are cramped and stuffy, especially while the windows are kept closed to protect occupants from shrapnel. Winter low temperatures in Southwest Asia are not frigid, but for troops who are in the field or manning observation posts on rooftops for days at a time, continuous exposure to near-freezing temperatures without shelter can be very stressful.

Dehydration and Wetness. Especially until one acclimates to the summertime heat in Southwest Asia, keeping up with fluid loss from sweating can be a challenge. And dehydration, even of mild degrees, significantly increases heart rate at rest and makes each physical and mental exertion a greater challenge. Thirst is a poor indicator of impending dehydration because combatants can quickly become somewhat numb to their own thirst, just as they do to most other forms of discomfort experienced in an operational theater. Dehydration can be deadly, but wetness can also be a physical and mental challenge. In the summer, uniforms stay soaked with sweat much of the time. And in the winter, heavy rains can contribute to hypothermia in those whose duties keep them out in the open.

Dirt and Mud. The Syrian Desert is covered not with sand but with a fine grit that in many places has the consistency of talcum powder. During the winter, this fine dirt can be suspended in the motionless air during "brown-outs" for days at a time. Brown-outs can be a serious hazard to visibility—a helicopter crash in Iraq on January 26, 2005, apparently caused by an unexpected brown-out, claimed the lives of 28 Marines and one Navy corpsman. But even when the dirt hanging in the air is not threatening lives, it makes breathing difficult and hygiene impossible. Heavy winter downpours convert every field of fine tan powder into a sea of thick brown mud, sometimes more than ankle-deep. For months during the rainy season, nearly everything everywhere is caked with mud.

Sleep Deprivation. Almost no one in an operational theater gets a full 6 to 8 hours of sleep every day. Combatants in the field must learn to function on no more than 4 hours sleep at a time—sometimes considerably less. During the push toward Baghdad during the 2003 invasion of Iraq, many warriors became so sleep deprived that they began having visual hallucinations and trouble thinking clearly. But even when the operational tempo is slower, warriors are often required to function for days or weeks with very little sleep. Sleep deprivation is an insidious hazard for several reasons. First, sleep deprivation significantly impairs cognitive mental functions, including memory, attention, and rational decision making. Second, tiredness and sleepiness are poor indicators of sleep need because, over time, individuals can adapt to sleep deprivation to feel less sleepy and tired, but they cannot make themselves think more clearly while sleep deprived (Van Dongen et al., 2003). Third, though still very little is known about the normal functions of sleep, there are compelling reasons to believe that sleep is required for restoration and healing, both physical and mental. The greater the overall stress level experienced by warriors in an operational environment, the more sleep they need to recover from their stress.

Noise and Blasts. In more secure forward operating bases, noise may usually be no more than a minor irritant, such as may be caused, for example, by the drone of diesel generators or malevolent-sounding Arabic chanting heard every night and morning coming from just outside the camp walls. But no camp or base is safe from indirect fire with mortars or rockets and occasional sniping with small arms. And in many areas, counter-fire howitzer batteries may be near enough to generate blasts and concussions almost equal to those caused by mortars falling nearby, especially when artillery shells are fired overhead. Still, those are sporadic sounds, which startle more than they overwhelm. During firefights, however, the noise of detonations and weapons being fired can be deafening and continuous for many minutes or hours. The intensity and raw power transmitted through a steady stream of such sounds can rattle one's bones and one's courage. Certain sounds, once heard, become particularly powerful stress producers; the sound of a rocket motor buzzing close overhead or of a mortar shell screaming down onto one's position are unforgettable. Worst of all, undoubtedly, are the sounds made by men and animals as they die.

Fumes and Smells. Individuals seem to differ in their sensitivity to various odors, but operational environments offer a number of uniformly offensive smells. The smell of human waste hangs around porta-johns, especially in areas where they are cleaned out relatively infrequently. In less established areas, human waste is burned daily in barrels of diesel oil. The smell of burning trash, especially plastic, can be particularly noxious. However, the most psychologically toxic odors are those generated directly by battle—the smells of blood, viscera, and burnt flesh. Recollections of such smells, even long afterward, can be so vivid as to border on hallucinations.

Bright Light or Darkness. The glare of bright sunlight in the Syrian Desert can certainly damage eyes and skin, especially if adequate protection is not available. And warriors can become sensitized to bright flashes of light after surviving nearby blasts, such as those caused by roadside improvised explosive devices (IEDs), since a light flash always precedes the concussion, noise, and shrapnel generated by powerful warheads. For many, however, greater day-to-day stress is generated by darkness. Because of the danger of drawing sniper fire at night, few military personnel use lights of any kind as they move around in the darkness. On patrols and convoys at night, vehicle headlights are left off both to make less of a target for the enemy and to keep from blinding our own gunners using infrared night vision goggles. High-speed maneuvering in the dark is certainly dangerous, as evidenced by the number of lives claimed in operational theaters by noncombat accidents at night. Darkness heightens the anxieties experienced by some combatants—on both sides of the fight—a fact that is freely exploited by our own military units who sometimes prefer to mount offensive operations at night.

Malnutrition. During the initial push toward Baghdad in March and April of 2003, many U.S. forces advanced so rapidly they could not be kept supplied with food of any kind, not even MREs (meals, ready to eat). Especially given the levels of exertion experienced by our troops, day after day, hunger became a significant stressor for many. A lack of adequate nutrition causes fatigue and heightened anxiety, and it interferes with sleep. For a few days in April 2003, some of our troops actually began to fear they would starve to death before supplies caught up with them. Though it is doubtful any of them were ever close to starvation, such a perception, in itself, can be a traumatic experience.

Illness or Injury. During the course of a 7- to 14-month deployment, it is inevitable that occasional minor illnesses would be weathered, such as seasonal viral upper respiratory infections. Such minor illnesses are not huge stressors for most people, of course, though while deployed to a war zone and weighed down by the cumulative effect of other operational stressors, a minor hassle such as a cold or flu can seem overwhelming. Besides causing physiological changes that enhance fatigue and reduce stress tolerance, even minor illnesses affect many people's feelings of wholeness, competence, and self-confidence. Worse in this regard than mild, transient illnesses, however, are the physical injuries that are not infrequent in war. As of June 23, 2006, the U.S. Department of Defense reported 2,511 combat deaths (KIAs) in Iraq since the war began on March 19, 2003, but more than 7 times that many—18,572—wounded in action (WIAs). More than half of those WIAs (10,064) were returned to duty (RTD), meaning that they returned to their units in Iraq soon after their injuries, usually while still recovering. Some of those injuries were minor, such as lacerations or perforated eardrums from IEDs or mortar blasts, but some were not so minor. Certainly, the greater the injury and the longer the time needed to recover, the more the injury added to individuals' stress loads, especially by affecting their confidence in themselves and their peers, leaders, and equipment. The alternative—to evacuate every injured person from an operational theater—is unthinkable. Evacuating warriors with minor injuries would not only add needlessly to the load borne by all the uninjured warriors left behind to do the job, but it would heap an additional burden of shame and guilt onto the shoulders of injured warriors being forced to abandon their buddies before the fight is over. It is for this very reason that few WIAs had to be coerced to return to their units. Some even eloped from aid stations and combat surgical hospitals before being officially released so they could rejoin their units as quickly as possible.

Cognitive Stressors

Lack of Information or Too Much Information. Uncertainty adds to everyone's stress load, especially when the uncertainty involves something that could soon threaten one's own life. Most strategic and tactical operational information never filters very far down chains of command except for those who really need to know something to get their job done. Rumors fly fast and furious among the troops to fill the void left by the lack of good intelligence filtered down to the lowest levels, but rumors often only add to individuals' stress levels by bombarding them with conflicting and rapidly changing information, much of it false. Good commanders make sure their troops are as well informed as they can be, but still—junior personnel on the ground seldom have much idea where they will be in the next few hours, let alone the next few days. Information from the outside world is a scarce commodity, too, though the Internet has helped a great deal to keep warriors in all but the most transient and isolated areas informed. It is not at all unusual for troops on forward operating bases in Iraq to listen through the course of a day or night to gun battles going on just outside the "wire" (the camp boundaries), but to not find out what happened—and that their camp was now safe—until they later read about it on the Internet or were lucky enough to catch a news broadcast on satellite television. Conversely, too much of the wrong kind of information can be stressful for troops, as well. Especially difficult to handle is information about a serious problem, whether in the operational theater or on the home front, that an individual is powerless to do anything about.

Ambiguous or Changing Mission or Role. After the end of official combat operations in Iraq on April 30, 2003, U.S. troops were hailed as liberators in most parts of Iraq, and they felt safe to move freely and in small numbers throughout the country. At that time, troops preparing to deploy to Iraq for Operation Iraqi Freedom II were trained in what was expected to be the peacetime mission of the U.S. military in Iraq—"Stability and

Support Operations" (SASO). SASO operations involve helping a country rebuild its own infrastructure through training, construction, health promotion, and interim civil peace-keeping assistance. During the remainder of 2003, however, SASO operations were increasingly derailed by the mounting insurgency. It became increasingly common for U.S. troops to be ambushed as they attempted to provide food, medicine, money, or other assistance to Iraqis in their impoverished and war-torn villages. Even more demoralizing and confusing for U.S. troops was to repeatedly learn that the very same village elders who graciously accepted their assistance during daylight hours supervised the firing of mortars and rockets at their encampments at night. Contractors and politicians who cooperated with the U.S. military, especially in Baghdad and in Ar Ramadi and Al Fallujah to the west, were increasingly assassinated or their family members kidnapped. Schools rebuilt by U.S. and Coalition SASO efforts one day would be blown back to rubble the next by the Mujahadeen. The mission to rebuild and provide peaceful assistance can seem to some combatants irreconcilable with their other mission—to destroy and kill.

Ambiguous or Changing Rules of Engagement. A second source of stress-producing ambiguity in recent conflicts involves the "rules of engagement" (ROE), the standards which determined when soldiers and Marines are permitted to fire their weapons, and at whom. Under nearly all circumstances, U.S. troops are prohibited from using deadly force unless a clearly armed adversary poses a clear and immediate threat to Coalition or civilian life. This is a laudable standard, one that all honorable warriors hope to meet at all times. But in the 3 years since the U.S. invaded Iraq, for example, a number of ambiguous situations have become almost commonplace for soldiers and Marines. One is the use by Mujahadeen of civilians, including women and children, as human shields. This was encountered in many areas of Iraq, particularly where fighting was the bloodiest and most contested, such as in An Nasiriyah during the initial push toward Baghdad and during the second battle of Al Fallujah in November 2004. But even when civilians were not being forced to stand between opposing sides in a firefight, they often found themselves too near the fighting to be clearly identified as civilian noncombatants. When a U.S. patrol or convoy is ambushed in an urban area, for example, every head that pops up in the windows or on the rooftops in the general direction from which fire is received has to be considered an armed adversary. As U.S. casualties mount during the course of an operation, it becomes increasingly understandable, if not imperative, for our troops to shoot first and ask questions later. But such impossible choices placed increasing cognitive stress burdens on our soldiers and Marines.

Loyalty Conflicts. Most soldiers and Marines deployed to operational environments have families of some sort back home, including parents, siblings, wives, husbands, and children. Though the urgencies of combat and operational military deployment eclipse most problems that might be brewing on the home front, it is difficult for some warriors to resolve the conflicts they experience due to divided loyalties between brothers-in-arms and others back home. Such conflicts are most stressful, of course, when a serious problem arises with family or friends in the United States, such as an illness, injury, or death. Deaths or injuries of family and close friends back home also greatly sensitize warriors to their own vulnerability on the battlefield.

Boredom and Monotony. Much of the day-to-day work of modern warfare is neither dramatic nor exciting, but rather tedious and boring. Soldiers and Marines stand guard in observation and guard posts and watchtowers for hours and days on end, scanning for threats and waiting for something to happen. Gunners and support personnel ride in convoys, mile after mile, looking for signs of roadside IEDs or, the greatest threat, explosives packed into a suicide bomber's unmarked vehicle—a suicide vehicle–borne improvised explosive device (SVBIED). Less tedious are foot patrols and "cordon and knock" sweeps through urban areas, searching for weapons caches and Mujahadeen sympathizers. But even patrols and sweeps, performed over the same areas over and over again, become monotonous. And everywhere in an operational environment, every day is much the same, without any chance for a day off or a vacation. Workdays are 24 hours, day after day, without predictable times for "liberty" or recreation. When recreational time is available, opportunities for novel and stimulating activities can be painfully limited. During operational deployment, seemingly endless repetition can be mind-numbing and (as for Bill Murray's character in the 1993 movie *Groundhog Day*) both disorienting and demoralizing.

Experiences That Don't Make Sense. Many experiences in an operational theater challenge warriors' belief systems, sometimes defying their attempts to find meaning and sense in them. Younger troops, in particular, deploy to a war zone for the first time holding on to somewhat unrealistic beliefs about "good" and "evil," and about their own mortality and personal importance. To regain their cognitive equilibrium after disrupting experiences, warriors must modify their belief systems to incorporate all the new facts and perceptions that do not fit in so well with their previous beliefs. Even when successful, this process of mental growth can be painful and confusing. Everyone is changed by belief-challenging experiences, but fortunately, not always in negative ways. The experiences of an operational deployment can catalyze rapid maturation in a young man or woman. It can force previously insecure, self-doubting individuals to be more self-confident after their successes in very trying circumstances. But too often, warriors have belief-challenging experiences in war that, try as they might, they cannot find any way to make sense out of and come to terms with.

Emotional Stressors

Losses of Friends to Death or Injury. The bonds that develop between comrades in an operational environment, especially in combat, are unlike any others in human experience. The degree of intimacy, trust, and life-and-death responsibility that warriors feel for each other while fighting and surviving a war together are unparalleled. As Jonathan Shay (1994) described so beautifully in *Achilles in Vietnam,* the bond between comrades-in-arms is closer in nature and intensity to that between a mother and child than that found in any mere friendship. Hence, the emotional impact of losing a close comrade in war is not unlike that experienced by a mother who loses a child, or vice versa. Shock, disbelief, guilt, shame, anger, and longing may be much the same for both. Unlike a grieving parent, however, a warrior who has lost a buddy in a war zone has little opportunity to really experience the normal but intense emotions that attend the loss, or to do the cognitive work necessary to make sense out of it and accept it. While deployed and still subjected to the very same dangers that just took the life of a close comrade, warriors cannot allow themselves to grieve. They must, instead, remain at least partially numb to their losses. They must store up their grief in emotional backpacks, to be neither put down nor emptied until the war is over and it has become safe to work through the losses.

Fear. When asked what they fear most during war, soldiers and Marines often give the easy answers first—they fear death, maiming, and losing friends. Nothing could be more obvious. The dangers to warriors' physical safety are constant in war, especially in counterinsurgency conflicts which lack battle lines behind which anyone can find safety. Fear of death and injury lies along a spectrum, ranging from the gnawing anticipation and dread of preparing to deploy or engage in combat action to the terror that follows being severely injured or nearly killed. But if death were truly the greatest fear of everyone in war, it would be impossible for military leaders to motivate their troops to stand and fight in spite of grave dangers to their personal safety. And warriors would never run *toward* danger instead of away from it. There would be no heroism. It takes only a few moments of reflection for warriors to acknowledge that their greatest fear is not death but failure, and the shame that accompanies failure. More than anything else, warriors fear letting themselves down and letting their leaders and friends down at a moment when it matters most. They fear most not losing their lives, but their honor. Fear of failure is ubiquitous and continuous, before, during, and after an operational deployment.

Shame and Guilt. Warriors suffer feelings of shame whenever they believe they have failed themselves or their comrades in some important way. But they sometimes suffer intense feelings of guilt, paradoxically, when they have *succeeded* in their jobs as warriors. Feelings of guilt can be experienced by combatants who have survived when their friends or leaders have not, or even for having killed others in battle. Survival guilt is common and "understandable" to those who suffer it. Warriors usually have no trouble recognizing their feelings of guilt for still being alive when others close to them were killed in combat. "It should have been me" are words spoken by many, if only to themselves. That is not to say that warriors can easily get over their survival guilt, but at least they can usually acknowledge it. Guilt over having killed others, especially guilt for killing unarmed men or women and children, may be more difficult for warriors to recognize and acknowledge. This may be because such acts also provoke feelings of shame that are very painful to admit into consciousness. Sometimes, the existence of feelings of intense guilt may be only apparent in warriors' nightmares or daytime fantasies, including self-destructive fantasies. Unconscious shame over perceived failure may be insidious and relentless, often banished from awareness by an unspoken pact among comrades-in-arms to avoid even speaking about shame-evoking events.

Helplessness. We all need to believe we are in active control of ourselves and our environments at all times. Our belief in our mastery over our situations makes them seem less chaotic and dangerous than they would otherwise be. The greater the chaos and danger we experience, perhaps, the greater our need to feel in control. So it is not surprising that warriors in operational environments avoid being placed in passive and helpless positions whenever possible. They prefer offense over defense, activity over passive waiting. Yet several features of modern asymmetrical warfare breed helplessness and passivity among U.S. and coalition troops. Since our adversaries in counterinsurgency operations neither wear uniforms nor set themselves apart from the rest of the population in any way, it is enormously difficult to mount sustained offensive operations against them. Usually, the only way to identify our adversaries is to wait for them to attack us in our convoys, patrols, and bases. Without clear battle lines or zones of control in an operational theater, no place is safe from at least intermittent and unpredictable indirect fire with mortars and rockets. Those who fire on our bases often hurry back to their homes long before being caught or identified. Ironically, the stress symptoms warriors can develop after exposure to intense or unrelenting stressors can, themselves, add to feelings of helplessness. To the extent stress-injured warriors' thoughts and feelings are not under their own control, as symptoms of their stress injuries, they often feel even more out of control and endangered.

The Horror of Carnage. We all also need to believe we are, more or less, secure and invulnerable (Janoff-Bulman, 1992). To some extent, the need to believe we are physically secure may spring from the impossibility of embracing our own mortality. Certainly, the experiences of life chip away relentlessly at our belief in our own invulnerability. But the fragility of life and the vulnerability of the human organism are lessons that cannot be learned too quickly without sustaining a stress injury. One of the most potently toxic experiences in war is the witnessing of human carnage. Seeing other people maimed, dismembered, or "turned into a pink mist" by a direct hit can be a highly traumatic experience, particularly when such carnage involves someone close, such as a friend or valued leader. The greater the identification with the damaged person, the greater the threat posed to one's own sense of security and invulnerability. The horror of gruesome scenes of carnage is one of the stress burdens placed most on ground combatants such as infantrymen. Because foot soldiers kill close up, they often cannot avoid seeing the bodies mutilated by their weapons. And worse, infantrymen are often right beside their comrades-in-arms when they are killed in battle. But no one deployed in a counterinsurgency operation is protected from witnessing carnage, since no base, vehicle, or city street, anywhere in theater, is secure from attack.

Killing. In his groundbreaking books, *On Killing* (1995) and *On Combat* (2004), Dave Grossman argued that the very act of killing another human being can be a significant, if not traumatic, stressor for many warfighters. Grossman suggests that all humans may have an instinctive aversion to killing members of their own species, an aversion that must be overcome on the battlefield in order to commit acts of interpersonal violence. McNair (2005) advanced the concept of perpetration-induced posttraumatic stress disorder as an entity distinct from stress disorders observed in the victims of natural or man-made trauma.

Social Stressors

Isolation from Social Supports. People vary greatly in how much they depend on others for emotional support, and deployed warriors vary greatly in how adapted they are to being away from their families, friends, and other loved ones. Certainly, the more deployment experience individuals have had, the more accustomed they are to being away from home, and the more closely they usually are bonded with their comrades. Younger, less experienced troops are often more vulnerable to experiences of loneliness and longing because of being away from home. Older, more experienced troops are often more comfortable with family separations, though sometimes only at great cost to their relationships. Also, individuals vary in their abilities to use telephones, e-mail, letters, and other means to stay connected with their loved ones. Some people just cannot communicate well on the phone or in writing. But all service members deployed overseas leave behind parts of themselves in their social units back home.

Lack of Privacy or Personal Space. Ironically, at the same time deployed warriors may be experiencing intense loneliness and social isolation, they are also almost always surrounded by a large number of comrades from whom they cannot possibly get away. This is largely positive, of course, since being closely surrounded by trusted peers and leaders is the best antidote to fear in war. Like mutts in a litter, most warriors in theater are seldom more than a few feet away from others whom they know more intimately than they know their own family members. But it also means an almost total absence of privacy and the need to share almost all spaces and equipment. Often the only items that can be considered as belonging to each person in theater, individually, are uniforms and weapons. Everything else is communal, to some extent.

The Media and Public Opinion. No one fully knows what motivates warriors to volunteer for military service and to willingly fight, suffer, and sacrifice in war. But love of country must certainly be one of their strongest motivations. Honor, pride, and patriotism are all highly valued by members of the military, especially at times when the greatest sacrifices are required of them. But how can warriors know whether their country values them and their efforts, in return? The spokespeople for the nation, in the minds of many warriors, are their own military leaders, the media, and public opinion, as perceived by them. Each of these three entities wields enormous power to validate or invalidate the sacrifices of warriors in an operational theater, as veterans of the Vietnam War and their families learned so painfully during the 1960s and 1970s. Fortunately, the people of the United States and its news media are far more supportive of those who serve in Iraq and Afghanistan in 2006 than those who served in Vietnam in 1970. But every criticism leveled at either the current war or the way it is being fought, whether publicly or privately, inflicts emotional wounds on the warriors who face death every day and have lost the closest friends they will ever have. Warriors who are unfortunate enough to see their own names or photographs in news stories critical of the war, especially while still deployed, suffer a great deal.

Spiritual Stressors

Loss of Faith in God. One belief that can be severely challenged for some warriors by the chaos and senselessness of war is their belief in God. Some find it difficult to continue believing in a benevolent, loving God after surviving the losses of operational deployment. Others cannot find a way to forgive their God for allowing the evils of war to exist. But others find, in the spiritual crucible of war, renewed and greatly deepened faith and religious conviction. Military chaplains returning from a tour in Iraq uniformly describe such experiences among their soldiers, Marines, and sailors. Lieutenant Carey H. Cash (2004), the Navy chaplain who accompanied the 1st Battalion, 5th Marines, during their fight toward Baghdad,

Iraq in 2003, wrote a moving account of such faith under fire in *A Table in the Presence.*

Inability to Forgive or Feel Forgiven. Some warriors return from an operational deployment feeling significantly disappointed and let down—by others whom they trusted, and by themselves. Awful things happen in war; they are often unavoidable. And even the bravest and strongest can be pushed to the point of acting in ways that later may be deeply regretted. Finding a way to forgive oneself, and others, for the weaknesses and failings war brings into focus can be a significant challenge.

CONCLUSIONS

The stressors generated by combat and operational deployment are many and varied. Compared to civilian life, military stressors can be uniquely powerful and uniquely unrelenting. All of them can contribute to stress problems before, during, and after deployment, including stress injuries. Appreciating the spectrum of stressors that warriors must endure in an operational theater is a necessary first step to helping them cope and heal.

But appreciating the stressors themselves is not enough. Helping professionals in the military, such as chaplains, medical personnel, and mental health professionals, must also understand the unique military culture in which operational stressors are experienced and endured, and in which stress symptoms are expressed. The set of beliefs and attitudes shared by personnel in military units forms the context of combat and operational stress. These cultural beliefs sometimes make it difficult for warriors to acknowledge their own stress, or to fully engage in a partnership with helping professionals to reduce operational stressors. A full conscious awareness of combat and operational stress might even detract from warriors' ability to perform their missions and get home safely. Doctors, medics, chaplains, and mental health professionals deployed with operational forces must always weigh the potential harm they may do any time they challenge warfighters' denial of their own stress and vulnerabilities. It must never be forgotten that military commanders bear full responsible for preventing stress behaviors from impacting on operational readiness, and for identifying and managing negative stress reactions when they arise. At most, medical and pastoral care professionals serve as educators, advisors, and adjuncts in the process of controlling combat stress. But helping professionals can never take responsibility for combat and operational stress control among military troops.

Empirical studies are urgently needed to broaden our understanding of the stressors of combat and operational deployment, as well as the military cultural contexts within which operational stress is experienced.

REFERENCES

Belenky, G. L. (1986). Military psychiatry in the Israeli Defense Force. In R. A. Gabriel (Ed.), *Military Psychiatry: A Comparative Perspective.* New York: Greenwood Press.

Bourne, P. G. (Ed.). (1969). *The Psychology and Physiology of Stress: With Reference to Special Studies of the Viet Nam War.* New York: Academic Press.

Cash, C. H. (2004). *A Table in the Presence.* Nashville, TN: W. Publishing Group.

Center for Army Leadership. (2004). *The U.S. Army Leadership Field Manual: Be, Know, Do (FM 22-100).* New York: McGraw-Hill.

Chairman of the Joint Chiefs of Staff. (2003). *Doctrine for Joint Psychological Operations (JP 3-53).* Washington, DC: U.S. Government Printing Office.

Clausewitz, C. von. (1982). *On War* (A. Rapoport, Ed., J. J. Graham, Trans.). New York: Penguin Classics. (Original published in 1832)

Copp, T., & McAndrew, B. (1990). *Battle Exhaustion: Soldiers and Psychiatrists in the Canadian Army, 1939–1945.* Montreal, Quebec: McGill-Queen's University Press.

Dean, E. T. (1997). *Shook Over Hell: Post-Traumatic Stress, Vietnam, and the Civil War.* Cambridge, MA: Harvard University Press.

Department of Defense. (2006). *Operation Iraqi Freedom (OIF) U.S. Casualty Status as of: June 23, 2006.* Retrieved June 25, 2006 from http://www.defenselink.mil/news/casualty.pdf

Department of the Army. (2003). *U.S. Army Combat Stress Control Handbook.* Guilford, CT: Lyons Press.

Figley, C. R. (Ed.). (1978). *Stress Disorders Among Vietnam Veterans.* New York: Brunner-Routledge.

Frankl, V. E. (1984). *Man's Search for Meaning* (I. Lasch, Trans.). New York: Simon & Schuster. (Original work published 1946)

Gilmore, D. (1991). *Manhood in the Making: Cultural Concepts of Masculinity.* New Haven, CT: Yale University Press.

Goldstein, J. S. (2001). *War and Gender: How Gender Shapes the War System and Vice Versa.* Cambridge, U.K.: Cambridge University Press.

Grinker, R. R., & Spiegel, J. P. (1945). *Men Under Stress.* Philadelphia: Blakiston.

Grossman, D. (1995). *On Killing: The Psychological Cost of Learning to Kill in War and Society.* Boston: Little, Brown.

Grossman, D. (2004). *On Combat: The Psychology and Physiology of Deadly Conflict in War and Peace.* St. Louis, MO: PPCT Research Publications.

Hendin, H., & Haas, A. P. (1984). *Wounds of War: The Psychological Aftermath of Combat in Vietnam.* New York: Basic Books.

Hoge, C. W., Castro, C. A., Messer, S. C., McGurk, D., Cotting, D. I., & Koffman, R. L. (2004). Combat duty in Iraq and Afghanistan, mental health problems, and barriers to care. *New England Journal of Medicine, 351,* 1798–1800.

Janoff-Bulman, R. (1992). *Shattered Assumptions: Towards a New Psychology of Trauma.* New York: The Free Press.

Jones, F. D. (1995a). Disorders of frustration and loneliness. In F. D. Jones, L. R. Sparacino, V. L. Wilcox, J. M. Rothberg, & J. W. Stokes (Eds.), *War Psychiatry.* Washington, DC: Borden Institute.

Jones, F. D. (1995b). Traditional warfare combat stress casualties. In F. D. Jones, L. R. Sparacino, V. L. Wilcox, J. M. Rothberg, & J. W. Stokes (Eds.), *War Psychiatry.* Washington, DC: Borden Institute.

Kulka, R. A., Schlenger, W. E., Fairbank, J. A., Hough, R. L., Jordan, B. K., Marmar, C. R., et al. (1990). *Trauma and the Vietnam War Generation: Report of Findings from the National Vietnam Veterans Readjustment Study.* New York: Brunner/Mazel.

McNair, R.M. (2005). *Perpetration-Induced Traumatic Stress: The Psychological Consequences of Killing.* New York: Authors Choice Press.

Moran, C. M. W. (1967). *The Anatomy of Courage.* Boston: Houghton Mifflin.

Schneider, R. (1986). Military psychiatry in the German army. In R. A. Gabriel (Ed.), *Military Psychiatry: A Comparative Perspective.* New York: Greenwood Press.

Shay, J. (1994). *Achilles in Vietnam: Combat Trauma and the Undoing of Character.* New York: Scribner.

Shay, J. (2002). *Odysseus in America: Combat Trauma and the Trials of Homecoming.* New York: Scribner.

Shephard, B. (2001). *A War of Nerves: Soldiers and Psychiatrists in the Twentieth Century.* Cambridge, MA: Harvard University Press.

Solomon, Z. (1993). *Combat Stress Reaction: The Enduring Toll of War.* New York: Plenum Press.

U.S. Marine Corps. (1995). *Leading Marines (MCWP 6-11).* Washington, DC: U.S. Government Printing Office.

U.S. Marine Corps. (1997). *Warfighting (MCDP 1).* Washington, DC: U.S. Government Printing Office.

U.S. Marine Corps. (2000). *Combat Stress (MCRP 6-11C).* Washington, DC: U.S. Government Printing Office.

Van Dongen, H. P. A., Maislin, G., Mullington, J. M., & Dinges, D. F. (2003). The cumulative cost of additional wakefulness: Dose-response effects on neurobehavioral functions and sleep physiology from chronic sleep restriction and total sleep deprivation. *Sleep, 2,* 117–126.

3

Combat/Operational Stress Adaptations and Injuries

WILLIAM P. NASH

You dance with the devil, you don't change him—the devil changes you.

Max California in the film 8 mm

Men and women who participate in combat or who deploy to military operations in support of combat have always been affected by these experiences. Persistent reactions to combat and operational stress are clearly identifiable in the literature of antiquity (Shay, 1994, 2002), and military surgeons have described characteristic stress reactions since at least the 18th century (Jones, 1995b). The specific reactions experienced by warriors have changed, somewhat, from generation to generation and war to war (Jones, 1995a), but a lot has not changed over time. Terror is still terror. Grief is still grief. Courage, honor, and self-sacrifice—the most venerable of reactions to stress—still play the role they always have in military operations. And the core features of the major adverse operational stress reactions are much the same now as they were in the American Civil War (Dean, 1997) and the many wars of the 20th century (Shephard, 2001; Solomon, 1993).

So why, then, have so many different labels been used over the past three centuries to describe the adverse reactions of warriors to combat and operational stress? And why now, in the 21st century, are we still not sure what to call stress reactions on the battlefield? Although it is not the goal of this chapter to answer these questions, it will nevertheless be useful to do a little reconnaissance of them before developing a framework for understanding and classifying combat and operational stress reactions.

PROBLEMS WITH LABELING COMBAT/OPERATIONAL STRESS REACTIONS

Two factors have shaped attempts in modern history to name and classify combat/operational stress reactions: (1) ethical dilemmas surrounding the labeling of wartime stress reactions because of the sometimes profound effects such labels can have on individual combatants as well as on the military units in which they serve; and (2) shifting and sometimes reductionistic theories about the nature and causes of mental and behavioral problems of all types, including operational stress reactions.

The Ethics of Labeling Combat/Operational Stress Reactions

Labels, especially psychiatric medical diagnoses, can have profound consequences, both for good and ill (Reich, 1991). Classifying a pattern of behavior or inner experience as a medical diagnosis can reassure troubled individuals that they are not alone and that their behavior and experience make sense. Diagnostic labels can also offer exculpation, or at least mitigation, to the extent they imply that individuals are afflicted with something outside of their control. Prior to the Age of Enlightenment in the 18th century, deviant behavior was often ascribed to demonic possession, sin, or fate (Porter, 2002), and the only label in wide use in the Middle Ages to describe the vice of failing to perform one's duties on a battlefield was "coward." Since both demonic possession and cowardice have historically been punishable by death, untold lives have been saved in recent centuries just by giving combat stress reactions a medical label. Medical diagnoses, to the extent they are based on standardized criteria, also allow scientific study of syndromes and disorders, including research on how they may be prevented and treated.

Diagnostic labels can also harm individual warriors and the military units in which they serve. Psychiatric labels of all kinds carry a heavy burden of stigma, particularly among warfighters whose profession requires them to remain calm, focused, and in control regardless of adversity. Psychiatric labels imply, for them, not only weakness but a failure in their core to live up to the warrior ideal. Like weapons found on close inspection to have defective components, psychiatrically labeled warfighters can lose the trust of their superiors and peers, and their own trust in themselves. Many modern warriors would rather be diagnosed with cancer than with depression, anxiety, or—worst of all—posttraumatic stress disorder. To the extent they explain and absolve stress symptoms on a battlefield, labels of any kind can also seem to give permission to individuals to give up trying to master their own stress symptoms. The popularization by the European media of the "shell shock" diagnosis during the First World War, for example, doubtless contributed to the droves of combat stress casualties sent home from the trenches of France (Moran, 1967/1945; Shephard, 2001). Military leaders strive to keep their troops on the frontlines focused outwardly, on their mission, instead of inwardly, on their stress symptoms. Labeling combat/operational stress reactions makes the job of military leaders more difficult because it invites introspection and because it blurs the line between what is a disciplinary problem and what is a health problem.

Given the problems associated with medicalizing and pathologizing operational stress problems, both for individual combatants and military leaders, a shift occurred during the 20th century toward normalizing stress reactions of all types—to see them as "a normal adaptive process of reaction to an abnormal situation" (Lifton, 1988, p. 9). Since WWI, one of the abiding principles of war psychiatry has been "expectancy"—a continuous attitude toward stress casualties that they are not ill or sick, but will soon recover and return to full duty (Wessely, 2005). Stress casualties have been kept separate from the physically wounded, and if given any label at all, they have been classified as having something benign like "battle fatigue," "exhaustion," or "combat stress reaction." The avoidance of labeling and a focus on normalization have also long been central to civilian crisis management efforts.

Normalization has proven effective at encouraging warriors to recover from their stress reactions and return to duty (Kormos, 1978; Wessely, 2005), but perhaps at the price of discouraging acknowledgment and treatment of stress reactions when they occur. After WWII, when as many as 10% of all combatants in heavily engaged armies were treated at some point for adverse stress reactions, the rates of diagnosed battlefield stress casualties declined to 3.7% in the Korean War and 1.2% in Vietnam (Bourne, 1969). Although stress casualty rates have not been published for United States troops deployed to Iraq, it is likely that no more than 2% of all soldiers and Marines deployed to Iraq have been diagnosed with a stress disorder in theater. There are many positive reasons for declining rates of diagnosed battlefield stress casualties. Much shorter operational tour lengths, all-volunteer forces, and advances in training and leadership have all contributed to the resiliency of the military. However, to the extent stress reactions in combat have been redefined as merely "normal," they have also become progressively less likely to receive any attention from anyone other than a buddy or small unit leader. Low rates of battlefield stress casualties do not necessarily predict low rates of eventually diagnosed stress problems. The mental health problems experienced by Vietnam veterans after their war ended attest to the gap between identified battlefield stress casualties in Vietnam and the true extent of operational stress reactions generated in that conflict. Hoge et al. (2004) found 17% of heavily engaged infantrymen to self-report significant stress symptoms 3 to 6 months after returning from Afghanistan or Iraq. But stigma and fear of negative career consequences prevented many of them from seeking care.

Having been told their stress symptoms are merely "normal," how can warriors with persistent stress symptoms ever admit to themselves or anyone else they need help? As Shalev (1996) pointed out, the "normal response" hypothesis implies that recovery from stress reactions should always be possible. How can warriors not blame themselves if they find themselves in the minority who fail to recover?

On one side of this ethical dilemma lies the danger of crippling normal warriors and depleting their ranks by pathologizing commonplace reactions to everyday military operations. On the other side is the danger of trivializing the moral, psychological, and biological damage that can result from severe and persistent stress, thereby discouraging the wounded from seeking care. Although this conundrum may be no better solved today than it was a century ago, we can at least try to keep both Scylla and Charybdis in full view as we navigate the ethical strait between them.

Shifting Theories About the Causes of Operational Stress Casualties

Since the 18th century, when the Age of Enlightenment first encouraged reason and empirical observation to replace superstition and irrationality in all human endeavors, a parade of labels and mutually exclusive theories have been used to describe and explain operational stress reactions. The first recorded label for a combat stress reaction was "nostalgia," which literally means homesickness, but which was often significantly more disabling than a mere longing for home. For example, the Austrian internist Josef Leopold Auenbrugger described nostalgia in 1761 (as cited in Jones, 1995a, p. 6) as follows:

> When young men who are still growing are forced to enter military service and thus lose all hope of returning safe and sound to their beloved homeland, they become sad, taciturn, listless, solitary, musing, full of sighs and moans. Finally, they cease to pay attention and become indifferent to everything which the maintenance of life requires of them.

The causes of nostalgia were believed to be largely psychological and social, including prolonged separation from home and family and loss of hope of ever getting back home (Jones, 1995a). During the American Civil War, the most common label for combat stress reactions continued to be nostalgia (Dean, 1997), but many other diagnoses were used, some reflecting a growing belief that adverse stress reactions could be caused by actual physical damage to the brain and body. Other operational stress diagnoses during the Civil War included insanity, sunstroke, and "irritable heart" or "trotting heart" (Dean, 1997), the latter two diagnoses referring to the paroxysms of rapid heart rate at rest that often accompanied what we now call panic attacks. This shift in theory and labeling in the 19th century followed the medical discoveries that mental illness could be caused by physical damage to the brain, such as by an infection with syphilis or heavy alcohol use.

These two divergent views of causation—psychological versus biological—found champions at the end of the 19th century in two physicians who studied mental trauma in civilians, Sigmund Freud and Pierre Janet. Both Freud and Janet described the phenomenon of dissociation—currently defined as a disruption in the usually integrated functions of consciousness, memory, identity, or perception (American Psychiatric Association, 2000)—in the immediate aftermath of a traumatic event (Breuer & Freud, 1957/1895; Janet, 1920). And both believed that dissociation was a key element in the development of psychopathology after a traumatic experience (Nemiah, 1998). But while Freud saw the fragmentation of consciousness in dissociation as a self-protective defense mechanism intended to keep overwhelmingly disturbing perceptions or feelings out of consciousness, Janet believed dissociation was due to an innate failure to integrate information in the brain under the impact of a "vehement emotion" (van der Kolk, Weisaeth, & van der Hart, 1996). In Freud's view, dissociation at the moment of trauma was a "deliberate and intentional" *choice* (Breuer & Freud, 1957/1895, p. 123), albeit an unconscious one. In Janet's view, on the other hand, dissociation was a symptom of a *breakdown* of brain function, a loss of adaptation (van der Kolk & van der Hart, 1989). This difference in theory of causation makes all the difference in treatment and prognosis. For Freud, conscious recall of repressed traumatic memories was curative; for Janet, attempts to recall traumatic memories before they were somehow detoxified would only again overwhelm the brain's integrative capacity and cause further breakdown (Nemiah, 1998).

The succession of labels used to describe operational stress reactions in the 20th century can be understood partly as an ongoing debate between those who believed such reactions were psychological in origin, and those who believed they had primarily biological causes (Shephard, 2001). "Shell shock" in the First World War conveyed the belief, at the time, that the varied and sometimes bizarre symptoms observed in the trenches of France were caused by physical damage to the brain from proximity to the explosion of artillery shells. Every attempt to find evidence of physical damage to the brain in shell shock cases failed, however, which steered theories of causation away from the biological and toward the psychological. Although the diagnosis of "neurasthenia"—which refers literally to an exhaustion of the nervous system—was used in both world wars, the purely psychological labels "traumatic neurosis" and "war neurosis" gained prominence in WWII. "Neurosis" was a concept which grew out of the Freudian psychoanalytic movement in the early 20th century, defined as symptoms produced by "emergency discharges" of psychic energy dammed up by unconscious conflict (Fenichel, 1945, p. 20; Nash and Baker, this volume, chapter 4). Shell shock and neurasthenia were considered "hardware" problems; war neurosis was thought to be a "software" problem. By the end of WWII, the most commonly used labels were "battle fatigue" and "exhaustion," both reflecting a psychological rather than a biological etiology. Citing war psychiatry experience in both world wars, Kormos declared in 1978, "fortunately, it is a relatively settled matter. All sources appear nowadays to be in agreement that we are dealing with a functional entity" (Kormos, 1978, p. 12).

After Vietnam, an explosion in research on persistent war-related stress disorders led to the official recognition in 1980 of posttraumatic stress disorder (PTSD) (American Psychiatric Association, 1980). At the same time, American psychiatry embraced the "biopsychosocial model," an integrative theoretical orientation based on the premise that all mental and behavioral problems have simultaneous causes in the biological, psychological, and social spheres (Engel, 1980). Since then, PTSD has become a paradigm of a true biopsychosocial disorder, with well-documented physical, mental, and interpersonal components (see chapter 4, this volume). But partly because of continued efforts to keep combat stress reactions demedicalized and distinct from mental disorders like PTSD, etiological theories regarding battlefield stress casualties have not kept pace. In his chapter on traditional warfare combat stress casualties in the U.S. Army's current *War Psychiatry,* Franklin Jones (1995b, p. 37) wrote:

> It is important to remember that most psychiatric casualties are soldiers who, because of the influence of negative psychological, social, and physiological factors, unconsciously seek a medical exit from combat.

This view of combat stress reaction as a choice rather than an affliction is still widely held. In what follows, reductionistic views of causation will be challenged, and an alternate system of description and classification will be offered. The central premise of this chapter is that although many reactions to the stress of war are adaptive choices, the worst reactions are truly injuries. And all combat/operational stress reactions have biological, psychological, and social components.

CONCEPTUAL FOUNDATION: DEFINING STRESS AND ADAPTATION

Over the past century, many definitions for the word "stress" have been offered, but none has encompassed all the usages of the term even in the scientific community (Lazarus & Folkman, 1984). Perhaps one reason for this is that stress is not a unitary concept, but rather a collection of many interacting variables and processes—in the body, in the mind, and in relation to the outside world. More than that, stress may best be understood as a *transaction* between each individual's unique biology and his or her environment, mediated by a multitude of psychological and social processes (Aldwin, 1994). In the course of adapting to stress, genes and chemical processes affect and are affected by conscious coping choices, personality styles, and interpersonal relationships.

Stress and Adaptation as Biological Processes

The modern study of stress began with the work in the 1930s of the Hungarian endocrinologist, Hans Selye, who discovered the mammalian biological stress response almost by accident. While attempting to isolate a new sex hormone by repeatedly injecting rats with ground-up extracts of rat placentas, Selye (1956) was excited to find that these injections of ground-up placenta provoked a consistent pattern of physiological response—hypertrophy of the adrenal glands, stomach ulcers, and atrophy of the thymus gland (which is involved in the immune response). To be sure that these physiological changes were really due to a new endocrine hormone contained in the extract, and not just nonspecific damage to the animals from the ground-up tissue he had injected, Selye then injected rats with a weak solution of formaldehyde, a chemical fixative that destroys living tissue. To his dismay, rats injected with formaldehyde also developed the same triad of physiological changes to their adrenal glands, stomach linings, and thymus glands. As a physician, Selye was also aware that human patients demonstrate similar physiological changes after suffering from chronic illnesses of many kinds. Perhaps, he reasoned, the triad of physiological changes he discovered represented the generic response of mammalian biology to *any* environmental demand or noxious agent.

Subsequent experiments by Selye and others confirmed this hypothesis. It has been found that the same hormonal and immunological changes occur in the bodies of laboratory mammals after being subjected to a wide variety of physical, mental, and social challenges, including cold, prolonged exposure to predators, forceful immobilization, overcrowding, and infection, among others (Selye, 1950). And humans have been found to have almost identical physiological reactions to life stressors of various kinds. Based on these findings, Selye (1956) came to the conclusion that "stress is the nonspecific response of the body to any demand" (p. 74). Stress, in Selye's view, was a biological process in response to any challenge, external or internal. To differentiate the process of stress as a reaction to a challenge from the challenge, itself, Selye coined the term "stressor" for the agent that provoked the stress response. The stressor is the challenge, and stress is the process by which the organism adapts to the stressor. Selye called the predictable pattern of biological response to stressors of all kinds the general adaptation syndrome (GAS).

Time Course of Adaptation

Having discovered the biological stress response—GAS—Selye went on to study how it evolved and changed over time. He found that GAS was a process that (a) took time to develop in response to a stressor, (b) consumed energy, and (c) could not, in most cases, be sustained indefinitely. A simple example will illustrate these three characteristics of GAS. Imagine someone dropping a moderate weight into the palm of your outstretched hand after asking you to hold your arm as steady and horizontal as possible. Initially, your arm would dip under the impact of the weight, and it would take a few moments for your nervous system to recruit additional muscle fibers to the task of trying to keep your arm horizontal. After this initial adjustment period, your arm and shoulder would steady and settle into the work of resisting the pull of gravity on the weight. But eventually, resources would be used up by your contracted muscle fibers, lactic acid would build up, and an increasingly painful exhaustion would force you to drop your arm.

Accordingly, Selye divided the time course of GAS into three phases, as illustrated in Figure 3.1. In the first phase, the *alarm* phase, the organism mobilizes its resources to respond to the challenge it faces. Initially, its performance worsens under the impact of the stressor. Then, as adaptive changes take place, the organism's performance improves and it develops a phase of *resistance* to the negative effects of that particular stressor. The final phase is fatigue or *exhaustion*, during which adaptive resistance to a stressor is lost and a period of recovery may be necessary before an adaptive response can again be mounted to the same stressor. Depending on the nature of individual stressors and the biological systems that respond to them, the three phases of adaptation may be short or long, and the time dimension in GAS may represent adaptive changes that occur over

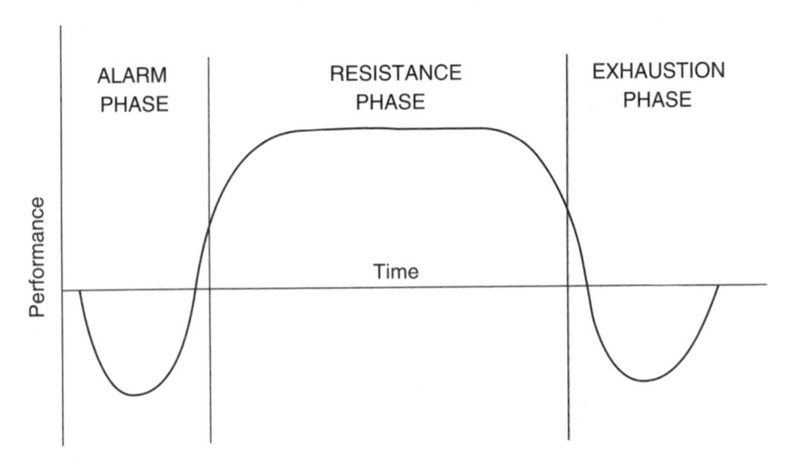

FIGURE 3.1 Three phases of adaptation to any stressor—the General Adaptation Syndrome.

minutes, days, or months. Whether one charted the response of the body to the stress of an all-out sprint during a firefight or the stress of a year-long operational deployment, the phases and their relationship would be much the same.

Selye noted that in most instances, adaptive responses do more than just return the functioning of organisms to their previous baseline. Physiological adaptation is more than just a homeostatic mechanism. In many cases, adaptation to a stressor results in *improved* functioning and performance in one way or another. In the example just given above, recruitment of muscle fibers would enhance strength for the moment. Over a longer time period, of course, repeated exposure to the stress of lifting the same fixed weight would build muscle mass and power. The fact that the performance of organisms or their component parts typically improves as they adapt to stressors led Selye to introduce the concept of "eustress," which he defined as the stress which is *necessary* for optimal functioning. Stress is necessary for the development of the body and the mind (Aldwin, 1994), and the benefits to warfighters of repetitive exposure to optimal stress underlie all training in the military.

Stress and Adaptation as Psychological Processes

The concepts of stress and coping have also been important organizing constructs in psychology over the past century (Lazarus & Folkman, 1984). In one guise or another, adaptation to stress has been a central theme in psychoanalysis, health psychology, behaviorism, and cognitive psychology. In all these theoretical systems, successful adaptation to the stressors of life has been declared to be crucial to mental health and well-being. However, compared to biological reactions to stress, adaptive responses in the cognitive, emotional, and behavioral spheres are much more diverse and variable, both between individuals and within each individual at different points in time. The responses of the body to stress are largely determined by fixed, inherited gene programs, while mental and behavioral responses to stress are the product more of learning and choice. As such, they are virtually unlimited in their variety and capacity to change over time.

Three components of psychological adaptation to stress are worthy of at least brief review: (1) reflexive and automatic responses to stress; (2) coping as conscious, volitional adaptation; and (3) appraisal as prelude to coping.

Reflexive and Automatic Responses to Stress. Reflexive behavioral responses to stress are those that lie at the end of the spectrum farthest from choice and free will because they are largely fixed and determined by gene programs. Examples include startle reactions to loud noises, and freezing, fleeing, or reflexly striking back in response to an external danger. Reflex behaviors operate entirely outside of conscious, voluntary control, as was noted by Charles Darwin (1915/1872) in his observations on emotional expression. In a demonstration of the independence of reflexive behaviors of willpower, Darwin attempted and failed to hold his face up against the glass of a terrarium containing a deadly viper as it repeatedly struck at him.

Less biologically determined responses to stress that seem, in their activation, almost as automatic as instinctive reflexes are what Lazarus (1999) and Lazarus and Folkman (1984) referred to as automatized responses. These are behavioral response patterns that have been learned through repeated practice to the point of no longer requiring conscious decision making. All complex skills, such as playing a musical instrument and driving a car, cannot be performed efficiently until they have been automatized through repeated practice. The goal of military training is to arm each warrior with a set of automatized response patterns that can be trusted to reliably occur in operational situations, regardless of the perceived danger or stress level. Automatized response patterns are, however, almost as fixed as biologically determined reflexive behaviors, since they cannot easily be modified without practice and retraining.

Another class of automatic psychological stress responses that are not learned through training and practice are the *defense mechanisms* that form part of the unconscious adaptive repertoire of every individual. Defenses are habitual but unconscious mechanisms of adaptation (Vaillant, 1977) that protect the individual against conscious awareness of painful thoughts, feelings, or perceptions. They range from the most mature and effective defenses such as altruism, suppression, and humor to the least mature defenses such as a splitting and projection. One particular defense mechanism that deserves special note in a discussion of adaptation to combat is *denial*. Denial can be defined as "the unconscious repudiation of some or all of the available meanings of an event to allay anxiety or other unpleasurable affects" (Dorpat, 1987, p. 24). Denial is not the avoidance of conscious awareness of the existence of an anxiety-producing stimulus; it is, rather, the avoidance of full awareness of the *meaning and significance* of that stimulus. Although once thought to be always pathological and maladaptive, denial has more recently been understood to be a necessary component of adaptation to severely stressful situations. To remain effective, warriors must remain aware of the existence of incoming fire at the same time they deny themselves a full appreciation of the danger posed by that fire (Grossman, 1995, 2004). Likewise, they must deny to themselves a full awareness of the effects on enemy combatants and civilian bystanders of their own fires.

"Coping" as Conscious, Effortful Adaptation. The responses to stress that are most characteristic of us as humans are those mediated by conscious decision making and effort. They are also the components of stress response that account for much of the variation in adaptive styles among individuals, including individual susceptibilities and vulnerabilities to stress. Furthermore, the conscious, volitional components of coping are

the ones that are the most amenable to modification through training, leadership, and esprit de corps in military units. For all these reasons, volitional coping deserves particular attention.

Lazarus and Folkman (1984) defined coping as "constantly changing cognitive and behavioral efforts to manage specific external and/or internal demands that are appraised as taxing or exceeding the resources of the person" (p. 141). The goals of coping, in their view, are not only mastery over the environment and problem solving, but also "managing emotions and maintaining self-esteem and a positive outlook, especially in the face of irremediable situations" (p. 139). The goal of coping is not merely to survive a severe stress, but to transcend it through courage, creativity, and growth. Effective coping not only manages suffering and adversity (Lazarus, 1999), but finds meaning in it (Frankl, 1984/1946).

Warfighters in a war zone can be incredibly creative in their development and use of coping strategies. Letter writing has long been an effective tool for deployed warriors to not only retain contact with loved ones back home, but to weave their experiences into coherent narratives in order to make sense out of them. Modern information technologies such as e-mail and instant messaging have raised the coping strategy of "letters from the front" to a new level of immediacy and impact. Digital cameras have also permitted warfighters to create photo journals of their experiences, sometimes even set to music selected to give the images the greatest meaning. Giving support to fellow combatants and receiving support from them continue to reign as the monarchs of battlefield coping strategies. Relationships forged in battle may be the most profound and honest of any that warfighters will ever have in their lives. Humor and play are everywhere in the war zone, even under the most dire of circumstances. And many deployed warfighters experience an epiphany of religious faith that can do much to neutralize the toxic effects of combat and operational stress. One of the most humanizing experiences possible in a war zone is the mere conscious awareness that, however much one may be buffeted by external factors outside of one's control, there are always still choices to be made. And these choices may not only save lives, but give meaning to otherwise chaotic experiences.

Appraisal as Prelude to Coping. The first step in the process of adaptation in humans is *appraisal* (Lazarus & Folkman, 1984). Sights, sounds, smells, and physical sensations do not have meaning for an individual until they are analyzed in their full context, including dangers posed and resources available to meet those dangers. On a forward operating base in Iraq, for example, the sound of nearby small arms fire would produce one response in individuals who appraised those sounds as coming from peers on a practice range, and another response in those who appraised them as coming from enemy forces attempting to breach the defenses of the base. It has long been a tenet of cognitive psychology that conscious appraisal is a crucial determinant not only of behavioral responses to stressors, but also of emotional responses to them (Lazarus, 1999). In their study of combat and operational stress among aircrews in World War II, Grinker and Spiegel (1945a) described the process of appraising and responding emotionally to threatened losses—whether the loss threatened was personal injury or death, harm to someone else who was loved, or failure to meet one's own expectations at a crucial moment.

> The emotional reaction aroused by a threat of such a loss is at first an undifferentiated combination of fear and anger, subjectively felt as increased tension, alertness, or awareness of danger. The whole organism is keyed up for trouble, a process whose physiological components have been well studied. Fear and anger are still undifferentiated, or at least mixed, as long as it is not known what action can be taken in the face of the threatened loss. If the loss can be averted, or the threat dealt with in active ways by being driven off or destroyed, aggressive activity accompanied by anger is called forth. This appraisal of the situation requires mental activity involving judgment, discrimination, and choice of activity, based largely on past experience. If on the basis of such mental activity it is seen that the loss cannot be averted, the situation is hopeless, and nothing can be done, then anxiety develops. (p. 122)

Individual differences in appraisal of stressful situations account for much of the difference in how individuals adapt to them (Lazarus & Folkman, 1984). This fact is important for understanding how to build resiliency in warfighters by modifying their appraisal of operational stressors through training, leadership, and unit cohesion. But it is also crucial for assessing individual risk for adverse stress reactions, since those most vulnerable will be those who appraise given situations as entailing the greatest personal loss.

Stress and Adaptation as Social Processes

The importance of social support in adaptation to extreme stress cannot be underestimated. Just as families, under ideal conditions, provide shelter, nurturance, and guidance for family members, relationships in cohesive military units are vital to the survival of each individual in them. Shared danger intensifies attachments, partly because each person's survival lies literally in the hands of his peers (Elder & Clipp, 1988). The resulting close social network buffers intense, negative emotions, and makes each dangerous encounter seem less threatening (Cohen, Gottlieb, & Underwood, 2000). As Boston psychiatrist Jonathan Shay (2002) so beautifully stated, the "human brain codes social recognition, support, and attachment as physical safety" (p. 210).

Effective military leaders can also promote adaptation in their subordinates to extreme stress, under ideal conditions. Grossman (1995) likened an effective military leader to a "well of fortitude" into which subordinates could repeatedly dip to restore their own flagging courage.

Of course, relationships can also have a negative impact on adaptation (Lazarus, 1999). Warfighters who are new to their units, such as replacements for combat losses, may have a particularly difficult time since they are initially excluded from the sustaining network of attachments in the unit. And to the extent warfighters depend on attachments in their units for their emotional survival, they are vulnerable to a catastrophic failure of adaptation if those attachments are abruptly lost (Elder & Clipp, 1988).

STRESS ADAPTATION SUMMED UP

Three additional aspects of stress adaptation may help integrate the above theory on the biological, psychological, and interpersonal components of adaptation, and make it easier to apply to real-life adaptive processes in a war zone. The first is a grouping of all adaptive processes into one of three tactical categories—*accommodate, neutralize,* or *disengage.* The second is another look at the time course of adaptation, this time with combat action and operational deployment in mind. The third is the reversibility of adaptive responses to stress.

Three Tactics of Adaptation: Accommodate, Neutralize, or Disengage

The point of all biological, psychological, and interpersonal adaptive processes—whether conscious or unconscious, voluntary or involuntary—is to restore lost homeostasis, to reduce alarm and anxiety, and to grow and develop through mastering challenges. However, moment to moment, it may be hard to see how specific adaptive tactics may lead to these strategic goals. Or, working backwards, it is hard to identify which thoughts, feelings, and behaviors are adaptive if one uses the strategic goals of adaptation as a yardstick. For this reason, it may be useful to conceive of all adaptive responses as falling into one of three tactical groups, the goals of which are much simpler. The three *tactics* of adaptation, moment to moment, are: (1) to change oneself to *accommodate* to the challenge faced, (2) to *neutralize* or eliminate the challenge, or (3) if neither of the first two tactics are possible, to *disengage* cognitively or emotionally from the source of the stress, in order to become numb to it (Nash, 1998). An example from civilian life may help illustrate these three tactics of adaptation.

Imagine a group of recreational runners deciding to enter their first-ever marathon race. Having never run 26.2 miles before, the runners would have to begin a program of tough training, running longer and longer miles at a faster and faster pace. The physical challenges of training would build power and endurance—force the runners' bodies to change to accommodate to the stress of distance running. But the training program would also promote adaptation in the runners' minds by building their self-confidence, focus, and will to endure. In addition to allowing themselves to develop and change in response to training, the runners would also seek ways to reduce the challenge of a marathon run—to neutralize the stress as much as possible. Of course, there is no way to shorten the course or make it flatter or downhill the whole way. The course is fixed. But runners can reduce the challenge of a marathon run in many smaller ways, such as by wearing optimal clothing, staying hydrated and nourished, and running in a pack to reduce wind resistance and the mental strain of running alone. No matter how hard runners may train, and regardless how clever they may be about reducing the impact of the challenges they face, however, it is impossible for them to make a 26.2-mile race easy. They will still suffer, both physically and mentally. To adapt to the challenge that is left after the runners have changed themselves as much as possible through training, and neutralized aspects of the challenge as much as possible through other actions, they simply have no choice but to make themselves numb to their own suffering—to disengage mentally from it.

In an operational theater, stressors come fast and furious, and they pile high and deep. Through training and experience, warfighters can accommodate to some of them—they can change themselves physically and mentally to be as suited as they can be to meet the challenges they face. And through the proper equipment, teamwork, and leadership, they can neutralize, or at least mitigate, a portion of the stress of war. But most of the danger, hardship, and ugliness of war cannot be removed by any amount of training or leadership. For the worst stressors of war, the only tactic available is controlled mental and emotional disengagement—to become as numb and unaware as necessary to endure and survive. Disengagement is partly a cognitive tactic involving denying the magnitude and significance of stressors. It is partly an emotional tactic involving dampening emotional reactions to stressors. And it is partly a physiological tactic, involving reducing the responsiveness of the nervous system through high levels of stress chemicals in the brain and body. Disengaging mentally from a severe challenge sufficiently to make it bearable while still maintaining focus and control is a skill that must be learned through repeated exposure to tough challenges. Toughness, a sine qua non for a warrior, is built on controlled and reversible mental disengagement from unavoidable stress.

The Stages of Adaptation: Dread, "in the Groove," and Rebound or Fatigue

Hans Selye's general adaptation syndrome, describing the time course of biological adaptation to stress, was discussed earlier and diagrammed in Figure 3.1. Selye knew his GAS applied to the nonbiological aspects of adaptation as well as the biological. Regardless of whether an adaptive

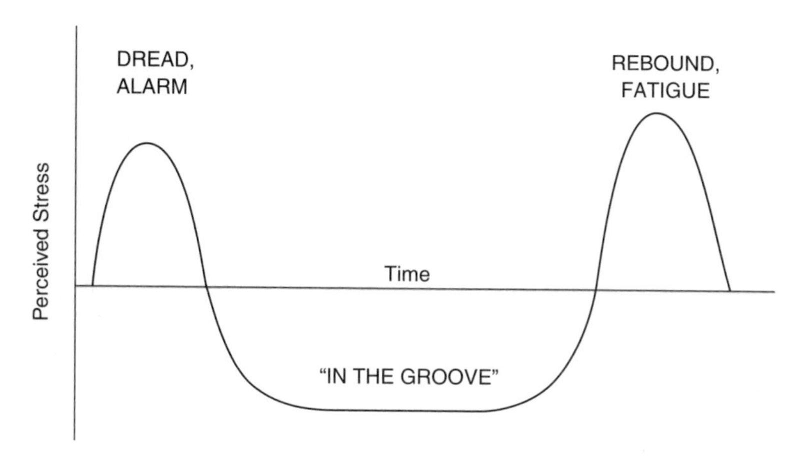

FIGURE 3.2 The time course of perceived stress in combat and other military operations, whether over minutes (e.g., a single firefight), days (e.g., a single mission), or months (e.g., an entire deployment).

response is mediated by the body, the mind, or social relationships, adaptation takes time and effort, and it cannot be sustained, in most cases, forever. Selye's GAS may be more easily applied to the adaptation of warfighters to the challenges of war, however, by flipping it upside down, as shown in Figure 3.2. Diagrammed in this way, the curve traces the time course of *perceived stress*, from the initial peak of stress caused by alarm or dread, through a period of reduced perceived stress while "in the groove," to a final reemergence of perceived stress either due to fatigue or rebound once a danger has passed. The timeline of Figure 3.2 can be minutes, hours, days, or months. The time course of adaptation is the same regardless of the duration of a stressor. Two examples will help illustrate.

Before a planned combat action, most warfighters experience a period of uneasiness and agitation because of the unknowns they face and because, before the action begins, there is little they can do to actively master their stress. With the commencement of combat, however, the pre-action dread dissipates quickly, especially for veterans of combat. Most warriors then quickly get into a groove—a period of exceptionally low-perceived stress, during which their thinking is clear, perceptions are sharp, and emotions are calm. The "in the groove" period may last the duration of a combat action, if it is not too long or too overwhelmingly stressful. Once the action ends, however, perceived stress shoots back up as warfighters emerge from their emotional and physical numbness and review in their minds and perceive in their bodies all the dangers and horrors they may have experienced. The veteran warfighter quickly masters this rebound stress, however, and perceived stress returns to baseline.

As a second example, consider the time course of the stress of an entire deployment to a combat theater. The dread of deployment, for some warfighters, begins long before they pack their duffle bags. Anticipation of dangers and losses, and of separation from family and friends, can progressively heighten pre-deployment anxiety. When warfighters finally set their boots down in the war zone, their initial dread may dissipate to some extent, but only to be replaced by the physiological and psychological alarm caused by the operational stressors encountered. It is no accident that the first 3 months of an operational deployment are when the majority of stress casualties may occur (Tischler, 1969). It is easier for those who have been there before, perhaps, but it takes time for warfighters to adapt to the hardships and dangers they face day in and day out—to get into the groove of deployment. The subsequent period of sustained adaptation during a deployment may last for many months, but since adaptation takes work and consumes resources, it cannot be sustained forever. If operational stressors are intense enough and last long enough, eventually warfighters become fatigued and their perceived stress level gradually rises. But even if the length and rigor of a deployment does not exceed the capacity of individual warfighters to adapt and cope while deployed, eventual redeployment out of the operational theater will cause a rebound increase in their perceived stress level much like that which follows short combat actions, only of longer duration. The new stressors associated with being home again—of returning to an environment that is familiar but to which the warfighter is no longer adapted—add to the perceived stress of homecoming.

The Reversibility of Adaptation

All adaptive responses, if they are truly mechanisms *chosen* by persons and their biological machinery to master challenges, must be reversible. However bent over a tree may grow to adapt to the wind blowing across a precipice, the tree retains the ability, at least theoretically, to grow back the other way if the wind direction changes. But just as adapting to a stress takes time and effort, readapting to the absence of that stress also takes work and cannot be accomplished instantaneously. After warfighters return home from fighting urban, counterinsurgency warfare, for example, they may continue for some time to scan strangers on the street for weapons or change lanes on the highway to avoid passing too close to a pile of rubbish that, in the war, might have hidden an improvised explosive device (IED). But unless warfighters' biological and psychological adaptive machinery has been damaged by the war, the adaptive changes that occurred during deployment should be reversible afterward.

STRESS INJURIES: BEYOND THE LIMITS OF ADAPTATION

There are compelling reasons to believe that the human mind, like the body, has a limited capacity to withstand external forces without suffering damage. Among them are the following:

- U.S. Army researchers during WWII found that the rates of stress casualties were proportional both to the intensity of combat (measured by the rate of physical wounding and death in battle) and to duration of combat. In the Italian campaign, for example, Appel and Beebe (1946) found that the breaking point for most individuals occurred after 88 days of combat in which at least one friendly casualty was sustained (210 calendar days, on average, in that theater). These and other observations on the epidemiology of combat/operational stress led to the military psychiatry principle of "ultimate vulnerability"—in other words, "everyone has a breaking point" (Jones, 1995a).
- Screening programs to weed out in advance those vulnerable to operational stress disorders have always failed to completely prevent such disorders (Copp & McAndrew, 1990; Shephard, 2001; Wessely, 2005). Ultimately, everyone is vulnerable.
- While adaptive strategies for coping with stress are almost infinitely variable, constrained only by opportunity and individual imagination, persistent adverse reactions to overwhelming stress tend to fall in one or more of a small number of remarkably predictable syndromic patterns. Combat stress casualties are fundamentally different in their nature from adaptive coping strategies; they are not choices, but afflictions.
- Research on the neurobiology and psychology of overwhelming stress has found a number of consistent patterns of persistent dysfunction in the brain and mind (see this volume, chapter 4).
- Individuals exposed to overwhelming stress have consistently described subjective experiences of feeling helpless to control or modulate their reactions to the stress. Rather, trauma is often experienced as a discontinuity or fracture of the self (Laufer, 1988; Lifton, 1988), and the subjective experience of helplessness during a traumatic experience typically provokes significant and persistent shame.
- Individuals exposed to overwhelming stress may be permanently changed by their experience, though not always negatively. Stress symptoms that develop during an operational deployment sometimes continue long after return from the war. Adverse stress reactions are often not reversible, although they can certainly heal.

The word "trauma" comes from the Greek word for wound. Although this term may have originally been intended metaphorically when first used in connection with stress, there is increasing reason to believe that overwhelming stress can inflict literal injuries in the brains and minds of warfighters and civilians. The term "injury" has significant advantages when communicating with warfighters about the nature of their reactions to severe stress and how best to care for them. Warriors understand that stress injuries, like sports injuries, may be unavoidable, at times—they are just part of the cost of doing what they do. And like sports injuries, most stress injuries heal up quickly, even without professional attention. But also like sports injuries, stress injuries are most likely to heal quickly and completely if warfighters monitor themselves for symptoms of injury and take proper care of those injuries that are sustained. The Canadian military has found that applying the term "stress injury" to persisting operational stress problems has helped destigmatize them (Grenier et al., this volume, chapter 13). "Injury" may also just happen to be the most accurate description of how overwhelming stress affects the mind and brain.

Several approaches have been used to categorize combat/operational stress injuries according to their phenomena (Grinker & Spiegel, 1945b; Kardiner & Spiegel, 1941; Solomon, 1993). But the shifting and polymorphic nature of stress injuries makes classification by symptoms difficult. The approach that will be used here, rather, will be to divide combat/operational stress injuries into three categories based on precipitating stressors—*traumatic stress,* caused by the impact of terror, horror, or helplessness; *operational fatigue,* caused by the wear and tear of accumulated stress; and *grief,* caused by the loss of someone or something that is highly valued.

Traumatic Stress: An Impact Injury

The core feature of psychological trauma is exposure to an event that is so toxic that a full and immediate adaptive response to it is impossible. Traumatic stress causes an impact injury to the mind and brain just as falling from too great a height will inflict an impact injury to the body. What makes a particular stressor toxic for a particular individual at a particular point in time is a bit more complex, though, than what causes a fall to break a bone. And the symptoms of traumatic stress injuries are more subjective than the swelling, bruising, and x-ray findings that signal a fracture. Nevertheless, a predictable symptom pattern has emerged in traumatic stress injuries of many types, including those common to combat. Although it is not without controversy (Bryant & Harvey, 2000; Marshall, Spitzer, & Liebowitz, 1999), the DSM-IV-TR diagnosis of acute stress disorder (ASD) best describes the features of traumatic stress injury (American Psychiatric Association, 2000, p. 471).

A. The person has been exposed to a traumatic event in which both of the following were present:
   (1) the person experienced, witnessed, or was confronted with an event or events that involved actual or threatened death or serious injury, or a threat to the physical integrity of self or others
   (2) the person's response involved intense fear, helplessness, or horror
B. Either while experiencing or after experiencing the distressing event, the individual has three (or more) of the following dissociative symptoms:

(1) a subjective sense of numbing, detachment, or absence of emotional responsiveness
(2) a reduction in awareness of his or her surroundings (e.g., "being in a daze")
(3) derealization
(4) depersonalization
(5) dissociative amnesia (i.e., inability to recall an important aspect of the trauma)

C. The traumatic event is persistently reexperienced in at least one of the following ways: recurrent images, thoughts, dreams, illusions, flashback episodes, or a sense of reliving the experience; or distress on exposure to reminders of the traumatic event.
D. Marked avoidance of stimuli that arouse recollections of the trauma (e.g., thoughts, feelings, conversations, activities, places, people).
E. Marked symptoms of anxiety or increased arousal (e.g., difficulty sleeping, irritability, poor concentration, hypervigilance, exaggerated startle response, motor restlessness).
F. The disturbance causes clinically significant distress or impairment in social, occupational, or other important areas of functioning, or impairs the individual's ability to pursue some necessary task, such as obtaining necessary assistance or mobilizing personal resources by telling family members about the traumatic experience.
G. The disturbance lasts for a minimum of 2 days and a maximum of 4 weeks and occurs within 4 weeks of the traumatic event.
H. The disturbance is not due to the direct physiological effects of substance (e.g., a drug of abuse, a medication) or a general medical condition, is not better accounted for by Brief Psychotic Disorder, and is not merely an exacerbation of a preexisting Axis I or Axis II disorder.

There are two parts to the DSM-IV-TR definition of a traumatic event. The first is a significant threat to the life or physical integrity of oneself or someone else in close proximity. The second is the person's response to the threat, which must involve terror, horror, or helplessness in order for the threat to qualify as traumatic. Threats vary in their ability to provoke terror, horror, or helplessness in the average person. In the author's experience with warfighters who had served in Afghanistan and Iraq, among the most potentially traumatic combat events were witnessing the violent death of a buddy or valued leader, being responsible for the death of unarmed children, failing to save a buddy from death or serious injury, friendly fire, witnessed atrocities, or surviving an unexpected assault in which many friendly casualties were suffered, such as a vehicle-born IED attack or a large ambush. But combat of any kind is potentially toxic. In their study of soldiers and Marines 3 to 6 months after Operation Enduring Freedom and Operation Iraqi Freedom I, Hoge et al. (2004) found a strong positive correlation between number of firefights and severity of traumatic stress symptoms.

One of the controversial features of the DSM-IV-TR diagnosis of ASD is the time course criterion. ASD cannot be given as a diagnosis to stress injured warfighters until their symptoms have persisted for more than 2 days after a traumatic event. DSM-IV-TR does not allow for any specific diagnosis to be made for traumatic stress symptoms occurring in the first 48 hours after an event, even though the symptoms and disability from them may be the same (or worse) on day 1 as on day 3 posttrauma. The rationale for this is the belief, expressed in the ICD-10 diagnostic criteria for acute stress reaction (WHO, 1992), that traumatic stress symptoms that appear within minutes of the impact of a traumatic event *usually* resolve within a few hours or a couple of days, at most. The committee that wrote the criteria for ASD wanted to avoid pathologizing traumatized individuals before their symptoms persisted beyond day two (Marshall, Spitzer, & Liebowitz, 1999). However, as clearly articulated in the *VA/DoD Clinical Practice Guideline for the Management of Post-traumatic Stress* (Department of Veterans Affairs & Department of Defense, 2004), traumatic stress symptoms deserve to be acknowledged and actively managed as soon as possible after they appear. Ignoring traumatic stress symptoms for the first 2 days would make as much sense as ignoring the signs and symptoms of an ankle fracture for the first 2 days just in case the injury might turn out to be only a sprain. These ethical entanglements may be side-stepped by using the term "traumatic stress injury" for all warfighters who experience significant posttraumatic stress symptoms, whether in the first 5 minutes or the first 5 days.

The symptoms of ASD listed above are merely that—observable symptoms. They offer little insight into the processes by which a traumatic event can injure a warfighter. Although existing research cannot explain everything that happens to the mind and brain at the moment of a traumatic impact, several consistent peritraumatic processes have been studied and reported. Four of these that may be especially useful in understanding traumatic stress injuries in combat are (1) physiological hyperarousal, (2) damage to core beliefs, (3) shame or guilt, and (4) dissociation.

Physiological Hyperarousal. All mammals normally react to imminent threats to their safety with an adaptive physiological "fight or flight" arousal, largely mediated by the neurotransmitters epinephrine and norepinephrine in the brain and body. Increases in the activity of norepinephrine circuits in the brain promote alertness and attention, while increases in the activity of norepinephrine neurons in the peripheral nervous system increase heart rate and blood flow to muscles and other vital organs. Arousal in response to a perceived threat is a universal adaptive mechanism. But there is evidence that *excessive* physiological arousal in response to a threat may be characteristic of traumatic stress injuries. Certainly, persistent hyperarousal in the form of difficulty sleeping, irritability, poor concentration, hypervigilance, exaggerated startle responses, or motor restlessness long after the threat has passed is a cardinal symptom of both ASD and PTSD (American Psychiatric Association, 2000). As markers of physiological hyperarousal, resting heart rates, and heart rate elevations in response to loud sounds or trauma reminders, have been found retrospectively to be greater in individuals with PTSD (Orr et al., 2003; Prins, Kaloupek, & Keane, 1995). Elevated heart rate soon after a traumatic event has also been found prospectively to be a predictor of who will go on to develop ASD or PTSD (Bryant et al., 2000, 2003; Kassam-Adams et al., 2005; Shalev et al., 1998).

Arousal is necessary to adapt to threats, but arousal beyond a certain optimal point is toxic. Hyperarousal reduces the efficiency of cognition and memory (van der Kolk, 1995; Yerkes & Dodson, 1908), and may make it more difficult to make sense out of and master a given situation. Excessive arousal can also promote physical damage to certain neurons in the brain, a process known as "excitotoxicity" (Stahl, 1996). Excitoxicity from excessive arousal has been implicated in the degeneration of the brain in several mental disorders. It may also be a mechanism by which traumatic stress damages neurons in the brain essential for overcoming fear and integrating traumatic experiences and memories (see this volume, chapter 4).

An important implication of the connection between arousal and traumatic stress injury is that anything that reduces arousal level at the moment of stress impact may mitigate or even prevent the resulting injury.

Damage to Core Beliefs. One of the recurring themes in the literature on traumatic stress is the ability of traumatic events to shatter necessary and deeply held beliefs (Davis, 2001; Janoff-Bulman, 1992; Kauffman, 2002). Everyone interprets life events and makes life decisions based on a set of core assumptions about the world and one's place in it. Janoff-Bulman (1992) proposed three fundamental assumptions common to all people at all times: (1) the world is benevolent, (2) the world is meaningful, and (3) the self is worthy. All people also need to believe that they are safe—that their lives will not be snuffed out in the next few seconds—and that a moral order exists in the universe that discriminates right from wrong. The importance of these core beliefs is easy to take for granted because they all operate beneath our radar screens, until something violates one of these beliefs. A feature common to all traumatic stressors, but particularly evident in combat operations, is their ability to betray one or more of these core assumptions. Young warriors experience death, chaos, and evil in an intimate way—and often not only once but repeatedly. Lifton (1988) called the radical intrusion of the reality of death into the minds of young warfighters "the death imprint." Because of "its suddenness, its extreme or protracted nature, or its association with terror of premature, unacceptable dying," Lifton (1988, p. 18) wrote, the death imprint may be very difficult for young warriors to assimilate and detoxify. Under almost no circumstance other than war would men and women in their teens and early twenties—barely adults—face the reality of their own mortality so brutally. Combat trauma also "destroys the capacity for social trust" (Shay, 1994, p. 33) because it shatters the illusion that people are basically benevolent and good. Shay also wrote eloquently, from his experience treating Vietnam veterans with traumatic stress injuries, how the betrayal in war of the moral order—of basic beliefs about right and wrong—can ruin the character of young warfighters. Grossman (1995) argued that for many young warfighters, the very act of killing another human being can shatter core beliefs, especially beliefs about one's own basic goodness.

Shattered beliefs are not beyond repair. Nor is the loss of the beliefs of youth necessarily a bad thing. Traumatic injuries of all kinds can promote positive growth that might otherwise not be possible (Calhoun & Tedeschi, 2001). Many warriors experience positive transformations in their self-esteem, life focus, and faith after a tour in a war zone. As Viktor Frankl (1984/1946) learned in the death camps of WWII, surviving unavoidable suffering can infuse life with a powerful new meaning. To prosper mentally, however, warfighters who have suffered damage to their core beliefs must construct new belief systems that transcend the old and incorporate, somehow, the brutal realities of war without sacrificing everything that is positive about human existence.

Shame or Guilt. Intense self-reproach in the form of shame or guilt may be a frequent consequence of traumatic stress injuries in combat (Figley, 1978; Grinker & Spiegel, 1945a; Grossman, 2004; Hendin & Haas, 1984; Shay, 1994; Solomon, 1993). Of course, shame and guilt can also accompany civilian traumatic stress injuries of many kinds (Herman, 1992; Horowitz, 1986). But combat may be uniquely toxic to self-esteem and self-image. To the extent warfighters go to war believing in their own strength and competence, they experience intense shame after becoming utterly helpless during a traumatic event. To the extent they believe they can protect their brother warriors from harm, they experience racking guilt after surviving their buddies' deaths. Civilians traumatized by natural disasters or crimes of violence may also feel intensely ashamed of their helplessness or guilty for surviving when others did not. But the shame of traumatized warriors is compounded by the fact that they all volunteered for military service knowing they would face the challenges of war. Since warfighters train and prepare themselves to withstand the traumas of combat, it is much harder for them to forgive themselves for failing in any way to triumph over it.

Recovering from traumatic shame and guilt requires the construction of a new set of beliefs about oneself and one's place in the world—beliefs that allow for very human weaknesses at sometimes crucial moments. Overcoming guilt and shame depends on forgiveness.

Dissociation. In recent years, there has been a resurgence of interest in the relationship between trauma and dissociation (Marmar, Weiss, & Metzler,

1998). Dissociation has been increasingly implicated in both the short-term and long-term disability that can follow a traumatic stress injury. On the battlefield, a sudden and profound disruption in the capacity of the brain and mind to process and integrate information can certainly make a warfighter a hazard to himself and his peers. A consummate movie portrayal of battlefield dissociation was that experienced by Tom Hanks' character in *Saving Private Ryan*, in which he went mentally blank while kneeling on the beach in Normandy, briefly becoming unable to respond to his surroundings or even to hear the explosions and screams around him. Since dissociation disrupts the processing of information, it is also implicated in the long-term failure of trauma-injured veterans to integrate their perceptions, feelings, and thoughts at the moment of traumatic injuries with the rest of their identities and memories. Many of the cardinal symptoms of ASD and PTSD involve a loss of authority over memory. Vivid images of traumatic experiences intrude, unwanted, into dreams and conscious awareness, and the individual struggles to avoid recall of these perpetually distressing memories. Yet, there is often a simultaneous amnesia for many of the details that might help put the puzzle pieces into some kind of order. Flashbacks in both ASD and PTSD are memories of trauma that provoke dissociation, again and again, in their recall. As long as traumatic memories are always accompanied by a breakdown in capacity to integrate and process information, such memories may forever remain independent of conscious control. Dissociation seems to be both a cause and effect of traumatic stress injury.

There are several unresolved controversies over the significance of dissociation during and after a traumatic stress, however. The first is whether dissociation truly is, as suggested by the DSM-IV-TR criteria for ASD, always a symptom of a traumatic stress injury. Peritraumatic dissociation is a strong predictor of future PTSD, but not everyone with PTSD reports having had symptoms of dissociation at the time of the trauma (Marshall, Spitzer, & Liebowitz, 1999).

Another controversy is whether dissociation is a normal and reversible adaptive process, or a symptom of irreversible (though capable of healing) injury. Seen as a normal adaptive process, dissociation has been cast as a defense mechanism, a form of self-hypnosis, or a genetically programmed reflex similar to the freezing or sham death behaviors of prey animals (Cardena, 1994). Supporting the idea that dissociation is normal and adaptive is the finding that dissociation occurs with great frequency in potentially traumatic situations. For example, a prospective study by Morgan et al. (2001) found that 96% of soldiers undergoing intense survival training experienced dissociation at some point in that training. But the frequency of dissociation in combat-like situations does not necessarily make it normal, any more than it should be considered nonpathological to have a black eye after being punched in the face. Janet's view of dissociation as a breakdown of adaptation, rather than as an adaptive choice, has gained renewed support in recent years (van der Kolk & van der Hart, 1989).

Advances in the neurobiology of traumatic stress injury have prompted one researcher to suggest that the subjective experience of dissociation may be just what it feels like to suffer an injury to integrative centers of brain during a traumatic stress (Bremner, 2002).

A third controversy involves which symptoms should be included in a list of trauma-induced disruptions "in the usually integrated functions of consciousness, memory, identity, or perception," as dissociation is defined by DSM-IV-TR (American Psychiatric Association, 2004, p. 519). The criteria for ASD, listed above, include only cognitive forms of dissociation, such as perceptions of unreality, numbing, being in a "daze" or on autopilot, or a loss of memory. However, as first described by Janet and others, traumatic dissociation could also include a host of physical forms of disrupted integration, including paralysis, blindness, deafness, shaking, stuttering, inability to speak, sleepwalking, and psychogenic pain (Janet, 1920; Nijenhuis, 2004). Physical (somatic) types of dissociation have been observed on the battlefield since the 18th century, made famous in the stress casualties of WWI and WWII. A survey of dissociation symptoms in policemen involved in combat includes similar symptoms of loss of physical control (Grossman, 2004). Besides cognitive and somatic forms of dissociation, a third type has been described both in warfighters and in civilians exposed to repeated or ritual abuse. This type has been termed "tertiary dissociation" by van der Kolk, van der Hart, and Marmar (1995, p. 316). Its characteristic feature is a shattering of the core self of the traumatized person into two or more partial selves, each with its own style and value system. In its most malignant form, tertiary dissociation underlies dissociative identity disorder (multiple personality disorder). But in a milder form, the fracture of self into partial personality fragments has been described in traumatized warfighters. Kind and gentle individuals can become abruptly cruel and vicious. Timid people can become fearless, even "berserk." Or strong and brave warriors can become childlike and regressed. Laufer (1988) termed the fractured component of the self that can develop in a traumatized warrior a "war self," which develops such different perceptions, values, and memories from the rest of the warfighter's identity that post-war integration into one whole person may be very difficult.

Prospective studies of warfighters engaged in combat will be necessary to resolve these controversies regarding dissociation. Prospective research will also clarify the role of physiological hyperarousal, damage to core beliefs, and shame and guilt in traumatic stress injuries on the battlefield. Meanwhile, to the extent these four components of traumatic injury may truly define combat trauma, they are exactly the places where we can intervene to prevent and mitigate traumatic stress injuries. Any tactic that reduces arousal level, protects or restores belief systems, relieves shame or guilt, or prevents or relieves dissociation should help keep warfighters healthy and ready.

Operational Fatigue: A Wear-and-Tear Injury

Not all stress injuries in operational settings are caused by trauma, just as not all post-deployment stress problems are PTSD. The National Vietnam Veterans Readjustment Study, funded by Congress in 1983, found that Vietnam veterans had high rates of persistent mental disorders other than PTSD, including depression, anxiety, substance use, and personality disorders (Kulka et al., 1990). The rates of these other mental disorders were higher in Vietnam veterans than in their civilian peers who had not served in Vietnam, and their rates were found to correlate directly with nearness to combat in theater. Similarly, Hoge et al. (2004) found significantly high rates of depression, anxiety, and substance use problems among warfighters 3 to 6 months after their return from Afghanistan or Iraq. But what do future depressive, anxiety, or substance use disorders look like on the battlefield? Are they identifiable among deployed warfighters? Or do these problems always spring from the ground in full bloom after return from a deployment?

These questions have not been well answered by prospective research to date. But observations over the past three centuries of the nature and causes of operational stress injuries suggest that there are characteristic patterns of symptoms that can arise in warfighters due merely to the wear-and-tear of accumulated stress, independent of specific traumatic events. Nostalgia cases in the 18th and 19th centuries were nontraumatic stress injuries. Jones (1995c) rediscovered the nostalgia concept in his explication of what he called "disorders of frustration and loneliness," which he believed could occur as easily in garrison as in a low-intensity operational deployment in a war zone. But others have noted the connection between accumulated stress in a war zone, even among support personnel not directly engaged in combat, and psychiatric breakdown.

Lord Moran, a battalion surgeon in France in WWI and later personal physician to Winston Churchill, observed that all warfighters had a limited supply of courage. And courage, like capital, was constantly paid out by warfighters in response the accumulated stressors of war (Moran, 1967/1945). Grinker and Spiegel (1945a) were among the first to describe the syndrome of "operational fatigue" in U.S. Army Air Forces personnel in WWII. Typical symptoms of operational fatigue included (p. 210):

- Persistent restlessness
- Irritability and anger
- Difficulty falling asleep
- Tension and subjective anxiety
- Depression
- Decreased appetite
- Decreased ability to concentrate
- Easy fatigue and low energy levels
- Tremor or other sympathetic hyperactivity

Appel and Beebe (1946) studied psychiatric breakdown in Italy near Cassino and Anzio, and found that exposure to danger took a relentless toll, even in the absence of specific, overwhelming events. They wrote that "men will break down in direct relation to the intensity and duration of their exposure. Thus, psychiatric casualties are as inevitable as gunshot and shrapnel wounds in warfare" (p. 1470). Older warriors have traditionally been found to be more susceptible than the young to operational fatigue injuries. Hence, the name that was given to these injuries in WWII—"old sergeant's syndrome" (Dean, 1997, p. 37).

The symptoms of operational fatigue—or nostalgia, for that matter—are indistinguishable from the symptoms of DSM-IV-TR depression and anxiety disorders. But further prospective research is needed to determine the relationship between operational stressors and the development of these symptoms, as well as under what circumstances such symptoms can become persistent and chronic.

Grief: A Loss Injury

Grief has been defined as a "reaction to the loss of a loved one through death" (Stroebe et al., 2001, p. 6). Grief is certainly a normal part of life since the loss of loved ones through death is inevitable for everyone who lives long enough. If the definition of grief is expanded to include reactions to the losses of non-human love objects, including valued states of mind and beliefs, then grief is an almost continuous process beginning in early childhood. Even though inevitable and necessary, however, grief may not quite qualify as an adaptive response to a manageable "eustress." Grief is not reversible. Nor is it a choice. The losses that provoke grief are afflictions. The symptoms of grief can be long lasting and, at times, disabling. Like trauma and operational fatigue, grief may best be conceived of as an injury. Like trauma, it is an impact injury, even if the death occurs over a long period of time. Of course, it is not being proposed here that grief is necessarily pathological. Rather, as with other combat/operational stress injuries, the point is to avoid minimizing or trivializing grief just because it is so common and "normal."

Many warfighters who participate in combat suffer the loss of someone who is loved, sometimes loved dearly. The attachments young warriors have for each other are infused with an intensity that has few parallels in civilian life, coming closer to the attachment a parent has for a child than the bonds of siblings or mere friends. Certainly, the responsibility warriors feel for each other is similar to that felt by a parent for a child. Hence, grief in war can be one of the most traumatic forms of grief.

The features of grief can vary greatly, but certain manifestations are typical, including symptoms in the following dimensions (Stroebe et al., 2001):

- Emotional: anger, hostility, depression, despair, anxiety, guilt
- Behavioral: agitation, fatigue, crying, social withdrawal
- Cognitive: preoccupation with thoughts of the deceased, self-reproach, memory loss, difficulties concentrating
- Physical: sleeplessness, energy loss, loss of appetite

In addition to the above manifestations, grief in reaction to losses that are particularly wounding can include some of the cardinal symptoms of traumatic stress injury, including the following (Jacobs, 1999):

- Dissociative symptoms, such as feeling stunned, dazed, shocked, or numb
- Intrusive, painful recollections about the deceased person
- Frequent efforts to avoid reminders of the deceased
- Damage to belief systems, including loss of security and trust

All the above grief symptoms are common among warfighters who have lost buddies and valued leaders. However, warriors deployed to an operational theater often do not experience the full impact of their grief until after they have returned to garrison, and their adaptive numbness and denial have worn off. Sometimes the reality of combat losses first begins to sink in for warfighters when they are on the airplane flying home, surrounded by too many empty seats.

SUMMARY AND CONCLUSIONS

Historical attempts to label and classify battlefield stress reactions have labored under several dilemmas. The first of these is that while medical labeling of significant combat stress problems has always carried a burden of stigma and may have interfered with the ability of leaders to keep their troops in the fight, avoiding medical labeling may have not only inadvertently discouraged seriously stressed warfighters from seeking needed help, but also added to their burden of shame by trivializing their persistent stress reactions. The second dilemma has arisen as a succession of biological, psychological, and social explanations for war-zone stress reactions has each failed to fully explain symptoms or to predict their course. Finally, existing taxonomies for combat/operational stress reactions have failed to draw clear distinctions between normal, reversible adaptations and irreversible repercussions to stress that have exceeded adaptive capacities.

In this chapter, normal processes of adaptation to stress have been examined from biological, psychological, and social perspectives, and their typical time course has been described. A nomenclature has been offered for the description of irreversible and involuntary stress reactions as "stress injuries," and a taxonomy for stress injuries has been suggested based on the nature of the causative stressor, dividing stress injuries into the broad categories of trauma, fatigue, and grief.

The potential advantages to the stress injury model of combat/operational stress reactions are several. First, it suggests guidelines for discriminating pathological stress symptoms from truly normal adaptive responses to stress. Second, it casts involuntary stress symptoms in terms already familiar to warfighters, suggesting that like physical injuries, stress injuries are sometimes unavoidable, part of the cost of doing their jobs. Stress injuries are no more to be considered the fault of the injured warfighter than are physical injuries during operational deployment. But also like physical injuries, stress injuries have the best chance of healing quickly and completely if they are acknowledged and given the proper care, even if that is no more than brief rest. The division of stress injuries into the three categories based on precipitating stressor increases the likelihood of recognizing and attending to those stress injuries that tend to get less attention—those *not* caused by an event involving overwhelming terror or horror. Additional research will determine whether the potential advantages of the stress injury conception of combat/operational stress reactions are realized, and whether it reduces stigma and promotes a broader and deeper understanding of how stress can damage body and mind.

REFERENCES

Aldwin, C. M. (1994). *Stress, Coping, and Development: An Integrative Perspective.* New York: Guilford Press.

American Psychiatric Association. (1980). *Diagnostic and Statistical Manual of Mental Disorders* (3rd ed.). Washington, DC: Author.

American Psychiatric Association. (2000). *Diagnostic and Statistical Manual of Mental Disorders* (4th ed., text rev.). Washington, DC: Author.

American Psychiatric Association. (2004). *Diagnostic and Statistical Manual of Mental Disorders* (4th ed., text rev.). Washington, DC: Author.

Appel, J. W., & Beebe, G. W. (1946). Preventive psychiatry: An epidemiologic approach. *Journal of the American Medical Association, 131*(18), 1469–1475.

Bourne, P. G. (1969). Military psychiatry and the Vietnam War in perspective. In P. G. Bourne (Ed.), *The Psychology and Physiology of Stress: With Reference to Special Studies of the Viet Nam War.* New York: Academic Press.

Bremner, J. D. (2002). *Does Stress Damage the Brain? Understanding Trauma-Related Disorders from a Mind-Body Perspective.* New York: Norton.

Breuer, J., & Freud, S. (1957). *Studies on Hysteria* (J. Strachey, Ed. & Trans.). New York: Basic Books. (Original published in 1895)

Bryant, R., Guthrie, R., Moulds, M., & Harvey, A. (2000). A prospective study of psychophysiological arousal, acute stress disorder, and posttraumatic stress disorder. *Journal of Abnormal Psychology, 109,* 341–344.

Bryant, R., Harvey, A., Guthrie, R., & Moulds, M. (2003). Acute psychophysiological arousal and posttraumatic stress disorder: A two-year prospective study. *Journal of Traumatic Stress, 16,* 439–443.

Bryant, R. A., & Harvey, A. G. (2000). *Acute Stress Disorder: A Handbook of Theory, Assessment, and Treatment.* Washington, DC: American Psychological Association.

Calhoun, L. G., & Tedeschi, R. G. (2001). Posttraumatic growth: the positive lessons of loss. In R. A. Neimeyer (Ed.), *Meaning Reconstruction & the Experience of Loss.* Washington, DC: American Psychological Association.

Cardena, E. (1994). The domain of dissociation. In S. J. Lynn, & J. W. Rhue (Eds.), *Dissociation: Clinical and Theoretical Perspectives.* New York: Guilford Press.

Cohen, S., Gottlieb, B., & Underwood, L. (2000). Social relationships and health. In S. Cohen, L. Underwood, & B. Gottlieb (Eds.), *Social Support Measurement and Intervention: A Guide for Health and Social Scientists.* New York: Oxford University Press.

Copp, T., & McAndrew, B. (1990). *Battle Exhaustion: Soldiers and Psychiatrists in the Canadian Army, 1939–1945.* Montreal, Quebec: McGill-Queen's University Press.

Darwin, C. (1915). *The Expression of the Emotions in Man and Animals.* New York: D. Appleton & Co. (Original published in 1872)

Davis, C. G. (2001). The tormented and the transformed: Understanding responses to loss and trauma. In R. A. Neimeyer (Ed.), *Meaning Reconstruction & the Experience of Loss.* Washington, DC: American Psychological Association.

Dean, E. T. (1997). *Shook Over Hell: Post-Traumatic Stress, Vietnam, and the Civil War.* Cambridge, MA: Harvard University Press.

Department of Veterans Affairs & Department of Defense. (2004). *VA/DoD Clinical Practice Guideline for the Management of Post-Traumatic Stress.* Washington, DC: Author.

Dorpat, T. L. (1987). A new look at denial and defense. *Annual of Psychoanalysis, 15,* 23–47.

Elder, G. H., & Clipp, E. C. (1988). Combat experience, comradeship, and psychological health. In J. P. Wilson, Z. Harel, & B. Kahana (Eds.), *Human Adaptation to Extreme Stress.* New York: Plenum Press.

Engel, G. L. (1980). The clinical application of the biopsychosocial model. *American Journal of Psychiatry, 137,* 535–544.

Fenichel, O. (1945). *The Psychoanalytic Theory of Neurosis.* New York: Norton.

Figley, C. R. (Ed.). (1978). *Stress Disorders among Vietnam Veterans: Theory, Research, and Treatment.* New York: Brunner-Routledge.

Figley, C. R. (Ed.). (1999). *Traumatology of Grieving: Conceptual, Theoretical, and Treatment Foundations.* New York: Brunner/Mazel.

Frankl, V. E. (1984). *Man's Search for Meaning.* New York: Simon & Schuster. (Original work published 1946)

Grinker, R. R., & Spiegel, J. P. (1945a). *Men under Stress.* Philadelphia: Blakiston.

Grinker, R. R., & Spiegel, J. P. (1945b). *War Neuroses.* Philadelphia: Blakiston.

Grossman, D. (1995). *On Killing: The Psychological Cost of Learning to Kill in War and Society.* Boston: Little, Brown.

Grossman, D. (2004). *On Combat: The Psychology and Physiology of Deadly Conflict in War and Peace.* St. Louis, MO: PPCT Research Publications.

Hendin, H., & Haas, A. P. (1984). *Wounds of War: The Psychological Aftermath of Combat in Vietnam.* New York: Basic Books.

Herman, J. (1992). *Trauma and Recovery: The Aftermath of Violence—From Domestic Abuse to Political Terror.* New York: Basic Books.

Hoge, C. W., Castro, C. A., Messer, S. C., McGurk, D., Cotting, D. I., & Koffman, R. L. (2004). Combat duty in Iraq and Afghanistan, mental health problems, and barriers to care. *New England Journal of Medicine, 351,* 1798–1800.

Horowitz, M. J. (1986). *Stress Response Syndromes.* Northvale, NJ: Jason Aronson.

Jacobs, S. (1999). *Traumatic Grief: Diagnosis, Treatment, and Prevention.* New York: Brunner/Mazel.

Janet, P. (1920). *The Major Symptoms of Hysteria: Fifteen Lectures Given in the Medical School of Harvard University* (2nd ed.). New York: Macmillan Co.

Janoff-Bulman, R. (1992). *Shattered Assumptions: Towards a New Psychology of Trauma.* New York: The Free Press.

Jones, F. D. (1995a). Psychiatric lessons of war. In F. D. Jones, L. R. Sparacino, V. L. Wilcox, J. M. Rothberg, & J. W. Stokes (Eds.), *War Psychiatry*. Washington, DC: Borden Institute.

Jones, F. D. (1995b). Traditional warfare combat stress casualties. In F. D. Jones, L. R. Sparacino, V. L. Wilcox, J. M. Rothberg, & J. W. Stokes (Eds.), *War Psychiatry*. Washington, DC: Borden Institute.

Jones, F. D. (1995c). Disorders of frustration and loneliness. In F. D. Jones, L. R. Sparacino, V. L. Wilcox, J. M. Rothberg, & J. W. Stokes (Eds.), *War Psychiatry*. Washington, DC: Borden Institute.

Kardiner, A., & Spiegel, H. (1941). *War Stress and Neurotic Illness*. New York: Paul B. Hoeber, Inc.

Kassam-Adams, N., Garcia-Espana, J. P., Fein, J. A., & Winston, F. K. (2005). Heart rate and posttraumatic stress in injured children. *Archives of General Psychiatry, 62*, 335–340.

Kauffman, J. (2002). *Loss of the Assumptive World: A Theory of Traumatic Loss*. New York: Brunner-Routledge.

Kormos, H. R. (1978). The nature of combat stress. In C. R. Figley (Ed.), *Stress Disorders among Vietnam Veterans*. New York: Brunner-Routledge.

Kulka, R. A., Schlenger, W. E., Fairbank, J. A., Hough, R. L., Jordan, B. K., Marmar, C. R., et al. (1990). *Trauma and the Vietnam War Generation: Report of Findings from the National Vietnam Veterans Readjustment Study*. New York: Brunner/Mazel.

Laufer, R. S. (1988). The serial self: War trauma, identity, and adult development. In J. P. Wilson, Z. Harel, & B. Kahana (Eds.), *Human Adaptation to Extreme Stress*. New York: Plenum Press.

Lazarus, R. S. (1999). *Stress and Emotion: A New Synthesis*. New York: Springer Publishing Company.

Lazarus, R. S., & Folkman, S. (1984). *Stress, Appraisal, and Coping*. New York: Springer Publishing Company.

Lifton, R. J. (1988). Understanding the traumatized self: Imagery, symbolization, and transformation. In J. P. Wilson, Z. Harel, & B. Kahana (Eds.), *Human Adaptation to Extreme Stress*. New York: Plenum Press.

Marmar, C. R., Weiss, D. S., & Metzler, T. (1998). Peritraumatic dissociation and posttraumatic stress disorder. In J. D. Bremner & C. R. Marmar (Eds.), *Trauma, Memory, and Dissociation*. Washington, DC: American Psychiatric Press.

Marshall, R. D., Spitzer, R., & Liebowitz, M. R. (1999). Review and critique of the new DSM-IV diagnosis of acute stress disorder. *American Journal of Psychiatry, 56*, 1677–1685.

Moran, C. M. W. (1967). *The Anatomy of Courage*. Boston: Houghton Mifflin. (Original published in 1945)

Morgan, C. A., Hazlett, G., Wang, S., Richardson, E. G., Schnurr, P., & Southwick, S. M. (2001). Symptoms of dissociation in humans experiencing acute, uncontrollable stress: A prospective investigation. *American Journal of Psychiatry, 158*, 1239–1247.

Nash, W. P. (1998). Information gating: An evolutionary model of personality function and dysfunction. *Psychiatry, 61*(1), 46–60.

Nemiah, J. C. (1998). Early concepts of trauma, dissociation, and the unconscious: Their history and current implications. In J. D. Bremner & C. R. Marmar (Eds.), *Trauma, Memory, and Dissociation*. Washington, DC: American Psychiatric Press.

Nijenhuis, E. R. S. (2004). *Somatoform Dissociation: Phenomena, Measurement, and Theoretical Issues*. New York: Norton.

Orr, S. P., Metzger, L. J., Lasko, N. B., Macklin, M. L., Hu, F. B., Shalev, A. Y., & Pitman, R. K. (2003). Physiological responses to sudden, loud tones in monozygotic twins discordant for combat exposure: Association with posttraumatic stress disorder. *Archives of General Psychiatry, 60*, 283–288.

Porter R. (2002). *Madness: A Brief History*. New York: Oxford University Press.

Prins, A., Kaloupek, D. G., & Keane, T. M. (1995). Psychophysiological evidence for autonomic arousal and startle in traumatized adult populations. In M. J. Friedman, D. S. Charney, & A. Y. Deutch (Eds.), *Neurobiological and Clinical Consequences of Stress: From Normal Adaptation to Post-Traumatic Stress Disorder.* Philadelphia: Lippincott-Raven.

Reich, W. (1991). Psychiatric diagnosis as an ethical problem. In S. Bloch & P. Chodoff (Eds.), *Psychiatric Ethics.* New York: Oxford University Press.

Selye, H. (1950). *The Physiology and Pathology of Exposure to Stress: A Treatise Based on the Concepts of the General-Adaptation-Syndrome and the Diseases of Adaptation.* Montreal, Quebec: Acta, Inc.

Selye, H. (1956). *The Stress of Life.* New York: McGraw-Hill.

Shalev, A. Y. (1996). Stress versus traumatic stress: From acute homeostatic reactions to chronic psychopathology. In B. A. van der Kolk, A. C. McFarlane, & L. Weisaeth (Eds.), *Traumatic Stress: The Effects of Overwhelming Experience on Mind, Body, and Society.* New York: Guilford Press.

Shalev, A. Y., Sahar, T., Freedman, S., Peri, T., Glick, N., Brandes, D., Orr, S. P., & Pitman, R. K. (1998). A prospective study of heart rate response following trauma and the subsequent development of posttraumatic stress disorder. *Archives of General Psychiatry, 55,* 553–559.

Shay, J. (1994). *Achilles in Vietnam: Combat Trauma and the Undoing of Character.* New York: Scribner.

Shay, J. (2002). *Odysseus in America: Combat Trauma and the Trials of Homecoming.* New York: Scribner.

Shephard, B. (2001). *A War of Nerves: Soldiers and Psychiatrists in the Twentieth Century.* Cambridge, MA: Harvard University Press.

Solomon, Z. (1993). *Combat Stress Reaction: The Enduring Toll of War.* New York: Plenum Press.

Stahl, S. M. (1996). *Essential Psychopharmacology.* New York: Cambridge University Press.

Stroebe, M. S., Hansson, R. O., Stoebe, W., & Schut, H. (2001). Introduction: Concepts and issues in contemporary research on bereavement. In M. S. Stroebe, R. O. Hansson, W. Stroebe, & H. Schut (Eds.), *Handbook of Bereavement Research: Consequences, Coping, and Care.* Washington, DC: American Psychological Association.

Tischler, G. L. (1969). Patterns of psychiatric attrition and of behavior in a combat zone. In P. G. Bourne (Ed.), *The Psychology and Physiology of Stress: With Reference to Special Studies of the Viet Nam War.* New York: Academic Press.

Vaillant, G. E. (1977). *Adaptation to Life.* Cambridge, MA: Harvard University Press.

van der Kolk, B. A. (1995). Trauma and memory. In B. A. van der Kolk, A. C. McFarlane, & L. Weisaeth (Eds.), *Traumatic Stress: The Effects of Overwhelming Experience on Mind, Body, and Society.* New York: Guilford Press.

van der Kolk, B. A., & van der Hart, O. (1989). Pierre Janet and the breakdown of adaptation in psychological trauma. *American Journal of Psychiatry, 146,* 1530–1540.

van der Kolk, B. A., van der Hart, O., & Marmar, C. R. (1995). Dissociation and information processing in posttraumatic stress disorder. In B. A. van der Kolk, A. C. McFarlane, & L. Weisaeth (Eds.), *Traumatic Stress: The Effects of Overwhelming Experience on Mind, Body, and Society.* New York: Guilford Press.

van der Kolk, B. A., Weisaeth, L., & van der Hart, O. (1996). History of trauma in psychiatry. In B. A. van der Kolk, A. C. McFarlane, & L. Weisaeth (Eds.), *Traumatic Stress: The Effects of Overwhelming Experience on Mind, Body, and Society.* New York: Guilford Press.

Wessely, S. (2005). Risk, psychiatry, and the military. *British Journal of Psychiatry, 186,* 459–466.

World Health Organization (WHO). (1992). *The ICD-10 Classification of Mental and Behavioral Disorders: F43.0 Acute Stress Reaction.* Geneva: Author.

Yerkes, R. M., & Dodson, J. D. (1908) The relation of strength of stimulus to rapidity of habit-formation. *Journal of Comparative Neurology and Psychology, 18,* 459–482.

4

Competing and Complementary Models of Combat Stress Injury

WILLIAM P. NASH AND DEWLEEN G. BAKER

On the eighth day of the invasion of Iraq, a squad of Marines sat silently in the cargo compartment of a lightly armored amphibious assault vehicle as it roared north toward Baghdad. The Marines were silent, but it was far from quiet inside their amtrac. Besides the whine of the engines and the growl of the tracks over the road surface, the Marines were nearly deafened by almost continuous incoming and outgoing automatic weapons fire, incoming small arms fire, and nearby rocket-propelled grenade explosions. Some of the Marines seemed very relaxed, even smiling, as they bounced along in their tiny steel box. A few were as tight as guitar strings, flinching as every 7.62-mm round ricocheted off the vehicle's nearly 2 inch-thick hull. The leader of this group of warriors, a gunnery sergeant with about 7 years on most of the others, stood up in the center of the amtrac with his head and shoulders poking through an open hatch, calmly munching on the cold entrée from an MRE (meal, ready to eat) as he watched the world go by. The gunny knew his Marines were studying him in the same way small children in uncertain situations study their parents, searching for clues about how great the danger was they faced. So it was with deliberate calm and good humor that the gunny bent down to give a bag of Skittles from his MRE to the youngest PFC in his squad, who not accidentally sat right at the gunny's feet. Everyone knew this particular PFC loved Skittles, and they knew that no one else ever got the gunny's candy. It was part of their bond, oldest to youngest, and both ignored the razzing they regularly received from the rest of the squad. After handing the bag of Skittles to the PFC, the gunny stood back up to look around through the open hatch. But almost as soon as his head was through the hatch, the gunny dropped back into the cargo compartment like a bag of rocks, his unrecognizable, exploded face plopping right into the lap of the young PFC, dousing him and his Skittles with sticky warm blood.

Theories of stress injury, by whatever name they are called, are tools crafted to answer fundamental questions raised by combat experiences like the one just described. Can we predict how individual Marines exposed to the terrors and horrors of combat will react to such extreme stressors? What might be the range of their responses? What factors of resiliency will determine who among them will be stunned but almost instinctively do exactly the right things to take care of themselves and others? And what factors of risk will determine who will "lose it" for a period of time, and shake or cry and become unable to act decisively? Then, once the immediate threat has passed, what factors will determine who will be troubled by no more than occasional painful memories of a terrible event, and who will develop lasting stress symptoms? In addition to answering questions about causation and individual risk and resiliency, models of stress and stress injury also lay the foundations on which valid treatment approaches are built. And, finally, all research on stress injuries contributes to scientific understanding by testing the hypotheses generated by these models.

The goal of this chapter is to survey the major theoretical models of stress and stress injury, including the traditional military model that places character and leadership at the core of risk and resiliency; the three psychological models that have developed from psychoanalytic, learning, and cognitive theory; and contemporary neurobiological models of stress and stress injury. Each of these models will be given a chance to answer the fundamental questions asked above. As will be apparent, no single model has ever satisfactorily answered all of them, though each has provided useful partial answers. Each model is like one of the blind men in the ancient Indian proverb trying to describe an elephant after feeling only one portion of the beast. To the extent all are accurate, the truth must include all their different views. It is impossible in one chapter to provide an encyclopedic review of all existing models of stress and stress injury. Nevertheless, it is hoped that the following "wave-top" review of major social, psychological, and biological theories of stress injury will provide the reader with a framework for understanding stress injury research and management—and hopefully, without too badly oversimplifying or trivializing centuries of scientific work.

CHARACTER-LEADERSHIP MODEL OF COMBAT/OPERATIONAL STRESS INJURY

> Psychologists, sociologists, and the like had not yet been invented so there was no pernicious jargon to cloud simple issues. Right was right and wrong was wrong and the Ten Commandments were an admirable guide.... A coward was not someone with a "complex" (we would not have known what it was) but just a despicable creature.
>
> *WWI Scottish Army officer, quoted by Shephard (2001, p. 25)*

The oldest theoretical model for understanding why warfighters respond to combat stress as they do may be called the "character-leadership model." This model is not a formal theoretical system that has been argued and developed scientifically over time so much as a set of assumptions that have permeated the beliefs of warriors for as long as war has existed. The basic premise of the character-leadership model is that the primary determinants of resiliency on a battlefield are the strength of each individual's character and the effectiveness of their leadership.

How "character" and "leadership" are defined has always been subjective and open to considerable interpretation (and prejudice). But the origins of the character-leadership model are apparent in the ancient Greek warrior ideal known as *arete*—the combination of strength, valor, and courage that was central to the character of the Greek aristocrat-warrior. For the ancient Greeks, *arete* was a defining characteristic of nobility, determined by parentage, education, individual prowess, virtuous behavior, and favor by the gods. The same warrior ideal is apparent in the samurai of ancient Japan, the knights of medieval Europe, and the leadership classes of modern armies. According to this ideal, warriors who are virtuous, noble, and strong—and who are guided by even more strong and noble leaders—should master even the most horrendous stressors in order to triumph over those who are weak, evil, or common. Conversely, according to the character-leadership model, failing to master the stress of combat or to achieve victory suggests possible moral weakness, selfishness, or vice—in individual warriors, their leaders, or both.

The assumptions of the character-leadership model are not merely relics of the distant past, of course. Nor are they without scientific merit. Experiences in World War I made it clear that certain individuals were more predisposed than others to developing "shell shock" and other stress disorders, particularly those who were intellectually challenged or already suffering from a serious mental disorder. This awareness led to the first concerted attempts by the military to screen out predisposed individuals—those judged to be "insane, feebleminded, psychopathic, or neuropathic" (Shephard, 2004, p. 41), or labeled as a "psychopathic inferior, feeble minded mental defective, moron, imbecile, neurotic, hysterical, sexual pervert, or psychotic" (Copp & McAndrew, 1990, p. 34). Shephard cites evidence that screening such individuals out of the American Army during the final years of WWI had a positive impact on American stress casualties in France, though we still had 69,394 psychiatric casualties by war's end (Shephard, 2001, 2004). Following WWI, it was postulated that a significant number of those veterans who did not recover from their war neuroses were impeded in their recovery by "character weakness" of one sort or another.

In World War II, vigorous preinduction screening was carried out by the British, Americans, and Canadians in an attempt to eliminate the vulnerable from the battlefield, but with nearly disastrous results. In the United States, the psychiatric rejection of nearly 1 in 3 conscripts caused an uproar in the press and the minds of Americans, who wondered how it was possible that so many of the nation's young men could be too immature or weak to serve effectively in the military. And what's worse, psychiatric screening of recruits in America during WWII did little to reduce the numbers of combat stress casualties in Europe, Africa, and the Pacific. The British and Canadians were so confident in their psychiatric preinduction screening that they landed at Normandy on D-Day with few plans or resources for managing acute combat stress casualties. However, British and Canadian psychiatric casualty rates were no lower than the American Army's, accounting for approximately 1 in every 4 nonfatal battlefield casualties (Copp & McAndrew, 1990).

On the other side of WWII, the Germans notoriously claimed to have had few if any combat stress casualties precisely because the character of their troops and the skill of their leaders were believed to be exceptional. However, retrospective analysis of Wehrmacht records shows that the German Army was no more immune to psychiatric breakdown than were allied forces (Schneider, 1986).

During and after the Vietnam War, preexisting character defects were often blamed for the signature stress problems from that war—drug and alcohol problems and the kinds of antisocial behavior immortalized in post-Vietnam movies like *Rambo* and *Taxi Driver*. The National Vietnam Veterans Readjustment Study in 1983 did, indeed, find slightly higher rates of substance use problems and antisocial personality disorder (the only type of personality pathology assessed in that study) in Vietnam veterans compared to control samples (Kulka et al., 1990). And a number of studies found a significant co-occurrence of personality pathology in Vietnam-era veterans suffering from posttraumatic stress disorder (PTSD) (e.g., Southwick, Yehuda, & Giller, 1993). Without prospective data, however, it is impossible to know to what extent personality pathology among Vietnam veterans preexisted their combat exposure, rather than being—like PTSD—an outcome of their wartime stress. A recent reanalysis of data collected by the Vietnam Veterans Readjustment Study seems to support the hypothesis that although premilitary experiences and behavior were the largest determinants of postmilitary antisocial behavior in Vietnam veterans, PTSD played a key role in converting war-zone stress into later antisocial behavior (Fontana & Rosenheck, 2005).

Obviously, much remains to be learned about the extent to which preexisting constitutional factors predict success or failure on the battlefield, or the subsequent development of stress disorders such as PTSD. There can be no doubt that preinduction screening has helped reduce numbers of psychiatric casualties since WWI. And as every warfighter knows, internal resources of fortitude and willpower are important components of resiliency. But studies of battle fatigue since WWII have shown convincingly that the primary cause of battle fatigue is not the warfighter's character but the intensity and duration of the fight (Appel & Beebe, 1946; Blood & Gauker, 1993; Jones & Wessely, 2001).

PSYCHOLOGICAL MODELS OF COMBAT/OPERATIONAL STRESS INJURY

Among the oldest theories devised to explain the relationship between stressful experiences and later symptoms are those crafted from a psychological perspective. The range and variety of psychological models of stress-induced dysfunction are too great to review in depth in this chapter. Rather, a very brief overview will be offered of three different approaches to the psychological understanding of combat/operational stress injuries—those based on psychoanalytic, learning, and cognitive theory.

Psychoanalytic Perspectives on "War Neurosis"

Sigmund Freud, the father of psychoanalysis, was a neurologist by training. His approach to understanding mental phenomena was very much rooted in physics and medicine, as they were understood at the turn of the 20th century. His most basic concept of mental dynamics—the "pleasure principle"—was based largely on the simple model of the neurological reflex arc. Just as the tap of a tendon by a physician's hammer elicits a reflex jerk of a limb, Freud theorized, life experiences elicit reflex impulses to act, which he called "drives." As infants, our drives to obtain physical and emotional sustenance are relatively unimpeded. If we are hungry or frightened, we cry. If we are content, we sleep. Obtaining pleasure and avoiding pain, according to Freud, is apparent in all behavior in the infant. As the infant grows, however, life becomes increasingly complicated. Prohibitions against freely acting on one's internal impulses are progressively internalized from parents, teachers, and peers; behavior becomes increasingly inhibited and controlled.

The central core of drives and impulses, present from birth, Freud termed the "id." The sum of all prohibitions later internalized from external authority figures he called the "superego." To mediate between the irreconcilable id and superego, according to Freud, each person slowly developed a set of executive mental functions that he collectively termed the "ego." The ego's job is to keep peace between id and superego while trying to meet the needs of both. One of the main tools the ego employs in its mediation between id and superego is "repression," which was the name given by Freud to the process of managing unacceptable or dangerous thoughts, feelings, and impulses by pushing them down into the realm of the unconscious mind. The ego maintains a barrier between the conscious decision-making functions of the mind and the unconscious storehouse of mental content on which the ego has decided the person had better not act. It is as if the ego says to itself, "what I don't know won't hurt me."

Under ideal conditions, the tension between impulses and prohibitions is kept to a minimum, and the barrier maintained by the ego between behavior and repressed, unconscious impulses remains intact. Under the impact of traumatic stress, however, unacceptable or dangerous emotions and impulses may be so intense as to be uncontainable beneath the repression barrier. On a battlefield, such overwhelming emotions may include terror or rage, for example, and be accompanied by intense impulses to flee or blindly attack. Because such intense impulses and the emotional reactions that generate them are unacceptable to the superego—or in direct conflict with other impulses, such as an urge to charge the enemy in a fit of rage may run hard against the wish to survive—the ego does its best to repress these impulses and their attendant emotions. But like too much water forced into a leaky barrel, these intense impulses and emotions spill or leak out in the form of "emergency discharges," which are the source, in Freud's view, of all neurotic symptoms (for further discussion, see Fenichel, 1945).

The symptoms that Freud saw as emergency discharges of excessive internal urges or emotions could be of many kinds, including losses of physical function ("conversion" of emotional urges into physical disability) such as blindness or paralysis, for example; irrational fears (phobias); or inexplicable feelings of anxiety or depression. Psychoanalytic theory contends that neurotic symptoms of all kinds represent not only emergency discharges of dammed up impulses, thoughts, or feelings, but also symbolic solutions to the unresolved unconscious conflicts associated with them. For example, warfighters who witness horrible scenes of carnage that they are subsequently unable to tolerate in consciousness may develop temporary blindness as a means of simultaneously convincing themselves they didn't really see anything horrible after all, protecting themselves against further visual horrors, and denying that they were unable to master an experience that others seemed to take in stride. As an example, a Marine who survives a roadside IED (improvised explosive device) blast may first become aware of total deafness in both ears at the moment his battalion sergeant major orders him to prepare to go immediately back out on the same stretch of road. Or an infantryman who cannot forgive himself for accidentally shooting and killing a young child or unarmed woman may begin to shake violently only when he picks up his rifle.

In the Freudian psychoanalytic view, the individuals who would be expected to be most susceptible to being injured by a particular stressor are those who have not yet gotten over neurotic conflicts from their pasts, whether stemming from prior traumas or other unhealthy early life experiences ("infantile neuroses"). To the extent a given individual's repression barrier is already leaking because of too great a quantity of unacceptable and dangerous impulses left over from childhood or past traumas, that person's ego is weakened against managing current life stressors.

One of the goals of aggressive preinduction screening carried out in the early years of WWII was to eliminate from military service all individuals judged to have preexisting neuroses. Of course, neuroses, as defined by the psychoanalytic model, lie on a continuum between health and disease; they are never so clearly either present or absent. And having neurotic symptoms before deploying to combat is not so clearly a predictor of vulnerability—one study of a group of so-called neurotic types in WWII showed that they were very capable on the battlefield, and had no higher rates of stress casualties than supposedly normal troops.

Treatments devised directly from the psychoanalytic model have typically focused on "abreaction"—of drawing unconscious, repressed memories and their associated thoughts and feelings into conscious awareness, to relieve the pressure on the repression barrier. During and after WWII, when psychoanalytic theories provided the dominant foundation for combat psychiatric treatment, hypnosis and drugs such as sodium amytal ("truth serum") were used to disinhibit stress-injured warriors to the point of reexperiencing what they feared to think and say. Psychoanalytic psychotherapy is a talking treatment which aims to gradually strengthen the ego over time as it reduces the power of repressed, unconscious thoughts and feelings to affect behavior by bringing them into conscious awareness.

Although there is little doubt that empathic listening and the recovery of the forgotten past can be powerful tools for healing, treatments based solely on psychoanalytic principles are no longer offered to stress-injured warriors because there has been little empirical evidence to support their long-term effectiveness. Furthermore, uncovering treatments can be very time consuming, and have sometimes made stress-injured clients worse rather than better.

Another possible shortcoming of the psychoanalytic model is its assumption that all stress injury symptoms are volitional—somehow chosen by individuals, albeit unconsciously, as solutions to otherwise unsolvable problems. This assumption is founded on the premise that all components of everyone's mental and neurobiological machinery can and should fall under their conscious control, and that stress injuries do not involve actual damage to this machinery. As will be discussed below, there are compelling reasons to doubt these premises in their entirety.

Learning Theory Perspectives on Conditioned Fear Responses

Learning theory, in its many present-day forms, is based on the discovery by the physiologist Ivan Pavlov of what he termed "conditioned reflexes" in experimental animals (Pavlov, 1960/1927). Pavlov started with the observation that hungry animals such as dogs naturally responded to the stimulus of the sight and smell of food by salivating. He called this an "unconditioned reflex" or "inborn reflex" because it was a natural response that required no learning or conditioning. By repeatedly exposing hungry dogs in his laboratory simultaneously to the sight and smell of food and the sound of a bell, however, Pavlov was eventually able to elicit a "conditioned reflex" of salivation to the sound of the bell alone. Pavlov's insight was that the temporal pairing of unrelated stimuli (food and bell) could cause an animal to learn to respond to both the same way.

The form of learning that Pavlov described has become known as "classical conditioning."

Although Pavlov used a rewarding stimulus (food) in his now-famous experiments, other researchers have since studied classical conditioning using aversive stimuli such as electric shocks or the sights or smells of predators. The form of classical conditioning that results from the pairing of an aversive stimulus and a neutral stimulus has become known as aversive conditioning or "fear conditioning." Classical fear conditioning is the theoretical paradigm that learning theorists have used to explain learned fear responses that result from traumatic experiences. Just as Pavlov's dogs learned to respond with salivation to the sound of a bell, even in the absence of food, warfighters exposed to terrifying situations in combat learn to respond with physiological arousal and subjective feelings of fear and anxiety when confronted with subsequent neutral reminders of those terrifying experiences. Such reminders (conditioned stimuli) could include loud noises, the sound of aircraft flying overhead, the sight or smell of vegetation similar to that experienced in the combat zone, the sight of crowds of strangers, or even a paper bag lying on the roadside that could conceal an IED. Recalling the vignette presented at the beginning of this chapter, it is not hard to imagine how the young PFC in whose lap the gunny fell would develop an intense reaction to the sight or taste of Skittles candy.

Classical fear conditioning is a normal process of learning, not necessarily a pathological process of injury. Most, if not all, individuals exposed to a terrifying or horrible event will later experience conditioned fear responses to neutral reminders of the event. But also normal is the process known as "extinction," in which repeated exposure to neutral stimuli without simultaneous experience of terror or horror causes the conditioned fear response to gradually fade. After passing enough roadside trash while traveling in the United States after returning from a tour in Iraq, for example, most warfighters find they eventually cease to respond with fear and arousal. Why, then, do certain individuals develop persistent, even worsening, fear responses to neutral cues after a traumatic stress injury? What prevents extinction from diminishing their conditioned fear responses over time?

To answer this question, Mowrer (1960) proposed a two-factor model for the development of posttraumatic stress symptoms. The first factor in his model was classical fear conditioning as described above. But the second factor in Mowrer's model was another type of learning altogether, one known since the work of B. F. Skinner as "instrumental" or "operant" conditioning. In operant conditioning, what is learned is not the pairing of a conditioned stimulus (such as the threat of an IED explosion) with an unconditioned stimulus (such as the sight of roadside trash), but the pairing of an environmental stimulus with a behavior that is devised through trial and error as a solution to the problem posed by the stimulus. For example, learning to drive on the centerline of the road or to swerve to avoid roadside trash are operant responses to the danger posed by roadside IEDs. How does operant conditioning contribute to the perpetuation of posttraumatic stress symptoms? Mowrer and subsequent theorists have proposed that trial-and-error learning is used by individuals to find ways to avoid the internal and external cues that provoke their anxiety just as laboratory animals in an experimental shock box can learn to jump off the electric grid beneath them to avoid being shocked if they are given warning of the shock to come. Thus, as long as combat veterans with PTSD continue to avoid the sights, sounds, and memories that trigger their conditioned fear responses, they can reduce their potentially painful experiences of anxiety and loss of control, which rewards their avoidance behaviors, but at the price of preventing themselves from going through the natural healing process of extinction. Thus, avoidance behaviors may prolong conditioned fear responses indefinitely.

Another explanation for the failure of individuals with PTSD to recover from their conditioned fear responses was proposed by Foa and Kozak (1986), who suggested that one of the factors preventing emotional processing (and extinction) of fear-related learning was the large number and variety of conditioned stimuli that develop as a "fear structure" under the impact of intense fear. In their model, conditioned fear responses could not diminish over time unless all the stimuli associated with a trauma—including the multitude of perceptions, emotions, thoughts, and physiological responses at the moment of the danger—were all simultaneously reexperienced in the absence of a real threat.

Treatments based on learning theory are among the most successful in controlled clinical trials for treatment of PTSD. They include systematic desensitization, stress inoculation training (SIT), and Edna Foa's prolonged exposure (PE) treatment (Foa & Rothbaum, 1998). Known collectively as exposure therapies, these treatments all make use of controlled reexperiencing of traumatic cues both in imagination and in real life in order to facilitate desensitization and extinction of conditioned fear responses. In PE, for example, traumatized individuals are asked to record on audiotape the story of their traumatic experiences in detail and in the present tense. They are then asked to listen to these tapes on a daily basis while practicing relaxation techniques. The theoretical basis for exposure treatments is related to the common sense approach to recovery from a frightening experience contained in the old aphorism about getting back on the horse that threw you. Virtual reality technologies are also now being studied as a potentially very effective method for exposure treatment of combat veterans (Rizzo, Rothbaum, & Graap, this volume, chapter 9; Spira, Pyne, & Wiederhold, this volume, chapter 10).

The strength of exposure treatments may also be one of their weaknesses, however. To the extent such treatments ask individuals to relive their traumatic experiences, they require a great deal of motivation and grit on the part of the client. Also, although exposure therapies have been used successfully with veterans with combat-related PTSD, such learning theory-based treatments have not been shown to be as effective with other types of combat/operational stress injury, such as those that are *not* caused by specific events involving terror or horror. And combat-related traumas, unlike civilian traumas such as rape or natural disasters, often involve some degree of perpetration of horrors and terrors, as well as passively experiencing them (MacNair, 2002).

Cognitive Theory Perspectives on Erroneous or Damaged Beliefs

Modern cognitive theories, which have also been immensely pertinent to the understanding and treatment of stress injuries, are all based on Jean Piaget's theories on the cognitive development of children. According to Piaget (1952), children learn about the world and their place in it by absorbing and storing knowledge in the form of cognitive "schemas." Each schema is a packet of information that the child—and later, the adult—uses to predict future events and to exert mastery over the environment. But schemas are much more than just packets of knowledge passively stored in the mind like so many words in a book. First of all, each new packet of knowledge is not absorbed in isolation but rather always in its relation to existing schemas and expectations about the world. It is a well-known fact that meaningless information—such as a string of random digits—is much more difficult to learn and remember than information that makes sense because it bears some relationship to already known facts. It is human nature to attempt to fit all new experiences into our existing beliefs. Second, our existing beliefs and expectations color our perceptions of the world and influence the way in which we record and remember our experience. Thus, it is also human nature to impose our beliefs on the world in an attempt to make each new encounter fit our preconceptions.

Over the course of a lifetime, schemas must continue to grow and evolve as the individual's knowledge of the world grows and becomes more accurate and sophisticated. Piaget postulated two different processes by which schemas could develop as a result of experience—*assimilation* and *accommodation*. Assimilation occurs when a new situation is encountered that *appears* to conform to the expectations generated by already existing schemas. As a simple example, imagine that a very young child has just seen an automobile for the first time in its life, and it just happened to be blue. The child will doubtless internalize information about the automobile, as much as could be perceived from its point of view, and it may establish the schema, "cars are blue." If the very next automobile it saw happened to be yellow, this new information could be *assimilated* into the existing schema, without changing it, by simply ignoring the color of the car, which would reassure the child that its basic car concept was not faulty. The child might also assimilate the experience of a yellow car by telling himself that what he had just seen was not a car after all, but some other unknown entity masquerading as a car. The experience of new information that conflicts with preexisting schemas provokes a form of discomfort that has been called "cognitive dissonance" (Festinger, 1957), and individuals of all ages tend to relieve this discomfort through all manner of filtering and distortion.

If, on the other hand, a child who held the schema "cars are blue" was willing and able to change his basic car concept after seeing a yellow car, he would be said to *accommodate* his car schema to fit the new experience. Thus, the accommodated new car schema may sound something like, "cars are yellow or blue." Similarly, if the first school bus the child saw was also yellow, the child might accommodate his car schema to include buses as follows: "vehicles are yellow or blue." The basic difference between assimilation and accommodation is that in the former, the input (perception) is modified to preserve the existing schema, while in the latter, the schema itself is modified to fit the external reality (Piaget & Inhelder, 1969).

Cognitive theories of stress and stress injury are based on the limitations and vulnerabilities of the cognitive processes described by Piaget. Given the complexity of the universe, on the one hand, and the limitations of human information processing capabilities, on the other, it should not be surprising that schemas are sometimes faulty or just plain wrong. One of the most successful and empirically validated forms of psychotherapy for depression, cognitive therapy, was developed by Aaron Beck (1967) based on the theory that dysfunctional emotions such as depression arise from erroneous and self-defeating cognitions (schemas) about the self and the world of which the individual is unaware. Current cognitive conceptions of trauma hold that traumatic experiences generate distorted negative and inappropriately generalized beliefs such as "I am helpless," "I am not safe anywhere," or "I deserved the bad things that happened to me" (Ehlers & Clark, 1999). As long as such negative and distorted beliefs remain unchallenged and largely unconscious, according to the cognitive model, the negative affects they generate can persist indefinitely.

But what happens when current life experience is so discordant with stored schemas and the assumptions associated with them that no immediate reconciliation is possible? Ronnie Janoff-Bulman (1992) was among the first to identify the central role of shattered assumptions in posttraumatic stress disorders. She postulated three fundamental and necessary assumptions that traumatic experiences, by their very nature, acutely damage: (1) the world is benevolent, (2) the world is meaningful, and (3) the self is worthy. Jonathan Shay (1994) has eloquently described the "moral injury" that is often at the heart of posttraumatic stress disorder in combat veterans. And many other clinicians and researchers have examined how both trauma and the loss of close attachments through death violate existing beliefs and meaning (e.g., Figley, 1999; Kauffman, 2002). One of the particular strengths of cognitive approaches to understanding trauma and loss is their focus on painful emotions other than fear—including shame, guilt, sadness, and rage—as consequences of violated belief systems.

Because the dissonance generated by the rift between old ways of seeing the world and the life-transforming experiences associated with trauma and loss tends to be so great, individuals often avoid thinking or feeling about those experiences, which perpetuates their continued existence apart from the rest of their inner world. Therefore, cognitive treatments for stress injury must gradually encourage, in an atmosphere of safety, processing and coming to terms with these areas of discordance with the aim of reconstructing damaged belief systems and finding meanings in traumatic experiences and their aftermath. To accomplish this, the reality of the traumatic experience and the individual's reactions to it must be integrated with the assumptions and world view that existed prior to the stress injury.

One potential shortcoming of cognitive approaches to the treatment of stress injuries such as PTSD is their focus on conscious and voluntary information processing, sometimes to the exclusion of the sort of unconscious and involuntary processing that is the focus of learning theory. To reconcile this dichotomy, Brewin, Dalgleish, and Joseph (1996) proposed a "dual representation" theory of PTSD, postulating that healing required both unconscious relearning and conscious restructuring. In current practice, most psychological therapies for stress injuries such as PTSD incorporate both exposure elements, based on learning theory, and cognitive restructuring elements, based on cognitive theory, with varying emphasis on one or the other. Two empirically and clinically successful treatments for PTSD that appear to emphasize cognitive elements, but include exposure, are eye movement desensitization and reprocessing (EMDR) therapy, developed by Francine Shapiro (1989, 2001), and cognitive processing therapy, developed by Resick and Schnicke (1992).

BIOLOGICAL MODELS OF STRESS AND STRESS INJURY

The explosion of research in biological aspects of stress and stress disorders has been so massive in recent years that an entire volume devoted to this subject would probably not do it justice. And new findings are added to this body of literature almost daily, so reviews of it may be obsolete soon after they are published. Nevertheless, we will attempt here to provide a brief and coherent overview of the current neurobiological conceptions of stress and stress injury that draws attention to major themes and areas of vigorous current research.

Patterns of Biological Response: Hard Wired and Fixed, or Learned and Plastic

> For all of its diversity, one can view neuroscience as being concerned with two great themes—the brain's "hard wiring" and its capacity for plasticity.
>
> *Kandel & Squire (2000)*

Brain "hard wiring," according to Kandel and Squire, refers to how connections develop between cells, how cells function and communicate, and how an organism's inborn functions are organized—its sleep-wake cycles, hunger and thirst, and its ability to perceive the world and respond to danger. These are ancient nervous system adaptations, evolved over thousands of millennia, which are too important to survival to be left to the vagaries of individual experience (Kandel & Squire, 2000). In contrast, brain plasticity refers to individuals' capacities for adaptation or change as a result of experiences encountered during their lifetimes. In comparison to other organisms in the animal kingdom, the larger cerebral cortices of human beings provide for a less rigid response to environmental dangers than that of lower animals; humans, therefore, have a greater capacity for innovation and an edge in the race for survival. However, our greater capacity for adaptation leaves greater leeway for adverse biological responses to the environment, including those associated with stress injuries such as PTSD. Although few biological response systems in the brain are entirely either hard wired or plastic, it might be helpful to begin by first describing those response systems that are primarily fixed by gene programs, followed by a description of those in which learning and memory play a larger role.

Hard-Wired Stress Response Systems

The simplest hard-wired stress response pattern is the reflex arc, consisting of an *afferent* neuron, bringing sensory information into the brain or spinal cord about a threat or opportunity, linked to an *efferent* neuron, sending action/response information from the central nervous system out to a muscle or glandular cell in the body. Examples of reflex behaviors include jerking one's hand off a hot stove or blinking when a puff of air strikes one's cornea. Reflexes seem superficially like simple electrical circuits—a button is pushed, and a buzzer sounds or a light flashes. However, even the simplest reflex arcs in the brain are capable of modulation through a number of mechanisms, almost all of them involving chemical messengers. Chemical messenger systems in the brain that control the amplitude of even the most hard-wired responses in the brain include neurotransmitters and hormones.

Major Neurotransmitters in the Brain. All neurons in the brain and body communicate with other neurons or effector cells in muscle or glandular tissue across microscopic gaps called *synapses*. Communication between neurons is transmitted across a synapse as small local clouds of neurotransmitter molecules, which diffuse across the tiny synaptic gap to initiate an excitatory or inhibitory electrical potential in the post-synaptic neuron or effector cell by binding with a *receptor*. Each neuron produces one or more particular types of neurotransmitter and/or neuromodulator chemicals that determine how that neuron affects the cells to which it carries electro-chemical impulses. Neurotransmitters involved in the stress response include the following, although others probably remain to be identified:

- Glutamate: the primary excitatory neurotransmitter in the brain. Glutamate neurons issue a "go" signal. One particular type of glutamate receptor, the NMDA (*N*-methyl-D-aspartate) receptor, has been found to be crucial to memory and cognitive processing systems in the brain, including those involved in classical fear conditioning (Fanselow & Kim, 1994; Miserendino et al., 1990).
- Gamma-aminobutyric acid (GABA): the primary inhibitory neurotransmitter in the brain. GABA neurons issue a "stop" signal. Substances, such as alcohol and benzodiazepine tranquilizers, that stimulate GABA receptors cause a reduction in perceived stress and anxiety, but at the expense of alertness.
- Norepinephrine: an activating neurotransmitter found in a network of neurons in the brain whose function is to modulate other neuron systems in the brain. Important hard-wired circuits that are activated by norepinephrine neurons include those that are part of the *sympathetic nervous system*, which accelerates heart rate, raises blood pressure, and prepares the body for action (Gold & Chrousos, 2002). Activation of norepinephrine neurons in the brain also causes increased arousal and focused attention, and it interferes with sleep. Norepinephrine has a biphasic effect on memory, enhancing memory at moderate levels, but inhibiting them at high levels (Cahill & McGaugh, 1998). Stress of any kind increases the activity of norepinephrine neurons in the brain, and prolonged, severe stress can deplete brain stores of norepinephrine. One of the most consistent biological findings in individuals with PTSD from any cause is an elevation and loss of normal modulation of norepinephrine activity in the brain (e.g., Geracioti et al., 2001). This is believed to account for much of the "hyper-arousal" cluster of PTSD symptoms.
- Dopamine: another activating neurotransmitter found in a limited network of neurons in the brain involved in modulating the activity of other neurons. Dopamine activation in a part of the brain called the *nucleus accumbens* has long been hypothesized to be central to motivation and pleasure, sometimes referred to as the "brain reward system" (Olds & Milner, 1954; Wise, 1996). There is substantial evidence

that addictive drugs and addictive behaviors act by significantly increasing dopamine activity in the nucleus accumbens (Koob & LeMoal, 1997). And depression has been hypothesized to involve dysfunction in the dopamine reward system in the brain (Tremblay et al., 2002). Low levels of stress increase dopamine activity in the circuit of the brain to which the nucleus accumbens belongs, while extreme stress in animals has been found to decrease dopamine activity in the nucleus accumbens (Charney, 2004; Horger & Roth, 1995).

- Serotonin: an inhibiting neurotransmitter found in the limited network of neurons making up the third major regulatory network of the brain (along with norepinephrine and dopamine). The functions of the serotonin system in the stress response are complex and not completely understood, but serotonin is known to inhibit components of the conditioned fear response, such as acoustic startle, as well as brain centers involved in the flight or flight response (Coupland, 2000). And in human subjects, low levels of serotonin activity have long been known to be associated with impulsivity, aggression, and suicidality (Lesch & Merschdorf, 2000).

Major Stress Hormones in the Brain and Body. Hormones in the brain and body differ from neurotransmitters mainly in the range of their activity—while neurotransmitters normally act only across a narrow synaptic gap, hormone messengers either diffuse through tissues or are carried systemically in the blood stream. The most important stress hormones in the brain and body are the following:

- Cortisol: a master regulator and facilitator of the stress response, secreted from the cortex of the adrenal glands. Levels of circulating cortisol in the body are controlled by a hormonal regulatory system collectively known as the HPA (hypothalamic-pituitary-adrenal) axis. The HPA axis is a cascade of hormonal secretions that begins in the hypothalamus—the link between electrical signals in the brain and hormonal signals in the body. Stress of any kind causes the hypothalamus to release CRF (corticotrophin-releasing factor) into the pituitary gland, which then releases ACTH (adrenocorticotropic hormone) into the bloodstream. Levels of cortisol secretion by the adrenal glands go up or down in direct relation to changing levels of ACTH from the pituitary gland. Cortisol is essential for life and has several important stress-related actions in the body, the most important of which is increasing the availability of glucose in the bloodstream to fuel the tissues of the body. Cortisol is also essential for turning off the stress response through feedback inhibition of ACTH release from the pituitary. But cortisol also has a number of potentially adverse effects on the body and brain. For example, cortisol inhibits the immune system, which can result in a number of stress-related illnesses. In persistent excess, cortisol promotes fat deposition and atherosclerosis. Finally, as will be discussed below, cortisol can be directly toxic to vulnerable neurons in the brain that also happen to be essential for coping with stress.

- Corticotrophin-releasing factor: a peptide chemical messenger that has two distinct roles in the brain, one as a hormone, the other as a neurotransmitter. As a hormone, CRF signals the pituitary gland to release ACTH into the bloodstream, ultimately inducing the release of cortisol from the adrenal glands. As a neurotransmitter found in many parts of the brain, CRF is a major mediator of the amplitude of the stress response. Acute and chronic stress increase CRF activity in the brain. Although receptors for CRF in the brain can either activate or inhibit stress-related brain circuits, CRF itself promotes stress-related emotions and behaviors. CRF injected directly into the central nervous systems of animals, for example, produces behaviors virtually indistinguishable from those known to be generated by stress and anxiety (Arborelius et al., 1999). CRF increases the activity of norepinephrine neurons in the brain (Page & Abercrombie, 1999; Van Bockstaele, Colago, & Valentino, 1998), thus directly promoting arousal in the brain and sympathetic nervous system. And high levels of CRF activity in the brain have been correlated with depression, anxiety, substance use disorders, and PTSD (Arborelius et al., 1999; Baker et al., 1999).
- Neuropeptide Y (NPY): a peptide hormone and neurotransmitter that appears to have significant antianxiety properties, perhaps by directly opposing the effects of CRF in the brain (Heilig et al., 1994). NPY injected directly into the brains of animals reduces anxious behaviors, and it reduces norepinephrine activity in the brain (Charney, 2004). Morgan and his colleagues (2000) found that high NPY levels were directly correlated with performance in warfighters undergoing SERE (survival, escape, resistance, and evasion) training. Combat veterans with PTSD have been found to have low levels of NPY compared to normal controls (Rasmusson et al., 2000).
- Endogenous opioids: peptide hormones in the brain and body that bind to the same receptors as narcotic analgesics such as morphine and heroin, with similar effects. Endogenous opioid peptides, such as the endorphins, cause "stress-induced analgesia," the relative insensitivity to pain that accompanies extreme stress (Terman et al., 1984). Stress-induced analgesia has been implicated in the emotional and physical numbing response that humans typically experience when exposed to severe stress. Endogenous opioids increase dopamine activity in the nucleus accumbens in the same way narcotic drugs of abuse do (Stout, Kilts, & Nemeroff, 1995).

Brain Stress Response Systems Involved in Learning

There are many different memory systems in the brain, each having separate neurocircuits that operate independently and in parallel (Squire, 2004). These memory systems may be divided into two broad groups, *explicit memory* and *implicit memory*. Explicit memory—also referred to as "declarative" memory because it can be consciously recalled and communicated at will—includes facts about the world and the contexts in which experiences occur. Implicit memory includes all types of memory that cannot be consciously recalled at will—although involuntary recall may be triggered by external or internal cues—including the emotional and physical components of fear conditioning. The part of the brain most central to the learning of explicit memories is the *hippocampus* and its associated circuitry in the brainstem and cerebral cortex (Kramer et al., 2005; Lepage & Richer, 2000; Moscovitch et al., 2005; Shimamura & Squire, 1987; Squire, 2004; Tulving & Markowitsch, 1998). The part of the brain most central to the type of implicit learning known as fear conditioning is the *amygdala,* along with its connecting neurocircuitry (Davis, 1992; Phelps & LeDoux, 2005; Squire, 2004). The hippocampus and amygdala lie side by side, one in each temporal lobe of the brain. Understanding how these two brain systems function and cooperate under normal conditions, and how they fail to cooperate under extreme stress, is central to biological models of stress injury, especially traumatic stress.

The Hippocampus. The function of the hippocampus in humans was first discovered through the study of individuals who had suffered damage to their hippocampi because of injury or surgery for epilepsy (Squire, 1987). Patients with damage to their hippocampi (like the character played by Drew Barrymore in the 2004 movie *50 First Dates*) are unable to remember explicit information beyond a few seconds or minutes. Their long-term memory is intact, and they can repeat and follow instructions, but they cannot remember new information from one day to the next. How the hippocampus mediates between experience and long-term memory is still unclear, but its crucial role in declarative memory is apparent from its position in the brain as a target of converging inputs from the many regions of the cortex that process sensory information, including from the eyes, ears, and the rest of the body. One current conception of how the hippocampus processes all that sensory information, as well as the cognitions and memories that are evoked by it, is to maintain an ongoing map of "memory space"—a continuous tracking registry of where the individual is currently located in time and space (Eichenbaum et al., 1999; O'Keefe & Nadel, 1978; Wallenstein, Eichenbaum, & Hasselmo, 1998). By associating sensory experiences that have a relationship in space or time—such as the flash of a muzzle, the sound of a gunshot, and the impact of a round, for example—the hippocampus helps to fit new experiences into long-term memory and belief systems. Through its reciprocal connections to cortex, the hippocampus gradually mediates the conversion of short-term memories into longer-term and more durable memories.

But the hippocampus seems to do much more than just record new experiences in explicit memory. The fact that drugs such as ketamine and PCP cause dissociative states and amnesia, very similar to stress-induced dissociation, by blocking NMDA glutamate receptors in the hippocampus suggests that the hippocampus is crucial to the moment-to-moment integration of thoughts, sensations, and feelings that is taken for granted in the absence of dissociation (Krystal et al., 1995). It also implicates the hippocampus in the production of dissociation symptoms during traumatic stress (Chambers et al., 1999). The hippocampus is also involved in fear conditioning in two ways. First, it records the details of the context in which a danger was experienced, all of which may later become triggers for the conditioned fear response. The hippocampus is also known to be important in turning off the stress response once it has been activated (Yehuda, 2000), and it may be involved in the extinction of fear conditioning (Falls & Davis, 1995; Vianna, Coitinho, & Izquiredo, 2004).

The functioning of the hippocampus under stress is modulated by many neurotransmitters and hormones. Norepinephrine in the brain and circulating epinephrine (adrenaline) in the bloodstream, for example, augment memory acquisition by the hippocampus up to a point, ensuring that experiences generating moderate levels of arousal are not soon forgotten. Excessively high levels of norepinephrine and epinephrine activity, however, impair memory acquisition (Coupland, 2000; Gold & McCarty, 1995). Activation of GABA inhibitory neurons, such as by alcohol or benzodiazepine tranquilizers, inhibits new memory acquisition through the hippocampus. Since the mechanism by which the hippocampus records information involves the actual growth of new cells (neurogenesis) and the remodeling and growth of new connections between cells in the hippocampus (Korte et al., 2005), hormones and neurotransmitters that increase or decrease neurogenesis and remodeling have a significant impact on hippocampal functioning under stress. One of these neurotransmitters is serotonin, whose activity is directly increased by antidepressant medications used to treat stress disorders such as PTSD. Serotonin activation stimulates neurogenesis in the hippocampus (Djavadian, 2004; Joels et al., 2004; Huang & Herbert, 2006). Another very important modulator of neurogenesis in the hippocampus is cortisol, which inhibits the growth of hippocampal neurons and their connections (Nacher & McEwen, 2006). As will be seen below, cortisol has other, more serious effects on the hippocampus under conditions of intense or prolonged stress.

The Amygdala. The amygdala, a small, almond-shaped nucleus in the temporal lobe on each side of the brain, is one of the most extensively studied parts of the brain (LeDoux, 1996; Phelps & LeDoux, 2005). The amygdala receives sensory input directly from the sensory organs (via sensory gateways in another brain center called the thalamus) rather than from the sensory processing regions of the cortex from which the hippocampus receives its sensory input. This means that circuits between sensory organs such as the eyes and ears and the amygdala are much shorter than—and conduct information in much less time than—those between sensory organs and the hippocampus. Outputs from the amygdala form the major triggers for many components of the acute stress response, including reflex behaviors such as freezing or startle reactions, and the activation of the norepinephrine arousal system and the HPA axis, which results in the release of cortisol from the adrenal glands. Just looking at inputs and outputs of the amygdala suggests that it may serve to monitor incoming stimuli in search of a pattern that appears to pose a threat, then responding instantly—and below the level of conscious awareness—with an automatic stress response. A multitude of studies with animals and humans confirm that this is exactly what the amygdala does. It is central to fear conditioning in all mammals.

Even brief exposure of an animal to a threat (such as a predator) causes protein synthesis and lasting changes in the neurocircuitry of the amygdala (Adamec, Blundell, & Burton, 2005; Adamec et al., 2006). And the only way to either prevent fear conditioning in animals from occurring or of erasing it once it does occur is by damaging or removing the amygdala. Evidence for the participation of the amygdala in fear learning in humans is provided by functional magnetic resonance imaging (fMRI) studies and by studies with humans who have brain lesions involving the amygdala. The fMRI studies show a positive association between fMRI signal intensity and fear conditioning, even when the subject is shown a neutral stimulus (conditioned stimulus), previously paired with the aversive one of which he is only subliminally aware (Buchel et al., 1998; Morris et al., 1998; Phelps & LeDoux, 2005). Studies in brain-damaged patients highlight the distinct roles of the hippocampus and amygdala in learning. Patients with amygdalar damage and an intact hippocampus fail to show physiological evidence of fear learning, but can report the events of the fear-conditioning procedures, whereas patients with bilateral hippocampal damage cannot describe the fear-conditioning procedures, but have normal physiologic expression of the fear response (Bechara et al., 1995; Phelps, 2004).

Modulators that enhance fear conditioning mediated by the amygdala are many, including norepinephrine, epinephrine, and CRF. Modulators that diminish conditioned fear responses mediated by the amygdala include GABA, serotonin, and neuropeptide Y. Significantly, whereas cortisol activity in the hippocampus *decreases* memory formation and consolidation under stress, the same cortisol activity in the amygdala has the opposite effect—it *increases* fear-conditioned memory formation and consolidation (Sapolsky, 2003).

Biological Models of Stress Injury

All the biological stress response systems described so far contribute to normal adaptations to stress, even severe stress. As normal adaptive responses, these biological mechanisms must be, by definition, reversible once the source of the stress has been removed. But what happens in the brain and body under the impact of a too-intense stressor such as a traumatic event or too-prolonged stressor such as many months of daily exposure to danger during an operational deployment? What models have been developed to explain how biological adaptive mechanisms can go awry in conditions of extreme stress to contribute to potentially irreversible stress injuries? Two neurobiological models of stress injury will be briefly described here to answer these questions. One model, termed "allostasis," involves the shifting of set points in biological systems under severe or prolonged stress, resulting in the inability of those systems to return to their previous baseline levels once the source of stress has been removed. This model has been used to explain how stress can lead to lasting depression or anxiety symptoms, including some of those characteristic of PTSD. The other model explains the symptoms of stress injuries as resulting from stress-induced damage to neurons in vulnerable centers of the brain, especially the hippocampus.

Allostasis. "Allostasis" is a term that was coined by McEwen and Stellar (1993) to contrast with "homeostasis," the body's natural tendency to return to baseline states after being disrupted by stressors. "Homeostasis" comes from the Greek roots *homos*, meaning "self," and *stasis*, meaning "standing still." In homeostasis, as first described by Cannon (1932), biological systems resist change by directly opposing the changes caused by external and internal stressors in order to remain at their original points of equilibrium. A simple metaphor for homeostasis is a beam balanced on a fulcrum. As weight is added to one side of the balance beam (stress), the body applies force to the other side of the balance beam to try to restore equilibrium (homeostasis). An example of homeostasis is the tendency of muscles to repeatedly, briefly contract (shiver) to generate heat to restore body temperature to its normal set point after exposure to cold.

Allostasis, on the other hand, comes from the Greek root, *allos*, which means "other" rather than "self." In allostasis, biological systems restore equilibrium by *shifting the set point* (moving the fulcrum) rather than solely by applying counterforce to the other side of the balance beam. By shifting the fulcrum in biological systems of adaptation, equilibrium can be maintained with less force being applied to counter that imposed by stress. More importantly, allostasis can maintain equilibrium under conditions of such extreme stress that biological systems would otherwise be unable to keep a balance. Of course, one consequence of allostasis is that once the disturbing stressor has been removed, the biological system that has been stressed no longer returns to its original prestress balance unless the set point can be shifted again back to its original position. An example of reversible allostasis is the shifting of the body temperature set point during a fever. But some allostatic changes in the brain as a result of severe stress are not so easily reversible, and may result in prolonged or even permanent symptomatology.

One example of potentially irreversible allostatic changes in the brain as a result of stress is contained in the "neurotransmitter receptor hypothesis" of depression and anxiety (Stahl, 2000). In this hypothesis, stress (interacting with genetic predisposition) causes the depletion of neurotransmitter chemicals involved in the stress response, most notably norepinephrine and serotonin. Depletion of these neurotransmitters, without sufficient rest and recovery time to replace their stores in the brain, threatens the brain's ability to maintain balanced and adaptive functioning. To maintain balance, neurons in the brain have the capacity to turn on genes controlling the synthesis of receptors for those depleted neurotransmitters, which then proliferate (or "up-regulate") on the surface of neurons. With more receptors available to interact with fewer neurotransmitters in synapses, neuronal activity levels can be maintained in spite of lower levels of chemical messengers available to carry signals from one neuron to the next. This is analogous to the volume control on a radio being turned up to compensate for an ever-weakening radio signal. But what happens after the stress is removed or the individual restores depleted neurotransmitter stores through much-needed rest? All those extra receptors on the surfaces of neurons can not instantly vanish—the volume control cannot instantly be turned back down. In fact, rather than neurons with up-regulated receptors inducing them to "down-regulate" by turning off the genes that created them in the first place, they may just make other changes to adapt to the new state of affairs. A new, maladaptive equilibrium may be maintained around the new, allostatic set point. The effects of persistently up-regulated receptors for neurotransmitters such as serotonin have been postulated to underlie many of the symptoms of persistent depression and anxiety disorders, including those that develop during or after operational deployments. And they have been postulated to explain the antidepressant and antianxiety effects of medications like sertraline and paroxetine that induce serotonin receptors to down-regulate. This model also explains why the therapeutic effects of these medications are not instantaneous, since receptor down-regulation is a process that takes weeks, at least, to accomplish.

Hippocampal Damage as a Biological Consequence of Stress. There is ample evidence from preclinical animal studies that neurons in the hippocampus are remodeled or even damaged by stress (Bremner, 2002; Sapolsky, 2000). It used to be widely held that neurons, once developed, were relatively fixed and incapable of significant change in size or shape. It was also widely believed that because neurons could not divide and multiply, nervous tissue could not regrow neurons that had died for some reason.

Although these principles still hold true for most of the nervous system, it appears not to apply to the hippocampus. Unlike in other parts of the nervous system, normal functioning in the hippocampus involves a continual process of shrinkage and death in some neurons balanced by the growth and expansion of others. Under conditions of severe or prolonged stress, however, the normal reversible process of remodeling can be shifted toward destruction that exceeds regrowth, resulting in the loss of hippocampal neurons and functional ability (Joels et al., 2004; Kaufer et al., 2004; Sapolsky, 1985; Sapolsky & Pulsinelli, 1985). This shift of the balance in the hippocampus between creation and destruction, caused by stress, has been identified as another form of allostasis—a potentially irreversible shifting of set-point or balance-point (McEwen, 2000a,b).

Stress-induced damage to hippocampal neurons in animals has been found to occur by several different mechanisms. First, chronic stress of many kinds has been shown to cause neurons in the hippocampus to atrophy and their connections to other neurons to shrivel. Second, stress of many kinds has been shown to inhibit the normal regrowth of neurons in the hippocampus. Third, and perhaps most significantly, acute and chronic stress in animals have been shown to induce the physical death of neurons in the hippocampus. These three mechanisms, together, produce not only demonstrable losses in hippocampal function in stressed animals, such as their explicit memory abilities, but measurable shrinkage in their hippocampal volumes (Sapolsky, 2000). The processes that produce these three mechanisms of damage to hippocampal neurons at a cellular and molecular level differ somewhat, but all have been found to implicate the stress hormone cortisol in one way or another.

Cortisol is toxic to hippocampal neurons in a couple of ways. First the regrowth of hippocampal neurons is dependent on the availability of a local hormone in the brain known as brain-derived neurotrophic factor, which has been shown to be present in reduced levels in animals exposed to high cortisol levels (Smith et al., 1995). Thus, high cortisol levels, such as are believed to result from both acute and chronic stress, slows the rate of regrowth of hippocampal neurons. More directly, though, cortisol has been shown to promote the destruction of neurons in the hippocampus through its interaction with the excitatory neurotransmitter glutamate and its NMDA receptors. Glutamate NMDA receptors are essential for complex learning to occur quickly, hence they are crucial to the functioning of the hippocampus. But neurons that use NMDA receptors for their function are vulnerable to a form of damage called "excitotoxicity." Excitoxicity, first discovered in the early 1970s, is the destruction of neurons that occurs when NMDA receptors are overstimulated by too much glutamate neurotransmitter (Olney, Sharpe, & Feigin, 1972). High levels of NMDA receptor stimulation by glutamate induces pores called calcium channels in the membranes of the neurons to open up, allowing calcium ions to pour into the cells, ultimately causing the cells to burst. Since neurons in the hippocampus contain glutamate, their destruction releases glutamate in their surrounding area, which further stimulates and pushes toward excessive stimulation of neighboring neurons. High cortisol levels make glutamate neurons in the hippocampus more vulnerable to destruction by NMDA receptor stimulation, so that cell death can occur with lower levels of stimulation than would otherwise be necessary to induce excitotoxicity (Sapolsky, 2000).

The ability of cortisol to promote death of hippocampal neurons has been demonstrated in numerous animal studies, including by direct application of cortisol to the hippocampi of animals (Bremner, 2002; McEwen, 1995; Sapolsky, 2000). And if similar, stress-induced hippocampal damage occurs in humans, this model might account for many features of PTSD, including the relationships between stress, dissociation, amnesia, and loss of control of traumatic memories, among others (Bremner, 2002). But has stress-induced hippocampal damage been observed in humans with stress injuries such as PTSD? And has the role of cortisol in the pathogenesis of stress disorders in humans also been observed? So far, the results of studies in humans have been inconclusive.

The findings of two meta-analyses of magnetic resonance imaging (MRI) studies of hippocampal volumes in adults with PTSD—the first, a meta-analysis of 9 studies and a second, a meta-analysis of 13 MRI studies comparing hippocampal volumes in adult patients with PTSD with those of well-matched controls—are that PTSD subjects have smaller hippocampal volumes than subjects without PTSD (Smith, 2005; Kitayama et al., 2005). In some of the imaging studies, subjects also show deficits in memory tasks, and these deficits are correlated with hippocampal volume (Bremner et al., 1995; Shin et al., 2004). However, studies of children with PTSD and a study of adults with recent-onset PTSD do not show smaller hippocampal volumes in comparison to normal subjects, suggesting that PTSD chronicity is a salient factor in hippocampal size (Bonne et al., 2001; Carrion et al., 2001; De Bellis et al., 2002; Thomas & De Bellis, 2004).

It has been hypothesized that perhaps a smaller hippocampus is not the result of PTSD, but predisposes to the disorder (Pitman, 2001). A study by Bonne et al. (2001), finding no correlation between hippocampal size and PTSD outcome, fails to support this hypothesis (Bonne et al., 2001). But authors of another study with findings of decreased hippocampal size early in the course of PTSD hypothesize that decreased hippocampal size either occurs earlier in the illness than originally thought, or is a predisposing factor (Wignall et al., 2004). And a study of two monozygotic twin-pair groups strongly supports the hypothesis that smaller hippocampal volume predisposes to PTSD. In the first twin pair group, one twin was exposed to trauma and developed PTSD, and the sibling co-twin had not been exposed to trauma. In another twin pair group, one twin had been exposed to trauma (at levels similar to the PTSD subject in the first group) but did not develop symptoms and the sibling co-twin had not been exposed. Although the twins with PTSD had smaller hippocampi than the trauma-exposed twins who did not develop the disorder, there was no difference in hippocampal volume between sibling co-twins, suggesting that the volume difference was not due to the trauma experience, but instead might represent a risk factor for PTSD development (Gilbertson et al., 2002). Prospective studies are needed to resolve this issue.

In addition to studies of hippocampal size, some studies of hippocampal function support the hypothesis of hippocampal abnormality in humans with stress injuries. Bremner et al. (2003), using positron imaging techniques, shows abnormalities in hippocampal blood flow in PTSD clients. These findings have not yet been replicated by others. Also, effective psychotherapeutic treatments of PTSD, such as exposure therapy, do not appear to impact hippocampal size. In PTSD subjects with smaller hippocampal volumes treated with verbal psychotherapy, posttreatment imaging studies show no difference in hippocampal volume, despite a positive response to therapy (Lindauer et al., 2005). There is, however, some evidence that the hippocampus responds to some types of pharmaceutical treatment, such as treatment with selective serotonin-reuptake inhibitors, that modulate serotonin availability. PTSD subjects who completed a paroxetine trial show a significant reduction in PTSD symptoms, improved verbal declarative memory, and a posttreatment increase in hippocampal volume (Vermetten et al., 2003).

Another uncertainty is the role of cortisol in possible damage to the hippocampi of humans with stress injuries. A preponderance of studies finding *low* rather than elevated cortisol levels in the blood and urine of PTSD clients spawned the model that insufficient cortisol levels, perhaps mediated by dysfunctions in ACTH feedback circuits, is associated with increased risk for PTSD (Yehuda, 2000, 2004). However, reduced peripheral cortisol has not been found in all studies (Heim et al., 2000; Lemieux & Coe, 1995; Pitman & Orr, 1990; Young et al., 2004). And, a single study of CSF cortisol concentrations shows higher than normal CSF cortisol in combat veterans with PTSD, despite normal urinary cortisol excretion and normal plasma cortisol concentrations (Baker et al., 2005).

CONCLUSION: "EVERYBODY HAS WON, AND ALL MUST HAVE PRIZES"

Much research remains to be done to promote a more complete understanding of how biology and experience interact, mediated by psychological and social processes, to produce stress injuries in vulnerable individuals. However, the brief survey presented in this chapter of the major psychological and biological models of stress and stress injury, along with some of the evidence supporting them, makes it clear that although all theoretical perspectives on stress injury answer some of the important questions raised at the outset, none has answered them completely. As the Dodo in Lewis Carroll's *Alice in Wonderland* said after a circular race with no start or finish lines, "everybody has won, and all must have prizes." In the effort to promote a full understanding of the causes and consequences of stress and the treatment of stress injuries, there may also be no finish line. But accomplishments toward that end are far from meaningless, especially for stress-injured warfighters and their families.

Further advancement of the science of stress and stress injury cries out for more integrative models that incorporate the insights of many different perspectives into a multidimensional construct. The more that is learned about the full nature of these injuries, the less room there remains for narrow, reductionistic models to explain them. Multidimensional, integrated theories require testing with multidimensional research that seeks to uncover not only individual cause-effect relationships but the interplay between multiple causes and effects at the levels of brain, mind, and society. The warfighters whose sacrifices are the focus of this work deserve no less.

REFERENCES

Adamec, R., Blundell, J., & Burton, P. (2005). Role of NMDA receptors in the lateralized potentiation of amygdala afferent and efferent neural transmission produced by predator stress. *Physiology & Behavior, 86,* 75–91.

Adamec, R., Strasser, K., Blundell, J., Burton, P., & McKay, D. W. (2006). Protein synthesis and the mechanisms of lasting change in anxiety induced by severe stress. *Behavioural Brain Research, 167,* 270–286.

Appel, J. W., & Beebe, G. W. (1946). Preventive psychiatry: An epidemiologic approach. *Journal of the American Medical Association, 131*(18), 1469–1475.

Arborelius, L., Owens, M. J., Plotsky, P. M., & Nemeroff, C. B. (1999). The role of corticotrophin-releasing factor in depression and anxiety disorders. *Journal of Endocrinology, 160,* 1–12.

Baker, D. G., Ekhator, N. N., Kasckow, J. W., Dashevsky, B., Horn, P. S., Bednarik, L., et al. (2005). Higher levels of basal serial CSF cortisol in combat veterans with posttraumatic stress disorder. *American Journal of Psychiatry, 162,* 992–994.

Baker, D. G., West, S. A., Nicholson, W. E., Ekhator, N. N., Kasckow, J. W., Hill, K. K., et al. (1999). Serial CSF corticotropin-releasing hormone levels and adrenocortical activity in combat veterans with posttraumatic stress disorder. *American Journal of Psychiatry, 156,* 585–588.

Bechara, A., Tranel, D., Damasio, H., Adolphs, R., Rockland, C., & Damasio, A. R. (1995). Double dissociation of conditioning and declarative knowledge relative to the amygdala and hippocampus in humans. *Science, 269,* 1115–1118.

Beck, A. T. (1967). *Depression: Causes and Treatment.* Philadelphia: University of Pennsylvania Press.

Blood, C. G., & Gauker, E. D. (1993). The relationship between battle intensity and disease rates among Marine Corps infantry units. *Military Medicine, 158,* 340–344.

Bonne, O., Brandes, D., Gilboa, A., Gomori, J. M., Shenton, M. E., Pitman, R. K., et al. (2001). Longitudinal MRI study of hippocampal volume in trauma survivors with PTSD. *American Journal of Psychiatry, 158,* 1248–1251.

Bremner, J. D. (2002). *Does Stress Damage the Brain? Understanding Trauma-Related Disorders from a Mind-Body Perspective.* New York: Norton.

Bremner, J. D., Krystal, J. H., Southwick, S. M., & Charney, D. S. (1995). Functional neuroanatomical correlates of the effects of stress on memory. *Journal of Traumatic Stress, 8,* 527–553.

Bremner, J. D., Vythilingam, M., Vermetten, E., Southwick, S. M., McGlashan, T., Nazeer, A., et al. (2003). MRI and PET study of deficits in hippocampal structure and function in women with childhood sexual abuse and posttraumatic stress disorder. *American Journal of Psychiatry, 160*, 924–932.

Brewin, C. R., Dalgleish, T., & Joseph, S. (1996). A dual representational theory of posttraumatic stress disorder. *Psychological Review, 103*, 670–686.

Buchel, C., Morris, J., Dolan, R. J., & Friston, K. J. (1998). Brain systems mediating aversive conditioning: An event-related fMRI study. *Neuron, 20*, 947–957.

Cahill, L., & McGaugh, J. L. (1998). Mechanisms of emotional arousal and lasting declarative memory. *Trends in Neurosciences, 21*, 294–299.

Cannon, W. B. (1932). *The Wisdom of the Body*. New York: Norton.

Carrion, V. G., Weems, C. F., Eliez, S., Patwardhan, A., Brown, W., Ray, R. D., et al. (2001). Attenuation of frontal asymmetry in pediatric posttraumatic stress disorder. *Biological Psychiatry, 50*, 943–951.

Chambers, R. A., Bremner, J. D., Moghaddam, B., Southwick, S. M., Charney, D. S., & Krystal, J. H. (1999). Glutamate and post-traumatic stress disorder: Toward a psychobiology of dissociation. *Seminars in Clinical Neuropsychiatry, 4*, 274–281.

Charney, D. S. (2004). Psychobiological mechanisms of resilience and vulnerability: Implications for successful adaptation to extreme stress. *American Journal of Psychiatry, 161*, 195–216.

Copp, T., & McAndrew, B. (1990). *Battle Exhaustion: Soldiers and Psychiatrists in the Canadian Army, 1939–1945*. Montreal, Quebec: McGill-Queen's University Press.

Coupland, N. J. (2000). Brain mechanism and neurotransmitters. In D. Nutt, J. R. T. Davidson, & J. Zohar (Eds.), *Post-traumatic Stress Disorder: Diagnosis, Management, and Treatment*. London: Martin Dunitz.

Davis, M. (1992). The role of the amygdala in fear and anxiety. *Annual Review of Neuroscience, 15*, 353–375.

De Bellis, M. D., Keshavan, M. S., Shifflett, H., Iyengar, S., Beers, S. R., Hall, J., et al. (2002). Brain structures in pediatric maltreatment-related posttraumatic stress disorder: A sociodemographically matched study. *Biological Psychiatry, 52*, 1066–1078.

Djavadian, R. L. (2004). Serotonin and neurogenesis in the hippocampal dentate gyrus of adult mammals. *Acta Neurobiologiae Experimentalis (Warsaw), 64*, 189–200.

Ehlers, A., & Clark, D. M. (1999). A cognitive model of posttraumatic stress disorder. *Behaviour Research and Therapy, 38*, 319–345.

Eichenbaum, H., Dudchenko, P. A., Wood, E. R., Shapiro, M., & Tanila, H. (1999) The hippocampus, memory, and place cells: Is it spatial memory or a memory space? *Neuron, 23*, 209–226.

Falls, W. A., & Davis, M. (1995). Behavioral and physiological analysis of fear inhibition: Extinction and conditioned inhibition. In M. J. Friedman, D. S. Charney, & A. Y. Deutch (Eds.), *Neurobiological and Clinical Consequences of Stress: From Normal Adaptation to PTSD*. Philadelphia: Lippincott-Raven.

Fanselow, M. S., & Kim, J. J. (1994). Acquisition of contextual Pavlovian fear conditioning is blocked by application of an NMDA receptor antagonist D,L-2-amino-5-phosphonovaleric acid to the basolateral amygdale. *Behavioral Neuroscience, 106*, 210–12.

Fenichel, O. (1945). *The Psychoanalytic Theory of Neurosis*. New York: Norton.

Festinger, L. (1957). *A Theory of Cognitive Dissonance*. Stanford, CA: Stanford University Press.

Figley, C. R. (Ed.). (1999). *Traumatology of Grieving: Conceptual, Theoretical, and Treatment Foundations*. New York: Brunner/Mazel.

Foa, E. B., & Kozak, M. J. (1986). Emotional processing in fear: Exposure to corrective information. *Psychological Bulletin, 99*, 20–35.

Foa, E. B., & Rothbaum, B. O. (1998). *Treating the Trauma of Rape*. New York: Guilford Press.

Fontana, A., & Rosenheck, R. (2005). The role of war-zone trauma and PTSD in the etiology of antisocial behavior. *Journal of Nervous and Mental Disease, 193*, 203–209.

Geracioti, Jr., T. D., Baker, D. G., Ekhator, N. N., West, S. A., Hill, K. K., Bruce, A. B., et al. (2001). CSF norepinephrine concentrations in posttraumatic stress disorder. *American Journal of Psychiatry, 158,* 1227–1230.

Gilbertson, M. W., Shenton, M. E., Ciszewski, A., Kasai, K., Lasko, N. B., Orr, S. P., et al. (2002). Smaller hippocampal volume predicts pathologic vulnerability to psychological trauma. *Nature Neuroscience, 5,* 1242–1247.

Gold, P. E., & McCarty, R. C. (1995). Stress regulation of memory processes: Role of peripheral catecholamines and glucose. In M. J. Friedman, D. S. Charney, & A. Y. Deutch (Eds.), *Neurobiological and Clinical Consequences of Stress: From Normal Adaptation to PTSD.* Philadelphia: Lippincott-Raven.

Gold, P. W., & Chrousos, G. P. (2002). Organization of the stress system and its dysregulation in melancholic and atypical depression: High vs. low CRH/NE states. *Molecular Psychiatry, 7,* 254–275.

Heilig, M., Koob, G. F., Ekman, R., & Britton, K. T. (1994). Corticotropin-releasing factor and neuropeptide Y: Role in emotional regulation. *Trends in Neurosciences, 17,* 80–85.

Heim, C., Newport, D. J., Heit, S., Graham, Y. P., Wilcox, M., et al. (2000). Pituitary-adrenal and autonomic responses to stress in women after sexual and physical abuse in childhood. *Journal of the American Medical Association, 284,* 592–597.

Horger, B. A., & Roth, R. H. (1995). Stress and central amino acid systems. In M. J. Friedman, D. S. Charney, & A. Y. Deutch (Eds.), *Neurobiological and Clinical Consequences of Stress: From Normal Adaptation to PTSD.* Philadelphia: Lippincott-Raven.

Huang, G. J., & Herbert, J. (2006). Stimulation of neurogenesis in the hippocampus of the adult rat by fluoxetine requires rhythmic change in corticosterone. *Biological Psychiatry, 59,* 619–624.

Janoff-Bulman, R. (1992). *Shattered Assumptions: Towards a New Psychology of Trauma.* New York: The Free Press.

Joels, M., Karst, H., Alfarez, D., Heine, V. M., Qin, Y., et al. (2004). Effects of chronic stress on structure and cell function in rat hippocampus and hypothalamus. *Stress, 7,* 221–231.

Jones, E., & Wessely, S. (2001). Psychiatric battle casualties: An intra- and inter-war comparison. *British Journal of Psychiatry, 178,* 243–245.

Kandel, E. R., & Squire, L. R. (2000). Neuroscience: Breaking down scientific barriers to the study of brain and mind. *Science, 290,* 1113–1120.

Kaufer, D., Ogle, W. O., Pincus, Z. S., Clark, K. L., Nicholas, A. C., Dinkel, K. M., et al. (2004). Restructuring the neuronal stress response with anti-glucocorticoid gene delivery. *Nature Neuroscience, 7,* 947–953.

Kauffman, J. (2002). *Loss of the Assumptive World: A Theory of Traumatic Loss.* New York: Brunner-Routledge.

Kitayama, N., Vaccarino, V., Kutner, M., Weiss, P., & Bremner, J. D. (2005). Magnetic resonance imaging (MRI) measurement of hippocampal volume in posttraumatic stress disorder: A meta-analysis. *Journal of Affective Disorders, 88,* 79–86.

Koob, G. F., & LeMoal, M. (1997). Drug abuse: Hedonic homeostatic dysregulation. *Science, 278,* 52–58.

Korte, S. M., Koolhaas, J. M., Wingfield, J. C., & McEwen, B. S. (2005). The Darwinian concept of stress: Benefits of allostasis and costs of allostatic load and the trade-offs in health and disease. *Neuroscience and Biobehavioral Reviews, 29,* 3–38.

Kramer, J. H., Rosen, H. J., Du, A. T., Schuff, N., Hollnagel, C., Weiner, M. W., et al. (2005). Dissociations in hippocampal and frontal contributions to episodic memory performance. *Neuropsychology, 19,* 799–805.

Krystal, J. H., Bennett, A. L., Bremner, J. D., Southwick, S. M., & Charney, D. S. (1995). Toward a cognitive neuroscience of dissociation and altered memory functions in posttraumatic stress disorder. In M. J. Friedman, D. S. Charney, & A. Y. Deutch (Eds.), *Neurobiological and Clinical Consequences of Stress: From Normal Adaptation to PTSD.* Philadelphia: Lippincott-Raven.

Kulka, R. A., Schlenger, W. E., Fairbank, J. A., Hough, R. L., Jordan, B. K., Marmar, C. R., et al. (1990). *Trauma and the Vietnam War Generation: Report of Findings from the National Vietnam Veterans Readjustment Study.* New York: Brunner/Mazel.

LeDoux, J. (1996). *The Emotional Brain: The Mysterious Underpinnings of Emotional Life.* New York: Touchstone.

Lemieux, A. M., & Coe, C. L. (1995). Abuse-related posttraumatic stress disorder: Evidence for chronic neuroendocrine activation in women. *Psychosomatic Medicine, 57,* 105–115.

Lepage, M., & Richer, F. (2000). Frontal brain lesions affect the use of advance information during response planning. *Behavioral Neuroscience, 114,* 1034–1040.

Lesch, K. P., & Merschdorf, U. (2000). Impulsivity, aggression, and serotonin: A molecular psychobiological perspective. *Behavioral Sciences and the Law, 18,* 581–604.

Lindauer, R. J., Vlieger, E. J., Jalink, M., Olff, M., Carlier, I. V., Majoie, C. B., et al. (2005). Effects of psychotherapy on hippocampal volume in out-patients with post-traumatic stress disorder: A MRI investigation. *Psychological Medicine, 35,* 1421–1431.

MacNair, R. M. (2002). *Perpetration-Induced Traumatic Stress: The Psychological Consequences of Killing.* New York: Authors Choice Press.

McEwen, B. S. (1995). Adrenal steroid actions on the brain: dissecting the fine line between protection and damage. In M. J. Friedman, D. S. Charney, & A. Y. Deutch (Eds.), *Neurobiological and Clinical Consequences of Stress: From Normal Adaptation to PTSD.* Philadelphia: Lippincott-Raven.

McEwen, B. S. (2000a). Allostasis, allostatic load, and the aging nervous system: Role of excitatory amino acids and excitotoxicity. *Neurochemical Research, 25,* 1219–1231.

McEwen, B. S. (2000b). Protective and damaging effects of stress mediators: Central role of the brain. *Progress in Brain Research, 122,* 25–34.

McEwen, B. S., & Stellar, E. (1993). Stress and the individual: Mechanisms leading to disease. *Archives of Internal Medicine, 153,* 2093–2101.

Miserendino, M. J. D., Sananes, C. B., Melia, K. R., & Davis, M. (1990). Blocking of acquisition but not expression of conditioned fear-potentiation startle by NMDA antagonists in the amygdala. *Nature, 345,* 716–718.

Morgan, C. A., Wang, S., Mason, J., Southwick, S. M., Fox, P., Hazlett, G., Charney, D. S., & Greenfield, G. (2000). Hormone profiles in humans experiencing military survival training. *Biological Psychiatry, 47,* 891–901.

Morris, J. S., Friston, K. J., Buchel, C., Frith, C. D., Young, A. W., Calder, A. J., et al. (1998). A neuromodulatory role for the human amygdala in processing emotional facial expressions. *Brain, 121*(Pt. 1), 47–57.

Moscovitch, M., Rosenbaum, R. S., Gilboa, A., Addis, D. R., Westmacott, R., Grady, C., et al. (2005). Functional neuroanatomy of remote episodic, semantic and spatial memory: a unified account based on multiple trace theory. *Journal of Anatomy, 207,* 35–66.

Mowrer, O. H. (1960). *Learning Theory and Behavior.* New York: John Wiley & Sons.

Nacher, J., & McEwen, B. S. (2006). The role of *N*-methyl-D-aspartate receptors in neurogenesis. *Hippocampus, 16,* 267–270.

O'Keefe, J., & Nadel, L. (1978). *The Hippocampus as a Cognitive Map.* Oxford, U.K.: Clarendon Press.

Olds, M. E., & Milner, P. M. (1954). Positive reinforcement produced by electrical stimulation of the septal area and other regions of the rat brain. *Journal of Comparative and Physiological Psychology, 47,* 419–427.

Olney, J. W., Sharpe, L. G., & Feigin, R. D. (1972). Glutamate-induced brain damage in infant primates. *Journal of Neuropathology and Experimental Neurology, 31,* 464–488.

Page, M. E., & Abercrombie, E. D. (1999). Discrete local application of corticotropin-releasing factor increases locus coeruleus discharge and extracellular norepinephrine in the rat hippocampus. *Synapse, 33,* 304–313.

Pavlov, I. P. (1960). *Conditioned Reflexes* (G. V. Anrep, Ed. & Trans.). Mineola, NY: Dover Press. (Original work published in 1927)

Phelps, E. A. (2004). Human emotion and memory: Interactions of the amygdala and hippocampal complex. *Current Opinion in Neurobiology, 14,* 198–202.
Phelps, E. A., & LeDoux, J. E. (2005). Contributions of the amygdala to emotion processing: from animal models to human behavior. *Neuron, 48,* 175–187.
Piaget, J. (1952). *The Origins of Intelligence in Children* (M. Cook, Trans.). New York: International Universities Press.
Piaget, J., & Inhelder, B. (1969). *The Psychology of the Child* (H. Weaver, Trans.). New York: Basic Books.
Pitman, R. K. (2001). Hippocampal diminution in PTSD: More (or less?) than meets the eye. *Hippocampus, 11,* 73–74.
Pitman, R. K., & Orr, S. P. (1990). Twenty-four hour urinary cortisol and catecholamine excretion in combat-related posttraumatic stress disorder. *Biological Psychiatry, 27,* 245–247.
Rasmusson, A. M., Hauger, R. L., Morgan, C. A., Bremner, J. D., Charney, D. S., & Southwick, S. M. (2000). Low baseline and yohimbine-stimulated plasma neuropeptide Y (NPY) levels in combat-related PTSD. *Biological Psychiatry, 47,* 526–539.
Resick, P. A., & Schnicke, M. K. (1992). Cognitive processing therapy for sexual assault victims. *Journal of Consulting and Clinical Psychology, 60,* 748–756.
Sapolsky, R. M. (1985). Glucocorticoid toxicity in the hippocampus: Temporal aspects of neuronal vulnerability. *Brain Research, 359,* 300–305.
Sapolsky, R. M. (2000). Glucocorticoids and hippocampal atrophy in neuropsychiatric disorders. *Archives of General Psychiatry, 57,* 925–935.
Sapolsky, R. M. (2003). Stress and plasticity in the limbic system. *Neurochemical Research, 28,* 1735–1742.
Sapolsky, R. M., & Pulsinelli, W. A. (1985). Glucocorticoids potentiate ischemic injury to neurons: Therapeutic implications. *Science, 229,* 1397–1400.
Schneider, R. (1986). Military psychiatry in the German army. In R. A. Gabriel (Ed.), *Military Psychiatry: A Comparative Perspective.* New York: Greenwood Press.
Shapiro, F. (1989). Eye movement desensitization: A new treatment for posttraumatic stress disorder. *Journal of Behavior Therapy and Experimental Psychiatry, 20,* 211–217.
Shapiro, F. (2001). *Eye Movement Desensitization and Reprocessing (EMDR): Basic Principles, Protocols, and Procedures* (2nd ed.). New York: Guilford Press.
Shay, J. (1994). *Achilles in Vietnam: Combat Trauma and the Undoing of Character.* New York: Scribner.
Shephard, B. (2001). *A War of Nerves: Soldiers and Psychiatrists in the Twentieth Century.* Cambridge, MA: Harvard University Press.
Shephard, B. (2004). Risk factors and PTSD: A historian's perspective. In G. M. Rosen (Ed.), *Posttraumatic Stress Disorder: Issues and Controversies.* Chichester, U.K.: John Wiley & Sons.
Shimamura, A. P., & Squire, L. R. (1987). A neuropsychological study of fact memory and source amnesia. *Journal of Experimental Psychology: Learning, Memory, and Cognition, 13,* 464–473.
Shin, L. M., Shin, P. S., Heckers, S., Krangel, T. S., Macklin, M. L., Orr, S. P., et al. (2004). Hippocampal function in posttraumatic stress disorder. *Hippocampus, 14,* 292–300.
Smith, M. A., Makino, S., Kvetnansky, R., & Post, R. M. (1995). Stress and glucocorticoids affect the expression of brain-derived neurotrophic factor and neurotrophin-3 mRNA in the hippocampus. *Journal of Neuroscience, 15,* 1768–1777.
Smith, M. E. (2005). Bilateral hippocampal volume reduction in adults with post-traumatic stress disorder: A meta-analysis of structural MRI studies. *Hippocampus, 15,* 798–807.
Southwick, S. M., Yehuda, R., & Giller, E. L., Jr. (1993). Personality disorders in treatment-seeking combat veterans with posttraumatic stress disorder. *American Journal of Psychiatry, 150,* 1020–1023.
Squire, L. R. (1987). *Memory and Brain.* New York: Oxford University Press.
Squire, L. R. (2004). Memory systems of the brain: A brief history and current perspective. *Neurobiology of Learning and Memory, 82,* 171–177.

Stahl, S. M. (2000). *Essential Psychopharmacology: Neuroscientific Basis and Practical Applications* (2nd ed). New York: Cambridge University Press.

Stout, S. C., Kilts, C. D., & Nemeroff, C. B. (1995). Neuropeptides and stress: Preclinical findings and implications for pathophysiology. In M. J. Friedman, D. S. Charney, & A. Y. Deutch (Eds.), *Neurobiological and Clinical Consequences of Stress: From Normal Adaptation to PTSD.* Philadelphia: Lippincott-Raven.

Terman, G. W., Shavit, Y., Lewis, J. W., Cannon, J. T., & Liebeskind. J. C. (1984). Intrinsic mechanisms of pain inhibition: Activation by stress. *Science, 266,* 1270–1277.

Thomas, L. A., & De Bellis, M. D. (2004). Pituitary volumes in pediatric maltreatment-related posttraumatic stress disorder. *Biological Psychiatry, 55,* 752–758.

Tremblay, L. K., Narajo, C. A., Cardenas, L., Herrmann, N., & Busto, U. E. (2002). Probing brain reward system function in major depressive disorder: Altered response to dextroamphetamine. *Archives of General Psychiatry, 59,* 409–416.

Tulving, E., & Markowitsch, H. J. (1998). Episodic and declarative memory: Role of the hippocampus. *Hippocampus, 8,* 198–204.

Van Bockstaele, E. J., Colago, E. E. O., & Valentino, R. J. (1998). Amygdaloid corticotrophin-releasing factor targets locus coeruleus dendrites: Substrate for the co-ordination of emotional and cognitive limbs of the stress response. *Journal of Neuroendocrinology, 10,* 743–757.

Vermetten, E., Vythilingam, M., Southwick, S. M., Charney, D. S., & Bremner, J. D. (2003). Long-term treatment with paroxetine increases verbal declarative memory and hippocampal volume in posttraumatic stress disorder. *Biological Psychiatry, 54,* 693–702.

Vianna, M. R., Coitinho, A. S., & Izquierdo, I. (2004). Role of the hippocampus and amygdala in the extinction of fear-motivated learning. *Current Neurovascular Research, 1,* 55–60.

Wallenstein, G. V., Eichenbaum, H., & Hasselmo, M. E. (1998). The hippocampus as an associator of discontiguous events. *Trends in Neurosciences, 21,* 317–323.

Wignall, E. L., Dickson, J. M., Vaughan, P., Farrow, T. F., Wilkinson, I. D., Hunter, M. D., et al. (2004). Smaller hippocampal volume in patients with recent-onset posttraumatic stress disorder. *Biological Psychiatry, 56,* 832–836.

Wise, R. A. (1996). Addictive drugs and brain stimulation reward. *Annual Review of Neuroscience, 19,* 319–340.

Yehuda, R. (2000). Neuroendocrinology. In D. Nutt, J. R. T. Davidson, & J. Zohar (Eds.), *Posttraumatic Stress Disorder: Diagnosis, Management, and Treatment.* London: Martin Dunitz.

Yehuda, R. (2004). Risk and resilience in posttraumatic stress disorder. *Journal of Clinical Psychiatry, 65*(Suppl 1), 29–36.

Young, E. A., Tolman, R., Witkowski, K., & Kaplan, G. (2004). Salivary cortisol and posttraumatic stress disorder in a low-income community sample of women. *Biological Psychiatry, 55,* 621–626.

Section II

Research Contributions to Combat Stress Injuries and Adaptation

As the first section laid the groundwork for a unified theory of combat stress injuries, Section II includes three chapters that summarize the research literature in demography, public health, sociology, traumatology, medicine, psychology, family sciences, and family therapy that confirm the validity of the models noted in the first section. Moreover, these chapters emphasize the importance of efforts to prevent and mitigate combat stress injuries and disorders. This latter area is the focus of the final section of this book. The first chapter in this section emphasizes the life and death nature of understanding and managing combat stress injuries. The second chapter notes the significance of physical injuries and its relationship to mental injuries. The final chapter in this section emphasizes the enormous human costs of such injuries in the lives of warfighter families, especially the children.

5

The Mortality Impact of Combat Stress 30 Years After Exposure: Implications for Prevention, Treatment, and Research

JOSEPH A. BOSCARINO

STUDY OVERVIEW

Military personnel exposed to combat service are known to be at greater risk for developing posttraumatic stress disorder (PTSD). While past research has suggested that PTSD is often associated with a broad range of psychological problems, a growing body of research now suggests that PTSD also is associated with later medical morbidity and premature mortality. To assess the long-term health impact of PTSD, we examined all-cause and cause-specific mortality among a national random sample of U.S. Army veterans with and without PTSD after military service. We examined the survival time and causes of death among 15,288 male U.S. Army veterans 16 years after completion of a telephone survey, approximately 30 years after their military service. These men were included in a national sample of veterans from the Vietnam War era. Our analyses adjusted for age, marital status, race, combat exposure, volunteer status, entry age, discharge status, illicit drug abuse, intelligence, and pack-years of cigarette smoking. Our findings indicated that the adjusted postwar mortality for all-cause, cardiovascular, cancer, and external causes of death (i.e., mortality due to suicides, homicides, drug overdoses, motor vehicle accidents, and injuries of undetermined intent) was associated with PTSD among theater veterans with Vietnam service (N = 7,924). For Vietnam era veterans without Vietnam service (N = 7,364), PTSD was associated with all-cause mortality and marginally associated with external-cause mortality. When level of combat exposure also was controlled among the theater veterans, PTSD was no longer significant for cardiovascular mortality, suggesting that other factors may be involved with this outcome, but remained significant for cancer and external mortality, suggesting that PTSD is central in these outcomes. Our study suggests that veterans with long-term PTSD are at risk of death from multiple causes. While the specific reasons for this increased mortality are unclear, these outcomes are likely related to biological, psychological, and behavioral phenomenon associated with PTSD. Although further research is warranted and issues related to barriers to care might need to be addressed, for many returning veterans today, the adverse health impact of PTSD is potentially avoidable, since effective treatments currently are available for this disorder and the sequelae related to this condition.

REVIEW OF EXISTING RESEARCH

Studies suggest that having a history of posttraumatic stress disorder or traumatic stress exposure is associated with later medical comorbidity (Armenian, Melkonian, & Hovanesian, 1998; Boscarino, 1997, 2004; Boscarino & Chang, 1999a; Cwikel et al., 1997; Schnurr & Green, 2004; Long, Chamberlain, & Vincent, 1992). Research among combat veterans, in particular, suggests both higher rates of postwar psychological problems and medical conditions than noncombat veterans or comparable nonveterans (Boscarino, 1997, 2000; Centers for Disease Control, 1988a,b; Kulka et al., 1990a). For example, when the postwar health status of Vietnam veterans was recently examined by whether they had PTSD, the PTSD-positive veterans had substantially higher postwar rates for many chronic conditions, including circulatory, nervous system, digestive, musculoskeletal, and respiratory diseases, even controlling for the major risk factors for these conditions (Boscarino, 1997). Since military personnel exposed to combat service are at significantly increased risk for PTSD (Boscarino, 2000, 2004; Centers for Disease Control, 1988a; Kulka et al., 1990a), the long-term health impact of these stress response syndromes are potentially great for this population.

In the medical field, the evidence linking traumatic stress exposures to cardiovascular disease is extensive (Boscarino, 2004). In addition to Vietnam veteran studies (Boscarino, 1997; Boscarino & Chang, 1999a), a population study involving World War II and Korean War veterans has found higher rates of physician-diagnosed cardiovascular disease among PTSD-positive veterans (Schnurr, Spiro, & Paris, 2000). Another study among Dutch Resistance Fighters found increased rates of reported angina pectoris among those with PTSD (Falger et al., 1992). In addition, studies related to the Beirut Civil War and the Croatia War found increases in arteriographically confirmed coronary heart disease, cardiovascular disease mortality, and increases in acute myocardial infarctions associated with population-level exposure to these conflicts (Miric et al., 2001; Sibai, Armenian, & Alam, 1989; Sibai, Fletcher, & Armenian, 2001). Other studies involving exposure to other types of traumatic events have also reported similar associations (Cwikel et al., 1997; Felitti et al., 1998). Finally, numerous studies have documented persistent increases in basal cardiovascular activity among PTSD victims that could result in the onset of disease (Buckley & Kaloupek, 2001).

Since coronary heart disease has been associated with PTSD and this condition is considered an "inflammatory" disease (Danesh et al., 1998), it also has been suggested that PTSD-positive veterans might also be at risk for autoimmune diseases (Boscarino, 2004). For example, some investigations have found that individuals who developed PTSD, particularly men exposed to combat, appeared to have lower plasma cortisol concurrent with higher catecholamine levels (Boscarino, 1996; Mason et al., 1988; Yehuda, 2002; Yehuda et al., 1992, 1995). One study reported that Vietnam veterans with current PTSD not only had lower cortisol, but that this had an inverse "dose-response" relationship with combat exposure (Boscarino, 1996, 2000). In addition, research has indicated that Vietnam veterans with current PTSD had clinically elevated leukocyte and T-cell counts (Boscarino & Chang, 1999b) and similar findings have been reported in nonveteran studies as well (Herbert & Cohen, 1993; Ironson et al., 1997). Consistent with these clinical findings, it has been reported recently that comorbid PTSD was associated with several autoimmune diseases, including rheumatoid arthritis and psoriasis (Boscarino, 2004), suggesting that the pathophysiology of PTSD may be linked to neuroendocrinological, immunological, and/or inflammatory disease mechanisms (Boscarino, 2004; Boscarino & Chang, 1999b).

Recently, investigators from the Centers for Disease Control and Prevention (CDC) ascertained the vital status and underlying cause of death among participants in the Vietnam Experience Study (VES), a longitudinal study of 18,313 male U.S. Army veterans from the end of their military service through December 31, 2000 (Boehmer et al., 2004). In the CDC study, all-cause mortality appeared higher among Vietnam theater veterans (i.e., those who served in Vietnam) compared with Vietnam era veterans (i.e., those who served elsewhere) during the 30-year follow-up period. The excess mortality among Vietnam veterans, however, appeared isolated to the first 5 years after discharge from active duty and appeared to have resulted from an increase in external causes of death, due to suicides, homicides, accidental poisonings, motor vehicle accidents, and unintended injuries. Cause-specific analyses revealed no difference in disease-related mortality by veteran status. Vietnam theater veterans, however, experienced an increase in unintentional poisoning and drug-related deaths throughout the follow-up period. Death rates from disease-related chronic conditions, including malignant neoplasms and circulatory diseases, did not appear to differ between the theater and era veterans, despite the increasing age of the cohort and the long follow-up period. In the current study, we examine vital status and underlying cause of death by posttraumatic stress disorder and combat exposure status, something not previously done.

Research suggests that the majority of Vietnam veterans with PTSD developed this disorder due to combat experiences in Vietnam (Boscarino, 1995, 1996; Koenen et al., 2003; Yehuda, 1999). In addition, available research suggests that a significant proportion of these veterans had this disorder for decades (Kulka et al., 1990a). Given these findings, if preliminary studies are correct, the long-term health consequences of PTSD among Vietnam veterans should be discoverable in a large population study of these veterans.

RESEARCH METHODS FOR CURRENT STUDY

The focus of this study was to examine the effects of PTSD among Vietnam veterans through an assessment of postservice mortality among those in the VES cohort who completed telephone interviews in the mid-1980s. Potential subjects for the current study included 17,867 U.S. Army veterans known to be alive in December 1983. Starting in January 1985 these men were contacted by Research Triangle Institute (RTI) to complete telephone interviews. Altogether, 15,288 of these men (86%) were located and completed this survey. The interviews included questions related to PTSD, health status, substance abuse, cigarette smoking, as well as demographic data and military history information. Data from the veteran's original military records also were available, including discharge status, aptitude test results at service induction, service rank, and other information.

The population for the current study was comprised of men who served in the U.S. Army during the Vietnam War. The cohort was identified through a random sample of 48,513 service records selected from the nearly 5 million records on file at the National Personnel Records Center. Of these, 18,581 veterans met the criteria for study eligibility, which were chosen to increase comparability between men who served in Vietnam and men who served elsewhere. These criteria included: entering the service for the first time between 1965 and 1971, serving only 1 term of enlistment, and having a pay grade no higher than E5. Participants were classified as Vietnam "theater" veterans, if they served at least 1 tour of duty in Vietnam, or as Vietnam "era" veterans, if they never served in Vietnam and served at least 1 tour in the United States, Germany, or Korea. The sample in the current study included 7,924 theater and 7,364 era veterans, for a total of 15,288 men who were known not to be deceased in 1983 *and* who completed the RTI telephone survey in 1985–1986. Further details regarding the study have been published elsewhere (Boehmer et al., 2004; Boscarino, 1997; Centers for Disease Control, 1987, 1988a,b).

Ascertainment of Veterans' Vital Status

For our study, we assessed veterans' vital status from the date of completion of the telephone interviews starting in January 1985 until December 31, 2000. Vital status was ascertained using three national mortality databases: the Department of Veterans Affairs Beneficiary Identification Record Locator Subsystem (VA BIRLS) death file, the Social Security Administration Death Master File (SSA DMF), and the National Death Index Plus (NDI Plus) (Boehmer et al., 2004). Investigators also manually reviewed the potential matches from each data source separately and classified the matches as true, false, or questionable (Boehmer et al., 2004). The final determination of vital status was obtained by combining information from all three sources. As needed, additional information, such as actual death certificates, were ascertained to confirm the veteran's status. Veterans who had a match on at least 1 of the 3 national databases were determined to be deceased. All veterans whose vital status was uncertain or who were not identified by any of the national databases were assumed to be alive on December 31, 2000.

Underlying cause-of-death classifications were obtained from the NDI Plus and coded according to the International Classification of Diseases (ICD) revision in place at the time of death: the ninth revision (ICD-9) for deaths between January 1, 1979, and December 31, 1998, and the tenth revision (ICD-10) for deaths between January 1, 1999, and December 31, 2000. For cases in which cause-of-death codes were not available from the NDI Plus, CDC investigators obtained official copies of death certificates, which were then coded by an experienced nosologist at the CDC's National Center for Health Statistics (NCHS) (Boehmer et al., 2004).

Assessment of Veterans' PTSD Status

In addition to our mortality study, we conducted an analysis using a subset of VES participants who completed *both* the telephone survey and personal interviews after the telephone survey. This was required because although the PTSD measure used in the RTI telephone survey had been used in previous studies (Centers for Disease Control, 1989a; Frey-Wouters & Laufer, 1986; True et al., 1993), it had not been clinically evaluated. Consequently, we compared the results of the RTI-PTSD scale to those obtained by the Diagnostic Interview Schedule–Version III (DIS-III) PTSD scale, based on the *Diagnostic and Statistical Manual of Mental Disorders*, 3rd edition (DSM-III) (American Psychiatric Association, 1980). DIS-III is a standardized questionnaire designed to assess the presence of psychiatric conditions consistent with DSM-III (Robins et al., 1981, 1987). In the VES, DIS-III PTSD diagnoses were available for the past 30 days and for lifetime (Centers for Disease Control, 1989b). For the personal interviews, a random subsample was selected from among the 15,288 interviewed by telephone. Altogether, 75% of the theater veterans selected (n = 2,490) and 63% of the era veterans (n = 1,972) completed the personal interviews (overall participation rate = 69%). The personal interviews were administered at Lovelace Medical Foundation, Albuquerque, New Mexico, between June 1985 and September 1986. On average, the time from combat exposure in Vietnam to the telephone surveys and personal interviews was about 16 years (Boscarino, 1997).

During the telephone survey, the veterans were asked to report 15 PTSD-related symptoms experienced in the past 6 months. Consistent with DSM-III, a veteran was classified as having current PTSD if he reported at least one criterion B symptom (reexperiencing), at least one criterion C symptom (avoidance), and at least two criterion D symptoms (hyperarousal). The DSM-III criterion A (exposure) was not explicitly used in the RTI-PTSD scale, but implicitly, since some of the symptoms included in the B and D criteria referred explicitly to Army experiences (e.g., "in past 6 month, had dreams or nightmares of Army experiences") (Centers for Disease Control, 1989a). Using the DIS-III criteria employed in the personal interviews, PTSD was diagnosed as *current* if the veteran met the criteria for criterion A through D in the past 30 days. However, because of the way RTI-PTSD was defined, for comparison purposes the A criterion for exposure in the DIS-III PTSD measure was defined in two ways: for combat experiences only and for *any* traumatic exposures. Finally, we compared the RTI-PTSD results for theater veterans to those of the combat exposure scale (CES) used in the personal interviews (Centers for Disease Control, 1989b). The CES has been shown to be a valid measure of combat exposure and has been used in several previous studies (Centers for Disease Control, 1989b; Janes, 1991).

The RTI-PTSD scale results indicated that 10.6% of the theater veterans and 2.9% of the era veterans were PTSD cases (odds ratio [OR] = 3.9; $p < 0.001$). When we compared these results among those who were reinterviewed and administered the DIS-III during the personal interviews, the results were as follows: Of those who met the DIS-III criteria for current PTSD in the past month for combat (n = 54), 61% were classified as having PTSD on the RTI-PTSD scale; of those classified as negative on the DIS-III for combat, 93% were classified as negative on the RTI-PTSD scale, for an OR of 22.3 (95% confidence interval [CI], 12.7–39.1). For those who met the DIS-III criteria for current PTSD in the past month for *any* trauma (n = 72), the results were similar (OR = 17.1; 95% CI, 10.6–27.6). Furthermore, for the combat exposure results for the theater veterans, we found a dose-response relationship between having low, moderate, high, and very high combat exposure (based by quartiles) and meeting the criteria on the RTI-PTSD measure, with 7%, 17%, 24%, and 52% positively diagnosed, respectively (chi-square trend test = 123.5; df = 1; $p < 0.0001$). In addition, Cronbach's alpha was high for the RTI-PTSD scale items (alpha = 0.92), suggesting internal reliability (Cattell, 1986). We suggest that these findings indicate that a positive diagnosis on the RTI-PTSD scale is generally consistent with a diagnosis of PTSD.

Controlling for Study Bias, Confounding, and Combat Exposure

The main research focus of this study was to determine if PTSD was associated with postservice mortality separate from combat exposure. To achieve this, we developed multivariate models predicting survival, which were controlled for obvious confounders and potential selection biases. To do this, we adjusted our models for age, race, marital status, volunteer status, entry age, discharge status, illicit drug use, and intelligence (Boscarino, 1997; Boscarino & Chang, 1999a,b). In addition, for cardiovascular and cancer mortality, we also controlled for pack-years of cigarette smoking. We controlled for these variables because we wanted to examine the association between PTSD and survival, unaffected by potential selection biases or confounders, such as race, intelligence, or Army volunteer status, which could obscure the results. Controlling for these variables is important because Vietnam veterans are reported to come from higher risk groups (Baskir & Strauss, 1978), factors often associated with poorer health outcomes (Syme, 1992). In addition, controlling for volunteer status is important, because those who volunteered for the U.S. Army or Vietnam service might also have different personality profiles that could affect health outcomes (Boscarino, 2004). With the exception of cardiovascular and cancer mortality, where we controlled for pack-years of cigarette smoking, we did not control for other behavioral risk factors such as postservice drug use or educational attainment. We did this because we wanted to avoid "overcontrolling" for behavioral variables potentially on the causal chain of events linking PTSD to mortality (Boscarino, 2004), which could bias our results. Finally, in all the models for theater veterans we also included level of combat exposure as a final covariate adjustment. Although PTSD and combat exposure are highly correlated (Boscarino, 1995), we did this because we wanted to determine whether the associations found were due to PTSD or something else connected with the combat exposure experience. If PTSD was still significant in our models then, more than likely, PTSD was a central cause for the relationships found, not factors associated with combat exposure per se, such as an altered sense of vulnerability or risk-taking.

For our study, *age* was based on the veteran's age at time of the interview and was used as a continuous variable. *Marital status* was based on whether the veteran was married or not when separated from the service and was taken from the military record. *Race* was based on the veteran's reported race (white 82%; black 11%; Hispanic 5%; other 2%) and coded as a two-category indicator variable, white vs. nonwhite. *Volunteer status* was based on whether the veteran volunteered for military service and was classified as "volunteer" vs. "draftee" and based on the military record. For theater veterans, *Vietnam volunteer status* was based on whether the theater veteran reported volunteering for Vietnam and used as a binary variable. *Entry age* was based on age at induction and was from the military record. *Discharge status* was classified as honorable vs. dishonorable/other discharge and was from the military record. *Illicit drug use* was classified as present if the veteran reported use of illicit drugs (e.g., narcotics, barbiturates, amphetamines, hallucinogens, or marijuana) while in the Army. *Intelligence* was based on the General Technical (GT) examination test administered at military induction and used as a continuous variable (Centers for Disease Control, 1989b). As noted, for our cardiovascular and cancer analyses, *pack-years of cigarette smoking* was included and based on an estimate of the number of cigarette packs smoked per day/per year and used as a four-category indicator variable (none, 1–9, 10–19, 20+ pack-years). For theater veterans, *combat exposure* was based on five combat-related questions asked in the telephone survey (e.g., frequency of exposure to snipers, mortar fire, ambushes, etc.). This measure was scaled in a similar manner to the longer combat scale used in the personal interviews (Centers for Disease Control, 1989b) and was highly correlated with this scale (Pearson's $r = 0.76$; $p < 0.0001$). In the current study, we used the RTI-combat scale as a binary measure, with Vietnam combat veterans scoring in the top quintile classified as high combat exposure cases based on previous research (Boscarino, 1996).

Statistical Methods

We use Cox regressions to calculate both crude (bivariate) and adjusted (multivariate) hazards ratios [HRs] using the control variables discussed for all-cause mortality, cardiovascular mortality, cancer mortality, and mortality due to external causes combined, which included homicide, suicide, accidental poisoning, and unintended injury. Since another study examined survival from Army discharge about 30 years previous (Boehmer et al., 2004), and we only included those who were alive and completed the 1985–1986 telephone interviews, our analyses examine survival time from interview completion stating in January 1985 through to December 31, 2000, a period of 16 years. For these analyses, we evaluated the main proportional hazards assumption (Hosmer & Lemeshow, 1999), controlled for confounding, and tested for effect modification. We also assessed the linearity assumption for covariates treated as continuous. Statistical analyses for our study were performed using *Stata 9.1* (StataCorp LP, College Station, Texas). For all-cause mortality, we included all deaths for the time period. For cause-specific mortality, we only included the specific death being considered. For example, with cancer mortality, if the veteran died of another cause of death other than cancer, then his survival time was counted until the time of death from the other cause and then he exited from the analysis, which is a conservative estimation method (Cleves, Gould, & Gutierrez, 2002). All *p*-values presented were based on the two-tail test.

STUDY RESULTS

Examination of major differences by veteran status indicates that theater veterans not only had higher rates of PTSD (10.6% vs. 2.9%), but also were younger at the 1985 follow-up survey (17.7% vs. 22.3%, > 40 years of age) and had nonhonorable discharges less often (1.9% vs. 6.3%, nonhonorable) (Table 5.1). Also noteworthy is that 21% of the theater veterans volunteered for Vietnam service. In terms of PTSD status, the major differences were even more striking. For example, not only were the PTSD-positive veterans more likely to have died since the 1985 survey (11.8% vs. 4.9%), but they also were different in terms of other measurement variables (Table 5.2). For example, PTSD-positive veterans were more likely to have been nonwhite (30% vs. 16.4%), in the lowest intelligence quintile (37.3% vs. 18%), and to have had entered the service at a younger age (25% vs. 12.6%). They were also more likely to have used illicit drugs in the Army (8.1% vs.

TABLE 5.1
Profile of Vietnam Theater Veterans vs. Vietnam Era Veterans

Variables	Vietnam Theater Veteran (%)	Vietnam Era Veteran (%)	p-value[a]
PTSD at interview	10.6	2.9	<0.001
Deceased at follow-up	5.5	5.2	0.39
Age 40+ at interview	17.7	22.3	<0.001
Nonwhite	16.8	18.0	0.054
Married at discharged	27.3	30.1	<0.001
Intelligence—lowest quintile	20.8	17.8	<0.001
Used illicit drugs in service	2.5	1.8	0.004
Drafted into military service	64.4	67.2	<0.001
Volunteered for Vietnam service[b]	20.5	0.0	<0.001
Very high combat exposure[b]	20.9	0.0	<0.001
Entered service at 18 or less	14.2	12.6	0.004
Less than honorable discharge	1.9	6.3	<0.001
N	7,924	7,364	

[a] 2-sided chi-square test, df = 1.
[b] Vietnam era veterans coded as "no/none" for this variable.

TABLE 5.2
Profile of PTSD-Negative vs. PTSD-Positive Vietnam Veterans

Variables	PTSD Negative (%)	PTSD Positive (%)	p-value[a]
Deceased at follow-up	4.9	11.8	<0.001
Age 40+ at interview	20.4	13.0	<0.001
Nonwhite	16.4	30.0	<0.001
Married at discharge	28.6	29.0	0.753
Intelligence—lowest quintile	18.0	37.3	<0.001
Used illicit drugs in service	1.7	8.1	<0.001
Drafted into military service	66.4	57.0	<0.001
Volunteered for Vietnam service[b]	9.6	25.0	<0.001
Very high combat exposure[b]	9.5	28.9	<0.001
Entered service at 18 or less	12.6	25.0	<0.001
Less than honorable discharge	3.6	8.7	<0.001
N	14,238	1,050	

[a] 2-sided chi-square test, df = 1.
[b] Vietnam era veterans coded as "no/none" for this variable.

1.7%), less likely to be drafted (57% vs. 66.4%), and more likely to have a less than honorable discharge (8.7% vs. 3.6%). As one would expect, PTSD was also associated with higher combat exposure (28.9% vs. 9.5% overall), as well as with volunteering for Vietnam service (25% vs. 9.6% overall).

Table 5.3 presents the unadjusted and adjusted mortality results, respectively, for all-cause and cause-specific mortality, as well as the number of deaths and the total person-years at risk for theater and era veterans. As can be seen, the unadjusted all-cause mortality for PTSD-positive veterans was higher for both the era and the theater veterans, with HRs of 2.6 and 2.5, respectively (both p-values <0.001). When adjusted for potential confounders and bias, these reduce the HRs to 2.0 for era veterans and 2.2 for theater veterans, both still significant. Furthermore, when we added high combat exposure to the all-cause mortality model for theater veterans the HR was only slightly reduced (HR = 2.1; $p < 0.001$). For cardiovascular mortality, neither the unadjusted nor the adjusted results for era veterans were significant, with HRs of 1.3 and 1.2, respectively. However, for the theater veterans, both the unadjusted and adjusted results were significant, with HRs of 1.8 ($p = 0.015$) and 1.6 ($p = 0.034$), respectively. Nevertheless, the introduction of combat exposure status in the final model reduced the HR for this outcome to 1.5, which was now marginally significant ($p = 0.087$). In terms of cancer mortality, the unadjusted and adjusted results were not significant for era veterans, with HRs of 1.1 and 1.2, respectively,

TABLE 5.3
Cox Proportional Hazards Regressions: Crude and Adjusted Hazard Ratios by Veteran and PTSD Status

	All-Cause Mortality (Total Deaths = 820)			Cardiovascular Mortality[a] (Total Deaths = 241)			Cancer Mortality[a] (Total Deaths = 188)			External-Cause Mortality (Total Deaths = 175)		
Veteran Status	HR	95% CI	*p-value*	HR	95% CI	*p-value*	HR	95% CI	*p-value*	HR	95% CI	*p-value*
Vietnam Era (N = 7,364) (Person risk years = 110,553) (Total PTSD cases = 214)												
PTSD—unadjusted	2.6	1.7–3.8	<0.001	1.3	0.5–3.6	0.57	1.1	0.4–3.6	0.84	2.9	1.3–6.7	0.012
PTSD—adjusted[b]	2.0	1.3–3.0	0.001	1.2	0.4–3.5	0.69	1.2	0.4–3.4	0.70	2.2	0.9–5.2	0.073
Vietnam Theater (N = 7,924) (Person risk years = 119,453) (Total PTSD cases = 836)												
PTSD—unadjusted	2.5	2.0–3.2	<0.001	1.8	1.1–2.8	0.015	2.2	1.3–3.7	0.003	2.6	1.6–4.1	<0.001
PTSD—adjusted[b]	2.2	1.7–2.7	<0.001	1.6	1.0–2.6	0.034	2.0	1.2–3.4	0.012	2.2	1.3–3.7	0.002
PTSD—also adjusted for high combat exposure[c]	2.1	1.6–2.7	<0.001	1.5	0.9–2.5	0.087	1.9	1.1–3.3	0.017	2.2	1.3–3.7	0.002

[a] Cardiovascular and cancer mortality also adjusted for pack-years of cigarette smoking.
[b] All models adjusted for age at interview, race, Army volunteer status, Army entry age, Army marital status, Army discharge status, Army illicit drug use, and intelligence. For theater veterans, models adjusted for the above variables, in addition to Vietnam volunteer status (see methods section for additional information).
[c] Adjusted for above variable, plus high combat exposure.
CI = confidence interval; HR = hazards ratio; PTSD = posttraumatic stress disorder.

but this was not the case for theater veterans. Here the unadjusted and adjusted cancer results for theater veterans were both significant, with HRs of 2.2 ($p = 0.003$) and 2.0 ($p = 0.012$), respectively. Furthermore, the addition of combat exposure status did not appreciably change this association (HR = 1.9; $p = 0.017$). For external-cause mortality, the unadjusted results were significant for the era veterans (HR = 2.9; $p = 0.012$), but the adjusted results were marginally significant for this outcome (HR = 2.2; $p = 0.073$). However, both the adjusted and unadjusted external-cause results were significant for the theater veterans (HR = 2.6; $p < 0.001$; and HR = 2.2; $p = 0.002$, respectively). Adding combat exposure to the external-cause mortality model basically had no impact on this association (HR = 2.2; $p = 0.002$). It should be noted that for the theater veterans combat exposure status was not generally significant when PTSD was also included in the model, which was not the case for PTSD and all-cause, cancer, and external-cause mortality. As shown, PTSD was significant in the models that also include combat exposure, with the exception of cardiovascular disease mortality.

Since cancer mortality was significant among the PTSD-positive theater veterans, we examined the detailed cancer results to determine if one specific cancer site was more prevalent than others. These results were mixed. PTSD-positive theater veterans tended to have a somewhat higher rate of death from lung as well as from all other cancers combined (both 108 cases per 10,000 persons). We also examined external-cause mortality in more detail for the veterans, since this outcome was also highly significant. While specific external-mortality classifications could not be modeled due to the small numbers, we qualitatively examined these by PTSD and veteran status. As can be seen in Table 5.4, since the overall prevalence is similar in each external death category, it appears that PTSD-positive theater veterans were more likely to be classified as a homicide or self-inflected death (i.e., suicide, firearm, drug, or alcohol-related mortality) for some reason compared with era veterans.

DISCUSSION OF STUDY FINDINGS

In the most recent CDC follow-up of Vietnam veterans, external causes, diseases of the circulatory system, and malignant neoplasms accounted for a substantial proportion of deaths (38.5%, 23.1%, and 17.5%, respectively), as would be expected for nonveteran men in the same age range (Anderson, 2002; Boehmer et al., 2004). Our current Vietnam veteran study found PTSD was associated with all three of these causes of death and is consistent with other preliminary studies of veterans (Fett et al., 1987; Watanabe & Kang, 1995). Compared to the CDC study and the other studies mentioned, however, our analysis yields more specific findings. PTSD was associated with an adjusted all-cause mortality for *both* era and

TABLE 5.4
Specific Cause-of-Death and Method-of-Death Classifications by PTSD-Positive and Veteran Status[a]

Specific Cause of Death	% PTSD Positive—All Veterans % All Deaths (n)	% PTSD Positive—Era Veterans% All Deaths (n)	% PTSD Positive—Theater Veterans% All Deaths (n)
Nonexternal cause of death	14.8 (648)	6.5 (306)	22.2 (342)
Transportation-related death	7.5 (53)	0.0 (23)	13.3 (30)
Nontransportation-related death	23.5 (34)	23.1 (13)	23.8 (21)
Suicide	15.3 (59)	3.4 (29)	26.7 (30)
Homicide	27.3 (22)	20.0 (10)	33.3 (12)
Death of undetermined intent	25.0 (4)	0.0 (2)	50.0 (2)
Total deaths (N)	— (820)	— (383)	— (437)
Method of Death	**% PTSD Positive—All Veterans % All Deaths (n)**	**% PTSD Positive—Era Veterans% All Deaths (n)**	**% PTSD Positive—Theater Veterans% All Deaths (n)**
Nonfirearm-, drug-, alcohol-related death	14.1 (708)	6.4 (327)	20.7 (381)
Firearm-related	13.0 (54)	7.7 (26)	17.9 (28)
Drug-related	19.0 (21)	0.0 (10)	36.4 (11)
Alcohol-related	35.1 (37)	15.0 (20)	58.8 (17)
Total deaths (N)	— (820)	— (383)	— (437)

[a] Results represent percentage of deaths in each category that are PTSD-positive cases by veteran status. For example, table shows that 6.5% of era veterans who died from nonexternal causes had PTSD at baseline, compared with 22.2% of theater veterans.

theater veterans. Among theater veterans, PTSD was also associated with cardiovascular, cancer, and external mortality, and for all-cause, cancer, and external mortality this was true *even* after controlling for combat exposure status. Nevertheless, for external-cause mortality, PTSD-positive era veterans also appeared to be at risk for death, which approached statistical significance (HR = 2.2; $p = 0.073$). The fact that the era veterans with noncombat PTSD were at risk for all-cause and potentially for external mortality is noteworthy, especially since for theater veterans combat exposure was *not* generally significant when PTSD was also included in the model. This suggests that it is likely PTSD, not combat exposure per se, which is associated with increased mortality, as has been suggested elsewhere (Boscarino, 2004). Furthermore, as noted, the specific cause-of-death classifications for external mortality suggested that PTSD-positive theater veterans were more likely to die from suicide, homicide, and from alcohol- and drug-related causes. Given these results and the lack of significance for combat status, our study suggests that it is PTSD that is associated with premature mortality, not combat exposure, but that type of external mortality (e.g., intended vs. unintended deaths) may differ between veteran cohorts for some reason.

This study has strengths and limitations. Use of multiple sources of vital status allowed for a more complete account of postservice mortality in the United States. However, investigators may have missed deaths that occurred elsewhere. Nevertheless, underreporting was likely to have occurred equally among Vietnam and non-Vietnam veterans. In addition, underlying cause of death, as reported on the death certificate, is known to underreport alcohol- and drug-related deaths and to overreport circulatory, ill-defined, and respiratory conditions (Boehmer et al., 2004). Furthermore, although our RTI-PTSD scale appeared to be a valid measure, this was an earlier version of the PTSD nomenclature (Center for Disease Control, 1989a), and it likely lacked specificity compared to a valid gold standard (e.g., a structured diagnostic clinician interview) with a comparable timeframe. It also should be mentioned that the DIS-III instrument used in the VES was an earlier version of current PTSD criteria, and has been found to be at variance with recent PTSD measures (Kulka et al., 1990b). However, given the consistency of our validation study discussed above, we conclude that the RTI-PTSD measure used in our study was generally consistent with the presence of PTSD among these men. Other limitations were that our study included only men and only those who survived to participate in the survey. Nevertheless, study strengths were that the research was based on a large population sample, not simply persons identified through medical clinics or treatment seeking, and it included controls for key biases and confounders.

IMPLICATIONS FOR CURRENT STUDY

As suggested, there is growing evidence that exposure to psychologically traumatic events is related to increased medical morbidity (Boscarino, 1997). We think that our study, together with others, suggests a link between long-term exposure to severe psychological distress (including combat- and noncombat-related stress) and premature mortality from multiple causes. More conclusive evidence will require additional research. A particular challenge in this research will be assessing the impact of behavioral risk factors that could be related to human trauma exposures, but which also could result in disease, such as alcohol abuse or tobacco dependence. However, we suggest that these behavioral aspects of disease also may prove especially promising, because acquiring health-enhancing behaviors could be protective for adverse outcomes along these specific causal pathways. For example, cognitive therapies are often recommended for treatment of PTSD and other anxiety disorders (Foa, Keane, & Friedman, 2000). If this therapy is effective in reducing PTSD symptoms, then the burden of disease may be decreased, possibly along several causal pathways, including psychological, behavioral, and/or biological ones (Boscarino, 2004; McEwen, 2000). Nevertheless, PTSD appears to have a biological foundation that exists below normal cognitive states (Boscarino, 1996, 1997), suggesting that purely psychotherapeutic approaches may be limited (Shean, 2001). As has been previously suggested (Boscarino, 1997), understanding both the physiological and the psychological aspects of traumatic phenomenon seems warranted to effectively treat the sequelae associated with these syndromes. Nevertheless, the behavioral risk-factor disparities shown for PTSD status in Table 5.2 clearly point to some of the potential challenges. In addition, it is also important to stress that our study population included *only* those who survived to participate in the survey, hence the force of morbidity related to PTSD is likely much greater than suggested here—485 men died *before* the baseline interviews ever occurred.

There is growing evidence that the development of PTSD may be related to alterations in neuroendocrine, immune system, and other psychoneuroendocrine-related functions. In particular, given the reduced cortisol levels often found among PTSD victims, it has been suggested that a down-regulated glucocorticoid system may result in elevations in leukocyte and other immune inflammatory activities (Chrousos, 1995). One biologic pathway often cited involves alterations in the hypothalamic-pituitary-adrenal stress axis in concert with sympathetic-adrenomedullary stress axis activation (Boyce & Jemerin, 1990; Chrousos & Gold, 1992; Kiecolt-Glaser & Glaser, 1995), which could result in a host of specific diseases (Boscarino, 2004). An increase in allostatic load as a result of these biologic alterations, or efforts to relieve their adverse psychological effects through substance use, or both, could contribute to the pathophysiologic process (McEwen, 1998). We recently reported that workers in New York City who received brief emergency mental health counseling following the World Trade Center attacks had significantly reduced mental health and substance abuse problems up to 2 years after this event (Boscarino, Adams, & Figley, 2005; Boscarino et al., 2006).

Figure 5.1 provides a schematic diagram of the potential causal nexus of factors that could affect health status that are likely related to past combat exposure. As can be seen from this diagram there are multiple causal pathways through which future health status could be related to combat stress exposures. The research presented suggests that it is likely the onset of PTSD that is the reason for the association between previous

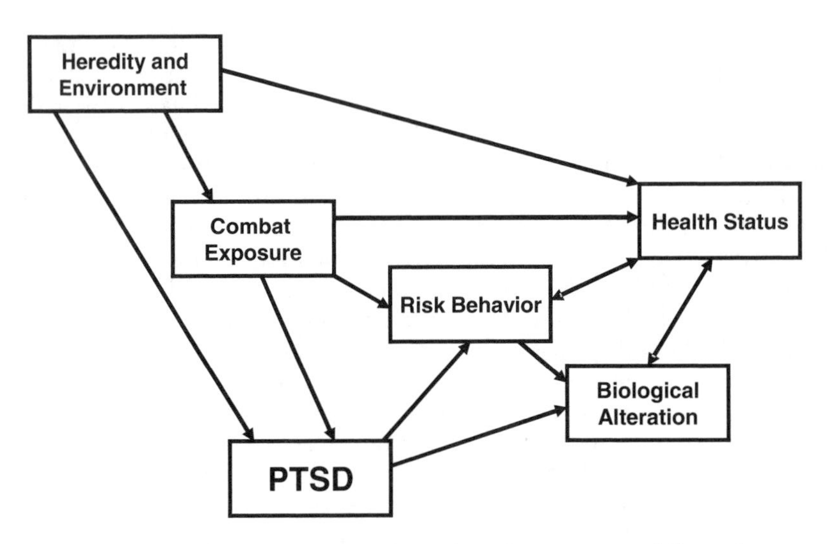

FIGURE 5.1 Multi-factor model for impact of combat exposure: different long-term causal pathways to disease outcomes.

exposure to traumatic stressors and advent of mortality and disease. If the PTSD can be successfully treated, however, then these outcomes should be averted.

Current clinical evidence suggests that the physiologic arousal often experienced during recollection of traumatic events by PTSD victims is associated with a cascade of neuroendocrine-related processes (Boscarino 1996, 1997, 1999a,b). Furthermore, while this response is initiated in the central nervous system, it is subsequently carried out by molecular mechanisms that have wide-ranging affects on human physiology. Although research suggests that these processes are not fully understood (Boscarino 1997; Boscarino 1996; McEwen 1998, 2000), it is likely that excessiveness of stress axis activation due to PTSD may directly or indirectly lead to disease among some individuals (Chrousos, 1995; McEwen, 1998). Furthermore, research suggests that there are genetic liabilities for PTSD, as well as exposure to psychological trauma (Gurvits et al., 2006; Stein et al., 2002; True et al., 1993). In addition, there is evidence that the effects of psychological stressors are mediated by gene and environmental interactions (Boomsma et al., 2002). Based on this new body of research, it seems likely that models of PTSD pathogenesis will emerge that incorporate molecular-level mechanism linked to specific genetic polymorphisms and environmental exposures. Finally, it is entirely possible that genes that make persons susceptible to PTSD (e.g., those related to adrenal functioning) may later also make them susceptible to certain diseases, such as inflammatory conditions and cardiovascular diseases (Boscarino, 2004). Given this growing body of new research, in the future we may likely see high-risk PTSD patients being treated not only with cognitive behavioral therapy and psychotropic drugs, but also with beta-blockers and anti-inflammatory medicines as well.

Preliminary reports related to military personnel returning from Afghanistan and Iraq suggest significant rates of psychological problems among these veterans (Friedman, 2004, 2005). In addition, consistent with our findings, anecdotal evidence suggests that these personnel may be at higher risk for suicide after return (Scott, 2005). One study that examined mental health problems among infantry units before and after deployment to Iraq or Afghanistan found significant increases in mental health disorders, including PTSD (Hoge et al., 2004). Another study tracking Department of Defense personnel, who had become eligible for Department of Veterans Affairs benefits during the Iraq and Afghanistan conflicts, showed an increase in treatment for mental health disorders and PTSD over time (Kang & Hyams, 2005). In addition, recent reports suggest that many of those wounded in these combat theaters are surviving injuries that would have previously proven fatal (Gawande, 2004; Peake, 2005). These medical advances may increase the prevalence of psychiatric casualties among wounded veterans returning from the current theaters of war (Jones, 2004). Recent reports that returning veterans may not be availing themselves of mental health treatment is worrisome (Hoge et al., 2004), and may lead to unnecessary long-term morbidity, increased healthcare costs, and premature mortality. Unfortunately, this is not an uncommon finding in the fields of disaster and war psychiatry (Boscarino, Adams, Stuber, & Galea, 2005). While additional research is needed, in light of the findings reported, barriers to mental health treatment need to be reduced as much as possible for all returning military personnel.

It is important to note that recognition of the multifactoral nature of PTSD is consistent with the observation that both pharmacotherapy and cognitive-behavioral psychotherapy are reported effective in treating PTSD (Boscarino, 2006; Boscarino, Adams, & Galea, 2006). In the case of pharmacotherapy, the pathophysiology of PTSD, in part, appears to involve the serotonergic and the noradrenergic systems; hence, drugs known to impact these mechanisms have been effective (Boscarino, 2006; Boscarino, Adams, & Galea, 2006). In the case of cognitive-behavioral therapy, this approach has been found effective in reducing PTSD-related symptomotology, by achieving psychological desensitization to stressful stimuli, by increasing control of aversive arousal, by enhancing anxiety management, and by other known behavioral-psychological mechanisms (Boscarino, 2006; Boscarino, Adams, & Galea, 2006; Foa, Keane, & Matthew, 2000). Thus, although the underlying causal mechanisms may differ for pharmacological (e.g., molecules) versus cognitive-behavioral therapy (e.g., cognitions), the outcomes would be similar—the psychopathology and underlying pathophysiology would be reduced and fewer adverse client symptoms manifested (Boscarino, 1997), hence, lowering the risk of such things as substance abuse (Boscarino, Adams, & Galea, 2006) as well as adverse physical health outcomes (Boscarino, 2006). Therefore, we would expect that as trauma-related symptoms are reduced through treatment, the risk for future negative health outcomes would also decrease. As noted, recent research following the World Trade Center disaster has confirmed this promising hypothesis (Boscarino, Adams, & Figley, 2005; Boscarino et al., 2006). Future research is now needed to confirm the long-term validity of this major finding for current warfighters.

REFERENCES

American Psychiatric Association. (1980). *Diagnostic and Statistical Manual of Mental Disorders* (3rd ed). Washington, DC: Author.

Anderson, R. N. (2002). Deaths: Leading causes. *National Vital Statistics Report, 50*(16), 1–85.

Armenian, H. K., Melkonian, A. K., & Hovanesian, A. P. (1998). Long term mortality and morbidity related to degree of damage following the 1988 earthquake in Armenia. *American Journal of Epidemiology, 148,* 1077–1084.

Baskir, L. M., & Strauss, W. A. (1978). *Chance and Circumstance: The Draft, the War and the Vietnam Generation.* New York: Vintage Books.

Boehmer, T. K., Flanders, W. D., McGeehin, M. A., Boyle, C., & Barrett, D. H. (2004). Postservice mortality in Vietnam veterans: 30-year follow-up. *Archives of Internal Medicine, 164,* 1908–1916.

Boomsma, D., Busjahn, A., & Peltonen, L. (2002). Classical twin studies and beyond. *Nature Reviews: Genetics, 3,* 872–882.

Boscarino, J. A. (1995). Post-traumatic stress and associated disorders among Vietnam veterans: The significance of combat exposure and social support. *Journal of Traumatic Stress, 8,* 317–336.

Boscarino, J. A. (1996). Post-traumatic stress disorder, exposure to combat, and lower plasma cortisol among Vietnam veterans: Findings and clinical implications. *Journal of Consulting and Clinical Psychology, 64,* 191–201.

Boscarino, J. A. (1997). Diseases among men 20 years after exposure to severe stress: Implications for clinical research and medical care. *Psychosomatic Medicine, 59,* 605–614.

Boscarino, J. A. (2000). Postwar experiences of Vietnam veterans. In G. Fink (Ed.), *Encyclopedia of Stress* (Vol. 3, pp. 656–662). New York: Academic Press.

Boscarino, J. A. (2004). Association between posttraumatic stress disorder and physical illness: Results and implications from clinical and epidemiologic studies. *Annals of the New York Academy of Sciences, 1032,* 141–153.

Boscarino, J. A. (2006). The postwar experiences and health outcomes of Vietnam veterans. In: G. Fink (Ed.), *Encyclopedia of Stress* (2nd ed.). New York: Academic Press (in press).

Boscarino, J. A., Adams, R. E., & Figley, C. R. (2005). A prospective cohort study of the effectiveness of employer-sponsored crisis interventions after a major disaster. *International Journal of Emergency Mental Health, 7,* 9–22.

Boscarino, J. A., Adams, R. E., Foa, E. B., & Landrigan, P. J. (2006). A propensity score analysis of brief worksite crisis interventions after the World Trade Center disaster: Implications for intervention and research. *Medical Care, 44,* 454–462.

Boscarino, J. A., Adams, R. E., & Galea, S. (2006). Alcohol use in New York after the terrorist attacks: A study of the effects of psychological trauma on drinking behavior. *Addictive Behaviors, 31,* 606–621.

Boscarino, J. A., Adams, R. E., Stuber, J., & Galea, S. (2005). Disparities in mental health treatment following the World Trade Center disaster: Implication for mental health care and services research. *Journal of Traumatic Stress, 18,* 287–297.

Boscarino, J. A., & Chang, J. (1999a). Electrocardiogram abnormalities among men with stress-related psychiatric disorders: Implications for coronary heart disease and clinical research. *Annals of Behavioral Medicine, 21,* 227–234.

Boscarino, J. A., & Chang, J. (1999b). Higher abnormal leukocyte and lymphocyte counts 20 years after exposure to severe stress: Research and clinical implications. *Psychosomatic Medicine, 61,* 378–386.

Boyce, W. T., & Jemerin, J. M. (1990). Psychobiological differences in childhood stress response. I: Patterns of illness and susceptibility. *Journal of Developmental and Behavioral Pediatrics, 11,* 86–94.

Buckley, T. C., & Kaloupek, D. G. (2001). A meta-analytic examination of basal cardiovascular activity in posttraumatic stress disorder. *Psychosomatic Medicine, 63,* 585–594.

Cattell, R. B. (1986). The psychometric property of tests: Consistency, validity, and efficiency. In R. B. Cattell, & R. C. Johnson (Eds.), *Functional Psychological Testing: Principles and Instruments* (pp. 54–78). New York: Brunner/Mazel.

Centers for Disease Control. (1987). Postservice mortality among Vietnam veterans. *Journal of the American Medical Association, 257,* 790–795.

Centers for Disease Control. (1988a). Health status of Vietnam veterans: I. Psychosocial characteristics. *Journal of the American Medical Association, 259,* 2701–2707.

Centers for Disease Control. (1988b). Health status of Vietnam veterans: II. Physical health. *Journal of the American Medical Association, 259,* 2708–2714.

Centers for Disease Control. (1989a). Health status of Vietnam veterans: Volume II. Telephone interview. Atlanta: Author.

Centers for Disease Control. (1989b). Health status of Vietnam veterans: Volume IV. Psychological and neuropsychological evaluation. Atlanta: Author.

Chrousos, G. P. (1995). The hypothalamic-pituitary-adrenal axis and immune-mediated inflammation. *New England Journal of Medicine, 332,* 1351–1362.

Chrousos, G. P., & Gold, P. W. (1992). The concepts of stress and stress system disorders: Overview of physical and behavioral homeostasis. *Journal of the American Medical Association, 267,* 1244–1252.

Cleves, M. A., Gould, W. W., & Gutierrez, R. G. (2002). *Introduction to Survival Analysis Using Stata.* College Station, TX: Stata Press.

Cwikel, J., Abdelgani, A., Goldsmith, J. R., Quastel, M., & Yevelson, I. (1997). Two-year follow up study of stress-related disorders among immigrants to Israel from the Chernobyl area. *Environmental Health Perspectives, 105*(Suppl 6), 1545–1550.

Danesh, J., Collins, R., Appleby, P., & Peto, R. (1998). Association of fibrinogen, C-reactive protein, albumin, or leukocyte count with coronary heart disease: Meta-analyses of prospective studies. *Journal of the American Medical Association, 279,* 1477–1482.

Falger, P. R. J., Op den Velde, W., Hovens, J. E., Schouten, E. G. W., Degroen, J. H. M., & Van Duijn, H. (1992). Current posttraumatic stress disorder and cardiovascular disease risk factors in Dutch Resistance veterans from World War II. *Psychotherapy and Psychosomatics, 57,* 164–171.

Felitti, V. J., Anda, R. F., Nordenberg, D., Williamson, D. F., Spitz, A. M., Edwards, V., et al. (1998). Relationship of childhood abuse and household dysfunction to many of the leading causes of death in adults: The Adverse Childhood Experience (ACE) Study. *American Journal of Preventive Medicine, 14,* 245–258.

Fett, M. J., Adena, M. A., Cobbin, D. M., & Dunn, M. (1987). Mortality among Australian conscripts of the Vietnam conflict era. I. Deaths from all causes. *American Journal of Epidemiology, 125,* 869–877.

Foa, E., Keane, T. M., & Friedman, M. J. (Eds.). (2000). *Effective Treatment for PTSD.* New York: Guilford Press.

Frey-Wouters, E., & Laufer, R. S. (1986). *Legacy of a War: The American Soldier in Vietnam.* Armonk, NY: Sharpe, Inc.

Friedman, M. J. (2004). Acknowledging the psychiatric cost of war. *New England Journal of Medicine, 351,* 75–77.

Friedman, M. J. (2005). Veterans' mental health in the wake of war. *New England Journal of Medicine, 352,* 1287–1290.

Gawande, A. (2004). Casualties of war: Military care for the wounded from Iraq and Afghanistan. *New England Journal of Medicine, 351,* 2471–2475.

Gurvits, T. V., Metzger, L. J., Lasko, N. B., Cannistraro, P. A., et al. (2006). Subtle neurologic compromise as a vulnerability for combat-related posttraumatic stress disorder. *Archives of General Psychiatry, 63,* 571–576.

Herbert, T. B., & Cohen, S. (1993). Stress and immunity in humans: A meta-analytic review. *Psychosomatic Medicine, 55,* 364–379.

Hoge, C. W., Castro, C. A., Messer, S. C., McGurk, D., Cotting, D. I., & Koffman, R.L. (2004). Combat duty in Iraq and Afghanistan, mental health problems, and barriers to care. *New England Journal of Medicine, 351,* 13–22.

Hosmer, D. W., & Lemeshow, S. (1999). *Applied Survival Analysis.* New York: John Wiley & Sons.

Ironson, G., Wynings, C., Schneiderman, N., Baum, A., Rodriguez, M., Greenwood, D., et al. (1997). Posttraumatic stress symptoms, intrusive thoughts, loss, and immune function after Hurricane Andrew. *Psychosomatic Medicine, 59,* 128–142.

Janes, G. R., Goldberg, J., Eisen, S. A., & True, W. R. (1991). Reliability and validity of a combat exposure index for Vietnam era veterans. *Journal of Clinical Psychology, 47,* 80–86.

Jones, S. (2004). Paying the price: The psychiatric cost of war. *Archives of Psychiatric Nursing, 18,* 119–120.

Kang, H. K., & Hyams, K. C. (2005). Mental health care needs among recent war veterans. *New England Journal of Medicine, 352,* 1289.

Kiecolt-Glaser, J. K., & Glaser, R. (1995). Psychoneuroimmunity and health consequences: Data and shared mechanisms. *Psychosomatic Medicine, 57,* 269–274.

Koenen, K. C., Lyons, M. J., Goldberg, J., Simpson, J., Williams, W. M., Toomy, R., et al. (2003). Co-twin control study of relationships among combat exposure, combat-related PTSD, and other mental disorders. *Journal of Traumatic Stress, 16,* 433–438.

Kulka, R. A., Schlenger, W. E., Fairbank, J. A., Hough, R. L., Jordan, B. K., Marmar, C. R., et al. (1990a). *Trauma and the Vietnam Generation: Report of Findings from the National Vietnam Readjustment Study.* New York: Brunner/Mazel.

Kulka, R. A., Schlenger, W. E., Fairbank, J. A., Hough, R. L., Jordan, B. K., Marmar, C.R., et al. (1990b). Appendix E: NVVRS prevalence estimate methodology and comparison with Vietnam Experience Study estimates. In *Trauma and the Vietnam Generation: Report of Findings from the National Vietnam Veterans Readjustment Study.* Tables of Findings and Technical Appendices (pp. E2–E25). New York: Brunner/Mazel.

Long, N., Chamberlain, K., & Vincent, C. (1992). The health and mental health of New Zealand Vietnam War veterans with posttraumatic stress disorder. *New Zealand Medical Journal, 105,* 417–419.

Mason, J. M., Giller, E. L., Kosten, T. R., & Harkness, L. (1988). Elevation of urinary norepinephrine/cortisol ratio in posttraumatic stress disorder. *Journal of Nervous and Mental Disease, 176,* 498–502.

McEwen, B. S. (1998). Protective and damaging effects of stress mediations. *New England Journal of Medicine, 338,* 171–179.

McEwen, B. S. (2000). Allostasis and allostatic load: Implication for neuropsychopharmacology. *Neuropsychopharmacology, 22,* 108–124.

Miric, D., Giunio, L., Bozic, I., Fabijanic, D., Martinovic, D., & Culic, V. (2001). Trends in myocardial infarction in Middle Dalmatia during war in Croatia. *Military Medicine, 166,* 419–421.

Peake, J. B. (2005). Beyond the Purple Heart: Continuity of care for the wounded in Iraq. *New England Journal of Medicine, 352,* 219–222.

Robins, L. N., Helzer, J. E., & Cottler, L. B. (1987). *The Diagnostic Interview Schedule Training Manual* (Version III-A). St. Louis, MO: Veterans Administration.

Robins, L. N., Helzer, J. E., Croughan, J., & Ratcliff, K. S. (1981). National Institute of Mental Health Diagnostic Interview Schedule: Its history, characteristics, and validity. *Archives of General Psychiatry, 38,* 381–389.

Schnurr, P. P., & Green, B. L. (Eds.). (2004). *Trauma and Health: Physical Health Consequences of Extreme Stress* (pp. 3–10). Washington, DC: American Psychological Association.

Schnurr, P. P, Spiro, A., & Paris, A. (2000). Physician-diagnosed medical disorders in relation to PTSD symptoms in older male military veterans. *Health Psychology, 19,* 91–97.

Scott, S. (2005). Wreath for those killed, even at their own hands. *New York Times.* June 3, A20.

Shean, C. (2001). A critical look at the assumptions of cognitive therapy. *Psychiatry, 64,* 158–164.

Sibai, A. M., Armenian, H. K., & Alam, S. (1989). Wartime determinants of arteriographically confirmed coronary artery disease in Beirut. *American Journal of Epidemiology, 130,* 623–631.

Sibai, A. M., Fletcher, A. F., & Armenian, H. K. (2001). Variations in the impact of long-term wartime stressors on mortality among the middle-aged and older population in Beirut, Lebanon, 1983–1993. *American Journal of Epidemiology, 154,* 128–137.

Stein, M. B., Jang, K. L., Taylor, S., Vernon, P. A., & Livesley, W. J. (2002). Genetic and environmental influences on trauma exposure and posttraumatic stress disorder symptoms: A twin study. *American Journal of Psychiatry, 159,* 1675–1681.

Syme, L. S. (1992). Social determinants of disease. In J. M. Last, R. B. Wallace, B. N. Doebbeling, et al. (Eds.). *Maxcy-Rosenau-Last: Public and Preventive Medicine,* 13th ed. (pp. 687–700). Norwalk, CT: Appleton & Lange.

True, W. R., Rice, J., Eisen, S. A., Heath, A. C., Goldberg, J., et al. (1993). A twin study of the genetic and environmental contributions of liability for posttraumatic stress symptoms. *Archives of General Psychiatry, 50,* 257–264.

Watanabe, K. K., & Kang, H. K. (1995). Military service in Vietnam and the risk of death from trauma and selected cancers. *Annals of Epidemiology, 5,* 407–412.

Yehuda, R. (Ed.). (1999). *Risk Factors for Posttraumatic Stress Disorder.* Washington, DC: American Psychiatric Press.

Yehuda, R. (2002). Current status of cortisol findings in post-traumatic stress disorder. *Psychiatric Clinics of North America, 25,* 341–368.

Yehuda, R., Kahana, B., Binder-Brynes, K., Southwick, S. M., Mason, J. W., & Giller, E. L. (1995). Low urinary cortisol excretion in holocaust survivors with posttraumatic stress disorder. *American Journal of Psychiatry, 152,* 982–986.

Yehuda, R., Southwick, S., Giller, E. L., Ma, X., & Mason, J. W. (1992). Urinary catecholamine excretion and severity of PTSD symptoms in Vietnam combat veterans. *Journal of Nervous and Mental Disease, 180,* 321–325.

6

Combat Stress Management: The Interplay Between Combat, Physical Injury, and Psychological Trauma

DANNY KOREN, YAIR HILEL, NOA IDAR,
DEBORAH HEMEL, AND EHUD M. KLEIN

Boom!...silence...I am unable to understand what people around me are saying. Everything moves in slow motion. I find myself struggling to breathe. I am unable to see because of the blood in my eyes. As I gasp for breathe, I realize that the bullet struck only an inch away from my heart. I try to breathe again, and again... I can't!

My soldiers run to help me. The medic starts working on me, and as I hear him shouting orders I realize that my leg is also injured and that I am losing a lot of blood from that wound. I did not even realize my leg was wounded because I was so preoccupied with trying to breathe. In the meantime, another medic joins in and they try to stop the blood flow. Then they lift me onto a stretcher and begin to run. Seconds later I fall from the stretcher, and as I fall, I see that my friend who was carrying the stretcher was hit in the leg...others are also hit. I am evacuated by helicopter to the hospital...I wake up the following morning and I am told that the doctors fought all night to keep me alive.

Thank God we all survived, but my battle is not over. Since that battle I suffer from nightmares: I see the blood, I hear the screams. I am frustrated, I feel like a failure; nothing seems to interest me any longer. I have lost my ability to be happy.

I wish that this battle would end soon so that I can return to being a normal 22-year-old kid.

From the "Personal Accounts" Column in NATAL's site
[a nonprofit organization treating Israeli victims of terror and war]

The account above, given by an Israeli commander who was injured during combat in September 2002, illustrates the suffering of combat survivors who not only experienced the horrors of a traumatic event but were also physically injured during the event.

The meaning of the Greek word *trauma* means "injury" or "wound." More specifically, trauma is defined as "an injury to living tissue caused by an extrinsic agent" (*Merriam-Webster's*, 1993). Accordingly, a psychological trauma is an emotional wound; an injury caused by a stressful extrinsic event that is perceived by an individual to be an actual threat to his or her life, or to his or her physical and emotional integrity.

In contrast to physical trauma, which is characterizable based on objective measures, emotional trauma depends on an individual's **subjective interpretation** of the objective characteristics of the trauma (Allen, 1995). In other words, the more traumatic an individual perceives the event to be, and the more helpless the individual feels, the greater the individual's risk of experiencing psychological trauma. Furthermore, a traumatic event is perceived as such only if the individual is aware of the threat. It is the subjective nature of psychological trauma that differentiates it from physical trauma, and this subjectiveness is key to understanding it.

Given the frequency with which warfighters suffer both physical and emotional injuries following combative events, a crucial question arises: How do the two traumas—the physical and the psychological—interact? How does physical trauma affect an individual's chances of developing a posttraumatic reaction? Is the risk of developing psychological trauma higher among those injured during a traumatic event, or might the injury act as a buffer? Does the injury enhance an individual's ability to cope with the psychological trauma? And in situations in which a physical injury results in permanent disability, is it possible to even determine when the physical trauma ends?

In this chapter we attempt to unravel the complexity of such questions. The chapter includes a review of the literature in the field of physical injury and psychological trauma, a description of the authors' research on soldiers injured during combat in Lebanon, and finally an examination of how such research will affect the ability of healthcare professionals to identify and treat emotional trauma among *physically* injured warfighters.

TRAUMATIC PHYSICAL INJURY: A RISK OR A PROTECTIVE FACTOR IN THE DEVELOPMENT OF PSYCHOLOGICAL TRAUMA?

During the past two decades, growing attention has been paid to the interplay between physical and psychological injuries; that is, to the psychological consequences of physical injury *caused by* a traumatic event (O'Donnell et al., 2003). Several reasons exist for this increasing interest in traumatic injury. First, there has been a steady increase in the number of individuals who suffer from nonlethal injuries following traumatic events such as traffic accidents, criminal assaults, and terrorist attacks. Second, improvements in emergency medical treatments have enabled doctors to save the lives of individuals who, in the past, would not have survived their injuries (O'Donnell et al., 2003). In fact, physical injury during a traumatic event has been one of the most frequent factors leading to posttraumatic reactions in the past few years. Lastly, from the moment injured trauma survivors are hospitalized, they are under the supervision of medical healthcare personnel; therefore, it is relatively easy to assess and observe the development of posttraumatic reactions over time (Mellman et al., 2001).

How do physical and psychological traumas interact? Does physical injury increase or decrease the risk for developing posttraumatic reactions? One might think the answer to this question would be obvious. After all, what could possibly be the psychological advantage of suffering a physical injury in addition to being exposed to a traumatic event? Interestingly, however, the answer is not straightforward.

Until recently, traditional views, particularly psychoanalytic ones, tended to regard bodily injury as a protective factor against the development of posttraumatic stress disorder (PTSD) (Ulman & Brothers, 1987). Several arguments support the traditional view. First, the focus on survival and healing after a physical injury provides a structured setting for coping, both physically *and* psychologically. In other words, the recuperative process becomes the center of attention, thereby shifting the injured individual's attention away from the negative emotional consequences of the trauma. In psychoanalytic terms, physical injury absorbs one's "free psychic energy," thus reducing the chances of developing anxious or conflicting feelings about the traumatic event. Second, unlike psychological wounds, physical injury typically engenders more sympathy from the environment. Finally, physical injury usually results in the removal of the individual from the stressful situation to a safe and protective setting, thereby drastically reducing environmental anxiety, especially in combat conditions.

Over the past two decades, numerous studies examining posttraumatic reactions among injured trauma survivors have challenged these traditional views. The most intuitive argument is provided by the dose-response model, based on classical learning theories (i.e., Pavlovian conditioning). According to the model, the more stressful the traumatic event is (unconditioned stimulus), the greater the intensity and severity of the posttraumatic reaction (conditioned reaction). The effect of the physical injury is similar to the effect of an electric shock used in experiments examining fear conditioning. Based on the intensity of the electric shock, researchers are able to predict the intensity of a rat's fear conditioning. Analogously, the severity of the physical injury (ranging from minor to severe) would be expected to be predictive of the intensity of the posttraumatic reaction. According to this argument, psychological trauma accompanied by physical injury increases the threat to one's life or physical integrity and therefore increases the stress.

TABLE 6.1
Summary of the Main Points Concerning the Role of Physical Injury as a Protective or Risk Factor in the Development of Posttraumatic Reactions

Risk Factor	Protective Factor
Increases the sense of horror and the perceived threat to one's life and physical integrity.	Free psychic energy is focused on survival and healing.
Medical procedures and pain may serve as secondary traumas.	Since the injury is visible it may engender sympathy from others.
Physical injury is a constant reminder (retrieval stimuli?) of the traumatic event.	The injury results in the removal of the individual from the stressful situation.

Undoubtedly, the dose-response model is based on straightforward assumptions. Yet, comprehensive reviews that directly examined its validity reveal that according to most studies, the relation between the severity of the injury and the severity of the posttraumatic reaction is not a linear one (Bowman, 1997, 1999).

Another argument supporting physical injury as a risk factor assumes that pain and discomfort, along with the medical procedures employed during treatment, are themselves causes for **secondary trauma**. Specifically, several studies have indicated that severe pain in itself can lead to the development of PTSD (Schreiber & Galai Gat, 1993). Some studies go so far as to argue that the term "posttraumatic" reactions might not be appropriate in cases of traumatic physical injury since the traumatic experience is not over (Noy, 2000). Finally, several theories emphasize the relationship between the memory of a traumatic event and the development and preservation of posttraumatic symptoms. According to these theories, physical injury decreases the likelihood that posttraumatic memories will fade as time passes because the injury itself acts as a retrieval cue for the trauma. In other words, it is harder for those injured during a traumatic event to forget the trauma because scars and pain may serve as constant reminders of the trauma. (See Table 6.1 for a summary of the theories' main points.)

POSTTRAUMATIC STRESS DISORDER FOLLOWING PHYSICAL INJURY: WHAT IS KNOWN?

What then is known about the relation between physical injury and the development of PTSD? Interestingly, there has been at least partial support for the view that physical injury may serve as a protective factor. For example, Merbaum and Hefez (1976) found that injured soldiers hospitalized during the Yom Kippur War showed minimal, if any, psychological

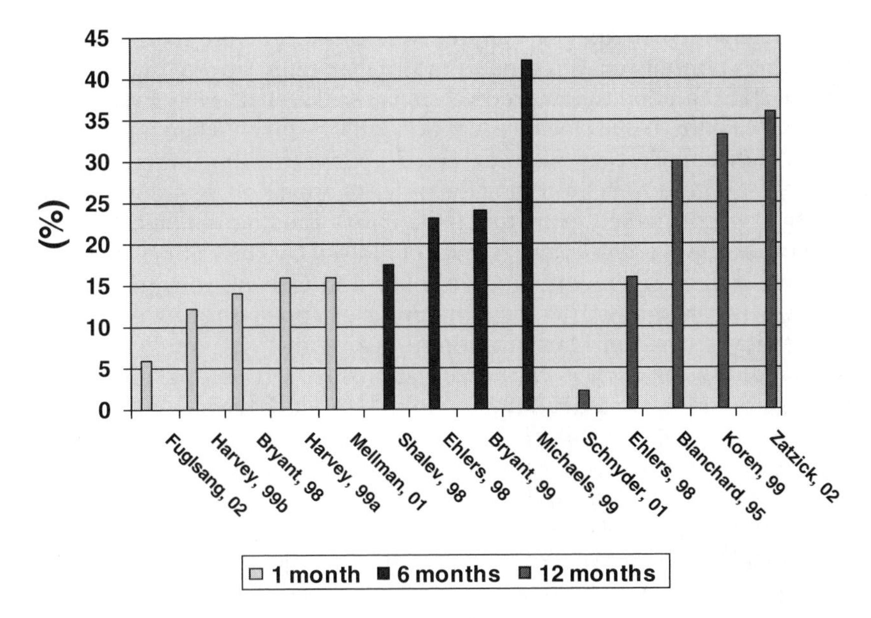

FIGURE 6.1 Frequency (%) of ASD or PTSD in injured populations by time.

disturbances even in cases of severe injuries, such as paralysis. Other studies were unable to find any relation between physical injury and the development of posttraumatic stress disorder (Foy & Card, 1987; Green et al., 1989).

In the past two decades, however, numerous studies have challenged traditional theories. Several studies found moderate to high rates of PTSD among injured survivors of other types of traumatic events such as traffic accidents (Blanchard et al., 1996; Ehlers et al., 1998; Koren et al., 1999; Mayou, Bryant, & Duthie, 1993; Ursano et al., 1999; Zatzick et al., 2002), terrorism (Shalev, 1992), criminal assault (Kilpatric et al., 1989; Zatzick et al., 2002), burn injuries (Perry et al., 1992), minor head injuries (Bryant & Harvey, 1998), and other types of injury (Michaels et al., 1999; Shalev et al., 1998). Although the prevalence of PTSD found in these studies varied considerably (2%–40%), the estimates were generally similar and were sometimes higher than those reported in samples of noninjured survivors. Similar findings were found concerning the prevalence of acute stress disorder in the first month after the trauma. Apart from one study which found a low prevalence of acute stress disorder (ASD) among traffic accident survivors (Fuglsang et al., 2002), most studies found ASD rates similar to those found in noninjured trauma survivor samples (Bryant & Harvey, 1998; Harvey & Bryant, 1999a,b; Mellman et al., 2001). Figure 6.1 summarizes the findings from the above studies according to the length of time that has elapsed from the traumatic event.

Studies which compared Vietnam veterans who were injured during the war to noninjured veterans found higher rates (approximately twofold) of PTSD among the injured veterans (Kulka et al., 1990; Pitman, Altman, & Macklin, 1989). However, caution must be taken when interpreting these studies since they were executed years after the traumatic event occurred and did not control for the order of appearance of the variables (in other words, which came first—the injury causing the posttraumatic reaction or the posttraumatic reaction followed by the injury). With this in mind, it is worth mentioning Palinkas and Coben's intriguing study (1987), which attempted to trace the order of appearance by examining the medical records of Marine veterans during the Vietnam War. On the one hand, the researchers found a relationship between physical injury and psychiatric hospitalization. On the other hand, they also found that in most cases the psychiatric hospitalization preceded the injury, thus implying that returning to combat after a psychiatric hospitalization may increase the chances for injury.

At present there exists a growing body of literature that indicates that other mental disorders, apart from posttraumatic stress disorder, may develop as a consequence of a traumatic event and that the probability of suffering from more than one mental disorder simultaneously is the rule and not the exception. Only a small number of studies have examined the prevalence of mental disorders such as phobias, depression, and drug abuse in relation to the prevalence of PTSD. However, the rates of comorbid disorders found in these studies were as high as those found in noninjured trauma survivor samples (Blanchard et al., 1995; Koren, Arnon, & Klein, 1999; Mayou, Bryant, & Ehlers, 2001; Schnyder et al., 2001; Zatzick et al., 2002). Comorbidity rates were highest in cases of depression, reaching a maximum of 50% (Blanchard et al., 1995; Koren, Arnon, & Klein, 1999).

In sum, the growing mass of research in the field of physical injury and posttraumatic reactions indicates that physical injury during a traumatic event is probably a risk factor, rather than a protective factor, in the development of PTSD. Furthermore, these studies indicate that traumatic physical injury may affect quality of life regardless of the development of full-blown PTSD. Although this growing body of literature contributes considerably to our understanding of the risk-elevating nature of physical injury with regard to PTSD, it reveals relatively little about the unique contribution of bodily injury to the subsequent development of PTSD (i.e., over and above the contribution of the trauma itself). One reason for this is that no study directly compared injured and noninjured survivors of the same trauma. In the following section, we will describe the research we conducted among warfighters who were injured during combat in Lebanon, in an attempt to estimate the unique contribution of physical injury to the development of PTSD (Koren et al., 2005). In order to do so we compared injured soldiers to their noninjured comrades who had participated in the same combat situations.

POSTTRAUMATIC REACTIONS AMONG WARFIGHTERS INJURED DURING COMBAT IN LEBANON

As mentioned above, the aim of the current study (Koren et al., 2005) was to determine the unique contribution of physical injury during traumatic events to the development of posttraumatic symptoms. An additional aim was to examine the relation between the nature of the injury, its severity, and the likelihood of developing posttraumatic reactions.

To accomplish these goals we employed a matched, injured-control design which enabled us to compare warfighters injured during combat to their noninjured comrades who participated in the same events. The research group consisted of soldiers admitted to the emergency trauma units of three major hospitals in the northern part of Israel, in the years 1998–2000, for medical treatment for combat-induced injuries. To ensure a minimal severity of injury criteria, soldiers hospitalized for less than 2 days were excluded from the research. Similarly, to minimize the chances that the results will be explained by mental disorders not caused by the trauma itself, soldiers who suffered head injuries or were treated for psychiatric disorders at the time of injury were also excluded. Based on these criteria, 172 warfighters were listed as suitable candidates for the study. An attempt to contact these candidates was made first by a letter informing them about the study and then followed by a telephone call to invite them to participate and to schedule an interview. Due to changes in address or journeys abroad we were able to contact 117 (68%) of the potential participants. Of these, 76 (65%) agreed to participate in the study. However, out of the 76, 16 (21%) could not keep their scheduled appointment, leaving the final sample at 60 injured warfighters, which represent a total response rate of 51.3%. The average time that elapsed between the injury and the interview was 15.8 months ($SD = 7.4$).

Next, consistent with the matched control method, we attempted to locate and match for every injured warfighter a noninjured comrade who participated in the same combat situation. To minimize possible alternative explanations for any differences between the study and comparison groups, the noninjured soldiers were selected according to their resemblance to the injured soldiers based on demographic variables (socioeconomic status, age, marital status, and ethnic origin) and military variables (rank, length of service, role in the unit, and pre-injury medical profile[1]). Taking these variables into account, we were able to trace 40 noninjured warfighters. Due to the stage-by-stage sampling method (i.e., first injured soldiers and then comparison noninjured ones) the average time that elapsed between the traumatic event and the interview in the comparison group was 18.3 months ($SD = 7.7$).

Suitable candidates who agreed to participate from both groups were invited to a one-on-one interview conducted by a graduate student in clinical psychology. Following a complete description of the study at the start of the interview, written informed consent was obtained. The first part of the interview focused on getting acquainted and obtaining personal background information such as family status, country of origin, level of education, and current occupation. Next, the participants were asked to answer questions regarding the details of the traumatic incident, such as when were they injured, how were they injured, how were they evacuated from the scene, and so forth. In addition, participants were asked to describe their emotions and thoughts during and immediately after the traumatic event. Lastly, an extensive battery of half-structured, self-report questionnaires were used to identify current and past psychiatric disorders (SCID, Structured Clinical Interview for DSM-IV), the frequency and intensity of clinical posttraumatic symptoms (CAPS, Clinician Administered PTSD Scale), anxiety, depression, dissociative tendencies, and a history of past traumas. The severity of the physical injury was assessed by a physician based on the medical charts at the time of the soldier's release from the hospital (ISS, Injury Severity Score; AIS, Abbreviated Injury Scale). The physician was blind to the soldier's mental state or diagnosis.

The findings were illuminating. First, the prevalence of PTSD among the injured group (16.7%) was approximately 7 times higher (!) than the prevalence of PTSD among the noninjured group (2.5%). In addition, three injured participants (5%), but none of the noninjured comparison group, suffered from partial PTSD (i.e., suffered from posttraumatic symptoms in 2 out of the 3 symptom clusters that classify PTSD according to the DSM). These differences were statistically significant. Similarly, although somewhat less dramatically, the prevalence of other psychiatric disorders (such as depression, drug abuse, and adjustment disorder) that developed after the traumatic event was 2 times higher among the injured participants (10%) than the noninjured ones (5%).

Second, the frequency of posttraumatic symptoms and the intensity of such symptoms, regardless of the formal PTSD classification, were significantly higher among the injured participants than among the noninjured comparison group. As shown in Figure 6.2, the difference in symptom frequency and intensity between the two groups is more or less consistent across the three clusters of posttraumatic symptoms defined by the DSM.

To determine whether these intergroup differences could be attributable to the larger number of participants with PTSD in the injured group, we excluded participants suffering from PTSD and then retested the effect of injury on all clinical symptoms. Interestingly, the frequency and intensity of posttraumatic symptoms remained significantly higher among the injured group as compared to the noninjured group.

Similar findings were obtained concerning the frequency and intensity of symptoms of depression and anxiety. As can be seen in Figure 6.3, injured participants as a group suffered from significantly higher levels of anxiety and depression. These differences remained significant even after the exclusion of participants with PTSD.

Since these findings indicate an unmistakable relationship between physical injury and the development of posttraumatic stress disorder, our

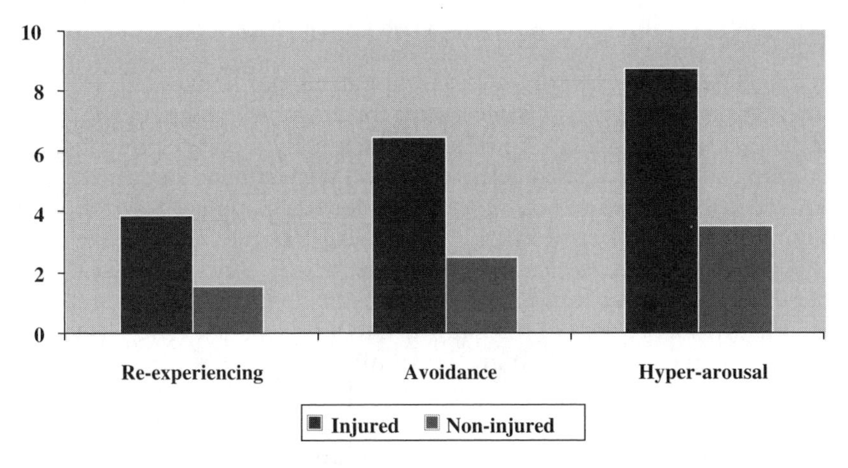

FIGURE 6.2 Level of PTSD symptoms in injured and non-injured warfighters.

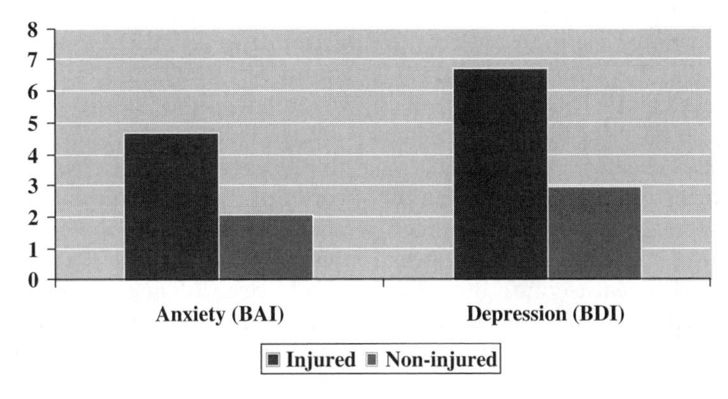

FIGURE 6.3 Level of anxiety and depression symptoms in injured and non-injured warfighters.

next question was whether this relationship is mediated by the severity of the injury or by the severity of the traumatic event. To answer this question we focused solely on the injured group and compared those with PTSD to those without based on several objective measures of the traumatic event. These included the duration of the traumatic event, elapsed time until medical help arrived, elapsed time until evacuation to the hospital, number of casualties, and severity of the injury. Surprisingly, we found no significant differences between injured participants with PTSD and injured participants without PTSD on each of the measures. In other words, among the injured participants we found no relation between PTSD and the severity of the injury or the severity of the traumatic event.

WHAT CAN (AND CANNOT) BE LEARNED FROM THE FINDINGS

Consistent with most of the recent literature, our findings clearly indicate that physical injury is a risk factor for the development of PTSD. It is important to note that our findings do not merely replicate previous findings; by using the matched-control method we went one step further and provided an approximation of the odds for developing PTSD following traumatic injury. More specifically, we found that the risk for developing PTSD following traumatic injury is approximately 8 times higher (!) than following injury-free trauma. In other words, physical injury is in itself a risk factor independent of the exposure to a traumatic event. In fact, our findings suggest that this high figure might even be an underestimate because approximately 35% of the originally approached injured soldiers, but none of the noninjured controls, refused to participate in the research. Based on the explanations given by the injured nonparticipants for their refusal, it was quite obvious that many of them did so for reasons that can be interpreted as posttraumatic avoidance.

Although we were able to establish a correlation between physical trauma and posttraumatic reactions, we were unable to determine how the severity of the injury affects this relationship. In other words, no differences in the severity of injury were found between injured participants suffering from PTSD and injured participants who did not develop PTSD. In fact, some of the injured participants who did not develop PTSD suffered injuries of greater severity than those who did develop PTSD. We obtained similar results when comparing the severity of the traumatic event.

How might the increased risk associated with injury be explained? The most straightforward answer is that traumatic injury increases the perceived threat to one's life or to one's physical integrity. According to the literature, a survivor's perceived level of threat is a better predictor of PTSD than the actual severity of the traumatic event (Kilpatric et al., 1989). Although we did not explicitly examine perceived threat, the nonlinear relationship between the risk for developing PTSD and the severity of the injury raises the possibility that perceived threat plays a mediating role in this relation. Previous studies indirectly support this hypothesis by implying that traumatic injury exerts its effect on perceived threat via its effect on the survivor's **sense of control** or ability to cope (Michaels et al., 1999; Pitman, Altman, & Mackli, 1989). In our study, one of the warfighters, who developed PTSD after suffering from a minor physical injury, explained that he did not panic during combat or after he was wounded; it was the evacuation that was traumatic. During the evacuation he was strapped to a stretcher and was unable to move while the ambulance was under attack.

Another potential mediating factor between physical injury and PTSD might be peritraumatic dissociation, which in previous studies has been implicated as a risk factor for developing PTSD (Shalev et al., 1996). It may be that the severe pain induced by the physical injury increases the odds for peritraumatic dissociation during the traumatic event. This possibility was supported in our study by injured participants, with or without PTSD, reporting significantly higher levels of peritraumatic dissociation than noninjured participants. Furthermore, controlling for peritraumatic dissociation remarkably decreased the differences between the injured and noninjured groups on all clinical measures, including PTSD. In other words, it may be that the degree of dissociation explains a large part of the variance in survivors' posttraumatic reactions. While dissociation is commonly perceived as a risk factor because it impairs the integration and processing of the trauma, an interesting alternative is that the relationship between physical injury and dissociation might be reversed. That is, people who tend to dissociate under stress might be at higher risk for getting injured. Palinkas and Coben's (1987) study indirectly supports this hypothesis by suggesting that mental illness precedes physical injury. Future studies should use prospective research methods in order to examine the relationship between physical injury and dissociation.

While not directly assessed in this study, noteworthy biological mechanisms may mediate the relationship between injury and PTSD. These hypothesized mechanisms relate to the complex interactions between the immune system and the stress regulating system, also known as the hypothalamic-pituitary-adrenal (HPA) axis. In addition to its key role in the regulation of the stress response, the HPA axis is also involved in the modulation of the immune system's response to inflammation and injury via proinflammatory mediators such as cytokines (Schobitz, Reul, & Holsboer, 1994). Alterations in the HPA axis have been suggested to be a risk factor for the development of PTSD (Yehuda, 2001). Therefore, situations which involve both emotional and injury-related stress may create an extra burden on an already compromised HPA axis. While this hypothesis has yet to be explored, numerous preclinical studies found that cytokines have adverse effects on memory, sleep, and mood, and that cytokines promote sickness behavior (Larson & Dunn, 2001). Indeed, a recent study found elevated cerebrospinal fluid (CSF) concentrations of inflammatory mediators in clients with PTSD versus normal controls (Baker et al., 2001).

While the current study significantly contributes to the field of physical injury and PTSD, several unsolved questions remain. First, posttraumatic symptoms were assessed relatively long after the traumatic injury occured. Thus, although our data provides a reliable estimate of chronic posttraumatic symptoms, our data does not examine acute stress reactions and is unable to determine the course of posttraumatic adjustment in relation to the physical recovery process. Additionally, severity of injury was assessed only at the time of hospitalization but not again at the time of assessment. Consequently, our data cannot clarify the degree to which current injury or physical disability may affect reported PTSD symptoms, nor can our data reveal the degree to which the pace of physical recovery may be influenced by PTSD. Finally, the event was defined as traumatic based on objective measures without taking into account the variance in soldiers' subjective perceptions of the event.

HELP-SEEKING BEHAVIOR AMONG INJURED WARFIGHTERS

Since the findings suggest that physical injury significantly increases the risk for developing PTSD, one of the questions they raise is how the injury affects help-seeking behavior. In other words, how does the physical injury affect the survivor's willingness to seek help in order to cope with the emotional consequences of the trauma? To examine this question we conducted a follow-up study (Hillel, 2003). We returned to participants in the original study and examined their attitude toward seeking help as well as their actual help-seeking behavior.

Out of 100 participants in the original study we were able to trace 85 (out of the 15 remaining participants, 9 changed address, and 6 were abroad at the time of the study). Eighty participants (94%) agreed to participate in the follow-up (4 out of the 5 refusing participants were from the injured group, out of which 2 were diagnosed as suffering from PTSD in the original study). We were able to interview 72 of the original (44 injured, 28 noninjured).

We obtained some interesting results in the follow-up study. While participants in both groups had moderately positive **attitudes** toward seeking help for emotional problems, behaviorally we found that injured participants seek more formal mental help (mainly from professionals from the Israel Ministry of Defense rehabilitation department and civilian professionals) and that they do so more frequently than noninjured trauma survivors. Noninjured trauma survivors mainly turn to informal sources of emotional support such as friends and family; rarely do they turn to professional sources. This finding is not consistent with previous studies which indicate that people with mental disorders in general, and war and traffic accident survivors in particular, find it difficult to turn to professional help (Bryant & Harvey, 1995; Green et al., 1993; Jeavons, Greenwood, & Horne, 1998; Solomon, Mikulincer, & Arad, 1991). One explanation for this disparity may be that for **injured** survivors, the status of being helpless and exposed when treated by medical staff paradoxically makes it easier for them to return to such a position when treated for emotional problems. Similarly, the fact that at least part of the medical treatment focuses on the emotional consequences of injury may legitimize injured survivors' need to seek help for other aspects of the traumatic experience. In other words, being treated for physical injuries may enable injured trauma survivors to overcome the social stigmas associated with psychological treatment. This hypothesis is supported by previous studies which found that the main factors preventing trauma survivors from seeking help are fear of social stigma, mild levels of posttraumatic symptoms (which enable reasonable daily functioning), and high levels of perceived competency (an individual's belief in his or her ability to cope in times of stress without the need for assistance from others) (Solomon, Mikulincer, & Arad, 1989).

Another significant factor which makes it easier for injured survivors to seek mental help is financial support. Psychotherapy is expensive, and its cost is an important consideration for those who are willing to seek therapy. The Israel Defense Ministry rehabilitation department fully funds mental help for injured or handicapped soldiers, thus making it easier for them to seek mental help. Noninjured warfighters are not recognized as handicapped and therefore are not entitled to receive financial assistance from the army when seeking mental help, unless they file a request to be recognized as mentally handicapped. In reality, most soldiers do not file such requests (in the follow-up study none of the participants filed such a request). In any case, it is important that follow-up studies focus on the factors enabling or preventing mental help-seeking behavior among injured survivors. In order to do so, a comparison should be made between injured survivors who seek help and injured survivors who do not.

CONCLUSIONS, UNRESOLVED ISSUES, AND FUTURE DIRECTIONS

The current study emphasizes the need for more attention to be directed at the emotional consequences of traumatic events. To that end, professionals specializing in diagnosis and early intervention in cases of emotional trauma should work alongside medical staff in treating trauma survivors. It should be stressed that early diagnosis and intervention by professionals in the field of trauma or by medical staff trained to perform therapeutic interventions in acute stress reactions may reduce trauma-related mental disorders.

Simple assessment tools used to screen for acute stress reactions immediately after a traumatic event should be implemented in hospitals. In recent years, several such assessment and screening questionnaires have been developed, validated, and found to predict accurately the development of chronic posttraumatic reactions (for example, Brewin et al., 2002). These questionnaires are short (for example, there are only 11 questions in Brewin's questionnaire), simple to complete, and easy to analyze. Therefore, it is relatively easy to train nurses in trauma units to use them. Based on a preliminary study conducted by our team, it seems that using such methods not only contributes to diagnosis and early intervention but that such methods also increase medical staff's awareness of posttraumatic emotional reactions and to the possible effects they may have on medical treatment.

This research emphasizes the need for a longitudinal study that would examine the developmental course of posttraumatic symptoms immediately after the injury and the interactions between the symptoms and physical recovery. Future studies would significantly contribute to the understanding of when and what type of intervention should be used in the course of physical recovery.

Due to the limited scope of this chapter, the effects of PTSD on recovery from physical injury will not be discussed. However, it is important to note that recent studies, including ones executed by our research group, suggest that regardless of the severity of the injury, the duration of hospitalization (whether in trauma wards or in rehabilitation centers) is significantly longer among injured survivors suffering from posttraumatic reactions than among injured survivors who do not suffer from PTSD (Arnon, 1997). Similarly, recent studies suggest that the severity of the posttraumatic reaction (Mason et al., 2002) and the perceived threat during the traumatic event (Schnyder et al., 2003) are better predictors of the average time it takes injured trauma survivors to return to work than is the severity of the injury. Therefore, we hypothesize that the effects of posttraumatic reactions on recovery from physical injury are mediated by the severity of the pain symptoms. Our study on traffic accident survivors supports this hypothesis since we found that regardless of the severity of the injury, injured survivors suffering from PTSD reported higher levels of subjective pain than injured survivors not suffering from PTSD (Arnon, 1997). These findings suggest that the relationship between physical injury and PTSD may be circuitous: physical injury increases the risk for developing PTSD, and PTSD in turn affects the physical healing process.

REFERENCES

Allen, J. G. (1995). *Coping with Trauma: A Guide to Self-Understanding*. Washington, DC: American Psychiatric Press.

Arnon, I. (1997). Development and prevalence of post traumatic stress disorder among vehicle accidents victims [Hebrew]. Sc.D. Thesis, Faculty of Medicine, Technion–Israel Institute of Technology.

Baker, D. G., Ekhator, N. N., Kasckow, J. W., Hill, K. K., Zoumakis, E., Dashevsky, B. A., et al. (2001). Plasma and cerebrospinal fluid interleukin concentrations in posttraumatic stress disorder. *Neuroimmunomodulation, 9,* 209–217.

Benbenishty, R. (1991). Combat stress reaction and changes in military medical profile. *Military Medicine, 156,* 68–70.

Blanchard, E. B., Hickling, E. J., Barton, K. A., & Taylor, A. E., (1996). One-year prospective follow-up of motor vehicle accident victims. *Behaviour Research and Therapy, 34,* 775–786.

Blanchard, E. B., Hickling, E. J., Taylor, A. E., & Loss, W. R. (1995). Psychiatric morbidity associated with motor vehicle accidents. *Journal of Traumatic Stress, 10,* 215–234.

Bowman, M. L. (1997). *Individual Differences in Posttraumatic Response: Problems with the Adversity-Distress Connection*. Mahwah, NJ: Lawrence Erlbaum Associates.

Bowman, M. L. (1999). Individual differences in responding to adversity with posttraumatic distress: Problems with the DSM-IV model. *Canadian Journal of Psychiatry, 44,* 21–33.

Brewin, C. R., Rose, S., Andrews, B., Green, J., Tata, P., McEvedy, C., et al. (2002). Brief screening instrument for post-traumatic stress disorder. *British Journal of Psychiatry, 185,* 158–162.

Bryant, R. A., & Harvey, A. G. (1995). Psychological impairment following motor vehicle accidents. *Australian Journal of Public Health, 19,* 185–188.

Bryant, R. A., & Harvey, A. G. (1998). Relationship between acute stress disorder and post-traumatic stress disorder following mild traumatic brain injury. *American Journal of Psychiatry, 155,* 625–629.

Ehlers, A., Mayou, R. A., & Bryant, B. (1998). Psychological predictors of chronic posttraumatic stress disorder after motor vehicle accidents. *Journal of Abnormal Psychology, 107,* 508–519.

Foy, D., & Card, J. (1987). Combat related posttraumatic stress disorder etiology: Replicated findings in a sample of Vietnam-era men. *Journal of Clinical Psychology, 43,* 28–31.

Fuglsang, A. K., Moergeli, H., Hepp-Beg, S., & Schnyder, U. (2002). Who develops acute stress disorder after accidental injuries? *Psychotherapy and Psychosomatics, 71,* 214–222.

Green, B., Lindy, J., Grace, M., & Gleser, G. C. (1989). Multiple diagnosis in posttraumatic stress disorder. *Journal of Nervous and Mental Disease, 177,* 329–335.

Green, M. M., McFarlane, A. C., Hunter, C. E., & Griggs, W. M. (1993). Undiagnosed post-traumatic stress disorder following motor vehicle accidents. *Medical Journal of Australia, 159,* 529–534.

Harvey, A. G., & Bryant, R. A. (1999a). Predictors of acute stress following motor vehicle accidents. *Journal of Traumatic Stress, 12,* 519–525

Harvey, A. G., & Bryant, R. A. (1999b). Acute stress disorder across trauma populations. *Journal of Nervous and Mental Disease, 187,* 443–446.

Hillel, Y. (2003). Post-traumatic reactions and mental help-seeking behavior following combat-related injury [Hebrew]. M.A. thesis, University of Haifa, Israel.

Jeavons, S., Greenwood, K., & Horne, D. (1998). Psychological symptoms in rural road trauma victims. *Australian Journal of Rural Health, 6,* 52–56.

Kilpatric, D. G., Saunders, B. E., Amick-McMuilan, A., Best, C. L., Veronen, L.J., & Resnik, H. S. (1989). Victim and crime factors associated with the development of crime-related post-traumatic stress disorder. *Behavior Therapy, 20,* 199–214.

Koren, D., Arnon, I., & Klein, E., (1999). Acute stress response and posttraumatic stress disorder in traffic accident victims: A one-year prospective, follow-up study. *American Journal of Psychiatry, 156,* 367–373.

Koren, D., Norman. D., Cohen. A., Berman. J., & Klein. E. M. (2005). Combat-related injury increases the risk for PTSD: An event-based matched injured-control study. *American Journal of Psychiatry, 162,* 276–282.

Kulka, R. A., Schlenger W. E., Fairbank J. A., Hough R. L., Jordan, B. K., et al. (1990). *Trauma and the Vietnam War Generation: Report of Findings from the National Vietnam Veterans Readjustment Study.* New York: Brunner/Mazel.

Larson, S. J., & Dunn, A. J. (2001). Behavioral effects of cytokines. *Brain, Behavior, and Immunity, 15,* 371–387.

Mason, S., Wardrope, J., Turpin, G., & Rowlands, A. (2002). Outcomes after injury: A comparison of workplace and nonworkplace injury. *Journal of Trauma, 53,* 98–103.

Mayou, R., Bryant, B., & Duthie, R. (1993). Psychiatric consequences of road traffic accidents. *British Medical Journal, 307,* 9–13.

Mayou, R., Bryant, B., & Ehlers, A., (2001). Prediction of psychological outcomes one year after a motor vehicle accident. *American Journal of Psychiatry, 158,* 1231–1238.

Mellman, T. A., David, D., Bustamante, V., Fins, A. I., & Esposito, K. (2001). Predictors of post-traumatic stress disorder following severe injury. *Depression and Anxiety, 14,* 226–231.

Merbaum, M., & Hefez, A. (1976). Some personality characteristics of soldiers exposed to extreme war stress. *Journal of Consulting and Clinical Psychology, 44*, 1–6.

Merriam-Webster's Collegiate Dictionary (10th ed.). (1993). Springfield, MA: Merriam-Webster.

Michaels, A. J., Michaels, C. E., Moon, C. H., Smith, J. S., Zimmerman, M. A., Taheri, P.A., et al. (1999). Posttraumatic stress disorder after injury: Impact on general health outcome and early risk assessment. *Journal of Trauma, 47*, 460–466.

Noy, S. (2000). *Traumatic Stress Situations* [Hebrew]. Tel Aviv, Israel: Shoken Press.

O'Donnell, M. L., Creamer, M., Bryant, R. A., Schnyder, U., & Shalev, A. (2003) Posttraumatic disorders following injury: An empirical and methodological review. *Clinical Psychology Review, 23*, 587–603.

Palinkas, L. A., & Coben, P. (1987). Psychiatric disorders among United States Marines wounded in action in Vietnam. *Journal of Nervous and Mental Disease, 175*, 291–300.

Perry, S., Difede J., Musngi, G., Frances, A., & Jacobsberg, L. (1992). Predictors of posttraumatic stress disorder after burn injury. *American Journal of Psychiatry, 149*, 931–935.

Pitman, R. K., Altman, B., & Macklin, M. L. (1989). Prevalence of posttraumatic stress disorder in wounded Vietnam veterans. *American Journal of Psychiatry, 146*, 667–669.

Shalev, A. Y. (1992). Posttraumatic stress disorder among injured survivors of a terrorist attack: Predictive value of early intrusion and avoidance symptoms. *Journal of Nervous and Mental Disease, 180*, 505–509.

Shalev, A. Y., Freedman, S., Peri, T., Brandes, D., Sahar, T., Orr, S. P., et al. (1998). Prospective study of posttraumatic stress disorder and depression following trauma. *American Journal of Psychiatry, 155*, 630–637.

Shalev, A. Y., Peri, T., Canetti, L., & Schreiber, S. (1996). Predictors of PTSD in injured trauma survivors: A prospective study. *American Journal of Psychiatry, 153*, 219–225.

Schnyder, U., Moergeli, H., Klaghofer, R., & Buddeberg, C. (2001). Incidence and prediction of posttraumatic stress disorder symptoms in severely injured accident victims. *American Journal of Psychiatry, 158*, 594–599.

Schnyder, U., Moergeli, H., Klaghofer, R., Sensky, T., & Buchi, S. (2003). Does patient cognition predict time off from work after life-threatening accidents? *American Journal of Psychiatry, 160*, 2025–2031.

Schobitz B., Reul, J. M., & Holsboer, F. (1994). The role of the hypothalamic-pituitary-adrenocortical system during inflammatory conditions. *Critical Reviews in Neurobiology, 8*, 263-291.

Schreiber, S., & Galai Gat, T. (1993). Uncontrolled pain following physical injury as the core-trauma in the post-traumatic stress disorder. *Pain, 54*, 107–110.

Solomon, Z., Mikulincer, M., & Arad, R. (1991). Monitoring and blunting: Implications for combat related post-traumatic stress disorder. *Journal of Traumatic Stress, 4*, 209–221.

Solomon, Z., Mikulincer, M., & Benbenishty, R. (1989). Combat stress reaction: Clinical manifestations and correlates. *Military Psychology, 1*, 35–4.

Ulman, R. B., & Brothers, D. (1987). A self-psychological reevaluation of posttraumatic stress disorder (PTSD) and its treatment: Shattered fantasies. *Journal of the American Academy of Psychoanalysis*, 15, 175–203.

Ursano, R. J., Fullerton C. S., Epstein, R. S., Crowley, B., Kao, T. C., Vance, K., et al. (1999). Acute and chronic posttraumatic stress disorder in motor vehicle accident victims. *American Journal of Psychiatry, 156*, 589–595.

Yehuda, R. (2001). Biology of posttraumatic stress disorder. *Journal of Clinical Psychiatry, 62*(Suppl), 41–46.

Zatzick, D. F., Kang, S. M., Muller, H. G., Russo, J. E., Rivara, F. P., Katon, W., et al. (2002). Predicting posttraumatic distress in hospitalized trauma survivors with acute injuries. *American Journal of Psychiatry, 159*, 941–946.

ENDNOTES

1. The medical profile is a global score used by the Israel Defense Forces (IDF) to characterize one's overall health status (both physically and mentally). Medical profile scores are assigned by a medical panel appointed by the Surgeon General of the IDF, and these scores range from a low of 21 to a high of 97. Assignment to frontline (profile >64) versus rearline (31< profile <65) units is determined by the medical profile scores. A person who receives a score of 24 or lower is considered unfit to serve. An explanation of this scale can be found in Benbenishty (1991).

This chapter reviews the literature on the secondary traumatization of wives of traumatized combat veterans. Secondary traumatization is one of several terms, including "compassion stress," "compassion fatigue," and "secondary victimization" (Figley, 1983), "co-victimization" (Hartsough & Myers, 1985), "traumatic countertransference" (Herman, 1992), and "vicarious traumatization" (McCann & Pearlman, 1989), that have been used to label the manifestations and processes of distress reported by persons in close proximity to victims of traumatic events that they themselves did not actually experience.

The term is used in this chapter, as in the literature, in both its narrow and broad sense. In the narrow sense, it refers to the transmission of nightmares, intrusive thoughts, flashbacks, and other symptoms typically experienced by traumatized individuals, to persons close to them. In the broad sense, it refers to any transmission of distress from someone who experienced a trauma to those around him or her and includes a wide range of manifestations of distress in addition to those that mimic post-traumatic stress disorder (PTSD) (Galovski & Lyons, 2004).

Some 15% to 40% of war veterans develop PTSD. By its nature, this is a disorder that puts tremendous difficulties in the way of the injured veteran's personal relations and functioning (Solomon, 1993). The avoidance symptoms of psychic numbing, withdrawal, detachment, constricted affect, and loss of interest in previously enjoyed activities are symptoms that undermine the individual's ability to maintain the intimacy of family life (Riggs et al., 1998). The hyper arousal symptoms include heightened irritability and hostility, which would make it difficult for the afflicted veteran to control his aggression. In addition, many PTSD casualties experience reduced sexual drive and problems in sexual functioning (Kotler et al., 2000; Letourneau, Schewe, & Frueh, 1997). Many have difficulty in functioning outside the home, both socially and at work, resulting in unstable employment and high unemployment rates (Solomon, 1993).

The first two sections of this chapter briefly summarize some descriptive accounts of being married to PTSD veterans and the empirical findings on the wives' secondary traumatization. The third section deals with separation and divorce among these couples. The fourth and fifth sections deal, respectively, with factors that have been found to predict secondary traumatization and with theoretical explanations of the phenomenon. The sixth section discusses treatment. The last section summarizes what is and is not known to date and makes recommendations for further study.

The chapter confines itself to what happens to the wives after their husbands return from service and does not address their experiences or state of mind when their husbands are away. It does not discuss the possible secondary traumatization of men whose wives served in the military and suffer from PTSD. Nor does it discuss the consequences of PTSD for partners of persons whose PTSD stems from other traumatic events, ranging from the Nazi Holocaust, through natural disasters, mass displacements, or personal traumas such as rape. Although all these issues are important, they are beyond the scope of this chapter.

DESCRIPTIVE ACCOUNTS OF LIVING WITH TRAUMATIZED VETERANS

The first accounts of the secondary traumatization of wives of traumatized veterans were clinical descriptions of living with veterans of the Vietnam War published in the 1980s. These accounts do not use the term secondary traumatization and do not mention PTSD, which was not well known yet. They do, however, provide clear testimony of the great distress the women suffered.

Williams (1980), based on group therapy sessions with veterans' wives, relates the "spread of effect" of the traumatized veterans' symptoms; the distress the women suffered as a result of their husbands' distancing and violence; and the "compassion trap" in which they found themselves when they sacrificed too many of their own needs in their efforts to improve their husbands' situation and to preserve their family life. Maloney (1988) describes six wives of Vietnam veterans with clear PTSD symptoms, including dreams of the war and panic attacks triggered by the same triggers as their husbands', such as the buzz of helicopters, sudden noises, gunfire, and the smell and sound of spring rain. In similar vein, Matsakis (1988) relates that the women in her support groups for veterans' wives told of dreaming of Vietnam, of suffering from insomnia and startle reactions, of being hypervigilant around their potentially violent husbands, and of feeling isolated and helpless in their marriages.

Patience Mason (1990), in a detailed and vivid book about life with her traumatized husband, describes, along with her PTSD symptoms, her fear of her husband's erratic explosions; her sense of "walking on eggshells"; the effort and energy she invested in trying to keep her husband from getting upset and flying into a rage; as well as her struggle to fulfill multiple roles of cook, provider, housekeeper, and of only parent to her children and her husband's only friend and rescuer.

These accounts convey a good sense of what living with a traumatized veteran can be like and the difficulties and distress it can cause. The details of these accounts are familiar today. At the time they were written, however, there was little awareness that trauma could be transmitted to someone who did not actually experience the traumatic event. Their importance for us today lies in the fact that they opened the door to the systematic research that followed.

EMPIRICAL FINDINGS OF SECONDARY TRAUMATIZATION IN WIVES OF PTSD VETERANS

The empirical studies of secondary traumatization of wives of PTSD veterans examine the wives' emotional distress and their perceptions of their marital relationships.

The Wife's Distress

The first large systematic study that examined the impact of PTSD on the emotional life of veterans' wives was Kulka et al.'s (1990) national epidemiological study of the impact of PTSD on the families of Vietnam War era veterans. The study reports indications of secondary traumatization in both the wife and children, though it does not identify it as such. In a subsample of 466 families, the authors found that the wives of PTSD veterans had lower subjective well-being and a greater sense that they were on the verge of a nervous breakdown than the wives of non-PTSD veterans. In a further analysis of this subsample, Jordan et al. (1992) found that the wives of PTSD veterans reported significantly lower happiness and life satisfaction and higher demoralization than the wives of veterans without PTSD.

In tandem, a more specific study was carried out by Solomon et al. (1992a) on 205 wives of Israeli combat veterans of the 1982 Lebanon War. This study, which focused solely on veterans' wives, examined a range of psychiatric symptomatology and social effects. The findings show that the wives of PTSD veterans reported significantly higher levels of somatization, depression, obsessive-compulsiveness, anxiety, paranoid ideation, interpersonal sensitivity, hostility, and somatic complaints than the wives of non-PTSD veterans, as well as greater loneliness and dissatisfaction with their wider social network. These findings, obtained 6 years after the war, provide further evidence of heightened emotional distress among wives of PTSD veterans. Similar findings were obtained on another sample of wives of Israeli war veterans of the Lebanon War some 20 years later (Ben-Arzi, Solomon, & Dekel, 2000).

Further evidence of secondary traumatization of the wives of traumatized veterans is provided by Dirkzwager et al.'s (2005) study of 708 partners of Dutch peacekeepers. Their findings showed that the partners of peacekeepers with PTSD reported more sleep problems, somatic problems, and negative social support than the partners of peacekeepers without PTSD, and judged their marital relationship less favorably.

In addition, these authors linked the wives' distress specifically to the marital situation. For the purpose of comparison, the authors also examined the responses of 332 of the peacekeepers' parents. No evidence of secondary traumatization was found among the parents, and no differences were found between parents of peacekeepers with and without PTSD. These findings suggest that it is the intimate nature of the marital relationship that makes the wife more vulnerable to secondary traumatization than members of the extended family.

The Marital Relationship

The first studies of the impact of PTSD on the marital relationship were carried out on the veterans themselves. These studies showed that the

PTSD veterans reported less marital satisfaction, less intimacy, and less self-disclosure and expressiveness than non-PTSD veterans, as well as more hostility and physical violence (e.g., Carroll, Rueger, Foy & Donahoe, 1985). While these findings point to difficulties in the marital relationships of PTSD veterans, it took another few years for researchers to realize that it was important to query the wives as well as their husbands.

The studies of the wives and couples provide evidence both of heightened marital distress among women married to veterans with PTSD and more dyadic problems in the couple relationship. Studies of wives of veterans of the Vietnam (Wilson & Kurtz, 1997) and Lebanon Wars (Mikulincer, Florian, & Solomon, 1995; Solomon et al., 1992a) show that wives of PTSD veterans report greater spousal conflict, less intimacy, less cohesion, and less marital satisfaction than wives of non-PTSD veterans, as well as more verbal and physical violence by their husbands (Frederikson, Chamberlain, & Long, 1996; Jordan et al., 1992; Rosenheck & Thomson, 1986). With respect to dyadic problems, Riggs et al. (1998) found that over 75% of Vietnam veterans with PTSD and their partners had Dyadic Adjustment Scale scores in the clinically significant range of marital distress, in contrast to 32% of couples where the veteran did not have PTSD. Using a different measure, Dirkzwager et al. (2005) found that 24% to 39% of the partners of peacekeepers with PTSD symptoms scored in the range of a clinically problematic relationship, compared with 16% of those whose man did not have PTSD symptoms. These findings suggest that heightened marital distress and impaired dyadic relationships are also manifestations of secondary traumatization.

These empirical findings were obtained with respect to different wars, at different times, and in different countries, including the United States, Israel, Australia, and the Netherlands. They provide solid evidence of the existence of secondary traumatization among veterans' wives. They also raise important questions: What keeps these women in their troubled marriages? What predicts which women are most likely to develop secondary traumatization and which ones are less likely? What treatment is available for the victims of secondary traumatization? The remainder of this chapter discusses the answers that the literature to date provides to these questions.

SEPARATION AND DIVORCE

The chronic strain that PTSD places on both partners in the marriage and on the relationship between them might lead us to expect heightened divorce rates among such couples. This expectation was well borne out in the aftermath of the Vietnam War. Reports on Vietnam veterans indicate much higher rates of divorce than among the rest of the population (Center for Policy Research, 1979). The expectation has not been borne out among Israeli couples, however. Data from in Israel indicate no upsurge in separation and divorce among traumatized veterans, whose divorce rates are no different from those of other Israelis.

Two questions arise. One is why women who remain married to PTSD veterans stick it out, despite their great distress and the problems in their relationships. The other is what enables them to stay in their troubled marriages. To our knowledge, the only study that has addressed these questions to date is a qualitative study of nine wives of Israeli veterans with PTSD (Dekel et al., 2005a).

With regard to the first question, most of the women who participated in the study told the researchers that they had considered divorce, and some that they had even discussed the possibility with her husbands but decided not to go through with it. As they explained it, they were stopped by a strong sense of moral commitment, stemming mainly from personal loyalty and internalized social norms, but, in some cases, also from fear for their husbands' lives. Most of the women told of good marital relationships prior to their husbands' developing PTSD. They described living happily with men who had been healthy, strong, and supportive. This, they said, reinforced their commitment to their husbands and their moral obligation to weather the difficult times with them. Their general sense was that one does not dump a man in time of hardship. For most of the women, this sense of obligation was reinforced by the conviction, strongly held throughout Israeli society, that one does not abandon an injured soldier in the field. As one put it:

> For me it's like abandoning an injured soldier on the battlefield or abandoning someone sick.... We created a family together; our relationship was established before; it's not like I'd break up the whole package because he's not pulling his weight. It doesn't work like that.

Some of the women who wanted to leave told that they were locked in their marriages by explicit threats their husbands made to commit suicide if they did. These women felt that they could not take the responsibility if their husband acted on his threat. All in all, the women who participated in the study did not feel that they really had the option of leaving their husbands. Boss (1999) noted a similar lack of choice for spouses of persons with other emotional or physical disabilities.

With regard to the second question, the women attributed their ability to cope with the many difficulties arising from their husbands' PTSD to several sources. One was the reservoir of good feelings they had from when their husbands were vibrant, healthy individuals and their marriages were happy. Another was watching their husbands struggle with PTSD day in and day out. This deepened their appreciation of his courage and determination, as well as their love for him, and also served as an example which encouraged them to struggle on as well. A third source of strength, reported by a few of the women, was an increased sensitivity on the part of their husbands. These women reported that their husband's difficulties and vulnerabilities seem to have made him more aware than he had previously been of the emotional difficulties that she and their children experienced. Finally, most of the women told that they gained a sense of strength and empowerment from their own struggle to help their husbands and to keep the family together. These initial observations suggest that there is some compensation or redress for the many difficulties and burdens of living with a PTSD husband.

The observations are consistent with the concept of posttraumatic growth proposed in recent literature (e.g., Tedeschi & Calhoun, 1996). This concept maintains that exposure to traumatic events may result not only in psychological distress but also in psychological and spiritual growth. The claim is that the struggle with the distress attendant on the traumatic exposure issues, among other things, in a heightened sense of power and mastery and a greater valuation of life and sense of meaning. These outcomes are very close to those described by the nine wives.

The question remains, however, of how generalizable the experiences of this small group of Israeli wives are to the wives of PTSD veterans elsewhere. In contrast to traumatized Vietnam War veterans, Israeli veterans with PTSD are not prone to self-medicate with alcohol and drugs, and are no more prone than other Israeli men to abuse their wives physically. These factors probably made living with the traumatized Israeli veterans less unbearable than living with their American counterparts. Moreover, despite rising divorce rates, Israeli society remains a traditional, family-oriented society, in which family unity is a central value (Cohen, 2003). We know that how families cope with a member's disability is affected by the unique social, historical, and cultural context of the society in which they live (Boss, 1987; Tubbs & Boss, 2000).

PREDICTORS OF SECONDARY TRAUMATIZATION AMONG WIVES OF PTSD VETERANS

The literature shows considerable variance in the adjustment of wives of PTSD veterans. Jordan et al. (1992), for example, found that although 42% of the wives of PTSD veterans reported feeling demoralized, 24% reported that they were satisfied with their lives. Such variability, which has been found in other studies of wives of PTSD veterans (Riggs et al., 1998; Solomon et al., 1992b), gives rise to the question of what predicts the different outcomes.

Broadly speaking, three main sets of predictors have been examined in the literature: predictors pertaining to the husband's PTSD, predictors pertaining to the wife, and predictors pertaining to the couple relationship.

Predictors Pertaining to the Husband's PTSD

These predictors include the severity of the PTSD, the avoidance symptoms, and the husband's violence.

PTSD Severity. Not surprisingly, several studies show that the more severe the husband's PTSD, the more severe the wife's distress (Beckham, Lytle, & Feldman, 1996; Riggs et al., 1998), and also that changes in the husband's PTSD severity predicted analogous changes in the wife's psychological distress and dysphoria (Beckham et al., 1996). Of particular interest is a study by Bramsen, van der Ploeg, and Twisk (2002), which suggests that the wife's distress is not only a function of her husband's PTSD, but also impacts on it. These authors examined the long-term psychological adjustment of 444 elderly Dutch couples drawn from a community sample some 50 years after the end of World War II. Both members of the couple had been exposed to the ravages of the war. These authors found that the level of each spouse's PTSD symptomatology was predicted not only by his or her own wartime experiences, but, beyond this, also by the severity of their spouse's PTSD. This finding brings home the spiralic interrelationship of posttraumatic distress in married couples. If the finding can be generalized to couples where only the man had direct exposure to the traumatic event, it suggests that the wife's secondary traumatization could exacerbate his PTSD, which in turn would exacerbate her own secondary traumatization.

Avoidance Symptoms. Several researchers went beyond the severity of the PTSD to examine the contribution of the three symptoms clusters to the quality of the marital relationship. These studies show that, of the three clusters, the avoidance symptoms make the greatest contribution to the quality of the marital relationship and, moreover, that within the avoidance cluster, symptoms of involuntary emotional numbing (e.g., emotional restriction, detachment from others) are more strongly related to the quality of the relationship than symptoms of effortful avoidance (e.g., attempts to avoid reminders and attempts to avoid thoughts and feelings) (Cook et al., 2004; Evans et al., 2003; Riggs et al., 1998). The damage caused by emotional numbing to the marital relationship has been attributed to the deficit it entails in the experience and/or expression of positive emotions (Litz, 1992). Since emotional expression plays an important role in the intimate exchanges integral to well-functioning relationships, the absence of or inability to express positive feelings toward a partner would diminish the quality of the relationship.

Husband's Violence. Several studies also show that the more frequent the PTSD husband's violence toward his wife, the higher the wife's distress (Calhoun, Beckham, & Bosworth, 2002) and the lower her martial satisfaction (Dekel & Solomon, 2006). These findings are similar to those found in the general population (Daniels, 2005; Woods, 2005). They are of special concern here because of the relatively high level of spousal violence perpetrated by men with PTSD (Byrne & Riggs, 1996; Jordan et al., 1992; Kulka et al., 1990).

Predictors Pertaining to the Wife

Two predictors pertaining to the wife have been examined: her caregiver burden and her separation-individuation.

Caregiver Burden. Caregiver burden is defined as the perception that one's emotional or physical health, social life, or financial status is adversely affected by caring for a relative who is ill or has special needs (Zarit, Todd, & Zarit, 1986). Hankin et al. (1992) describe PTSD as a long-term condition that places a heavy burden on the caregiving partner, similar to that borne by partners of other persons suffering from chronic impairments. Findings in Israel (Bleich, Solomon, & Dekel, 1997) show that the burden borne by wives of PTSD veterans is similar in its intensity and components (Novak & Guest, 1989) to that borne by caretakers of the elderly (Caserta, Lund, & Wright, 1996). In both cases, the most onerous components of the overall burden were the time pressures stemming from the demands of caretaking and the sense that the caretaker's personal development was sacrificed to the needs of the person being cared for. These were followed by chronic fatigue and physical effort attendant on caretaking. In both groups, the conflicts between the obligations of caretaking and other obligations (e.g., children, work) and the negative feelings (e.g., frustration, resentment) evoked by the caregiving ranked at the bottom of the scale.

Several studies conducted on American and Israeli wives of PTSD veterans found that the greater their sense of caregiver burden, the greater their emotional distress (Beckham, Lytle, & Feldman, 1996; Ben-Arzi, Solomon, & Dekel, 2000; Calhoun, Beckham, & Bosworth, 2002). The last study also found that changes in the wives' sense of burden over time predicted analogous changes in their psychological distress, dysphoria, and state anxiety. The role of caregiver burden in wives' secondary traumatization is further highlighted by findings that this variable completely mediated the associations between veterans' psychiatric symptoms and their wives' symptoms, and between veterans' functioning and wives' marital adjustment (Dekel, Solomon, & Bleich, 2005b).

It is also of note that findings also show that wives' sense of caregiver burden was associated both with the severity of their husband's PTSD (Beckham, Lytle, & Feldman, 1996) and with the degree of impairment in his day-to-day and occupational functioning (Dekel, Solomon, & Bleich, 2005b).

Wives' Separation-Individuation. Separation-individuation refers to the establishment of a distinct sense of self, through the achievement of emotional autonomy and independence from the mother (Mahler, Pipe, &

Bergman, 1975). Findings show that individuals with high levels of separation-individuation enjoy a good sense of mastery, increased coping ability, and reduced anxiety and inner conflict, which facilitate their coping with difficulties in the marital system (Blos, 1979; Bowen, 1978). The importance of separation-individuation to the emotional adjustment of wives of PTSD veterans is suggested by the qualitative study discussed above of nine Israeli women married to such veterans (Dekel et al., 2005a). These women described a constant struggle to lead a life of their own even as they were drawn into fusion with their needy husbands. They described the struggle to set boundaries and maintain individuation, as their caregiving encroached on their private space. Those women who did not succeed in maintaining a degree of autonomy and independence from their PTSD husband reported feeling that they were drowning in his experiences and demands.

Only one study to date has empirically examined the relationship between the adjustment of wives of PTSD veterans and their separation-individuation (Ben-Arzi, Solomon, & Dekel, 2000). This study shows that women with higher levels of separation-individuation reported lower sense of burden and less psychological distress. The explanation may lie in Bowen's (1978) claim that high levels of separation-individuation enable one to maneuver between intimacy and autonomy in the family system. If this is the case, high levels of separation-individuation might enable wives of PTSD veterans to act as caregivers and supportive partners without being emotionally overwhelmed or losing their autonomy and identity in the process.

Predictors Pertaining to the Marital Relationship

The association between the marital relationship and the secondary traumatization of women married to PTSD veterans has been examined in several ways. Solomon et al. (1991) examined the contribution of the wives' relationships with key members of their family and social networks. Their findings show that of all the relationships examined—between the wife and her father, mother, mother-in-law, siblings, husband, children, and friends—the only one that made a significant contribution to reducing the wives' distress was their relationship with her husband. In particular, a good marital relationship contributed significantly to reducing the wives' depression, anxiety, and hostility.

Waysman et al. (1993) examined the role of family environment. Their findings showed that wives in "conflict-oriented families" reported the highest levels of psychiatric symptomatology and loneliness, while those in "expressive families" reported the lowest levels of both. Further evidence of the emotional benefits of expressiveness come from Solomon et al.'s (1991) findings that high expressiveness, defined as the degree to which the marital relationship allowed for open and direct expression of feelings, was significantly associated with positive psychological adjustment among wives of traumatized combat veterans, whereas marital cohesion, intimacy, and conflict were not.

These findings all point to the key role that the nature and quality of the marital relationship plays in the wife's secondary traumatization. They indicate that a good marital relationship, and especially an expressive one, can mitigate the wife's distress, while a poor relationship, especially a conflicted one, can exacerbate it.

The findings are problematic in two respects, however. None of the studies cited controlled for the severity and manifestations of the husband's PTSD. This makes it impossible to distinguish between the impact of the PTSD and the impact of the marital relationship on the wives' secondary traumatization. As pointed out above, findings show a close relationship between PTSD severity and the quality of the marriage.

Moreover, since the studies are cross-sectional, we cannot be sure that the wife's secondary traumatization does not affect the quality of the marriage more than the quality of the marriage affects her secondary traumatization. Indeed, several scholars have suggested that there is a bidirectional impact (Fals-Stewart & Kelly, 2005)

Theoretical Explanations

The literature offers a variety of theoretical explanations of secondary traumatization among wives of traumatized men. The explanations vary in the quality and amount of empirical support they have.

Identification and Empathy. The first theoretical explanation was offered by Maloney (1988) in an effort to understand the resemblance of the symptoms reported by wives of traumatized veterans and those of the veterans themselves. Maloney suggested that the resemblance may stem from the wives' tendency to identify with their husbands, to internalize their experiences, and to experience in fantasy the same kinds of traumatic events that they had experienced.

Figley (1989, 1995, 1998) uses the term "empathy" rather than identification; but, in essence, his account seems to be an attempt to explain how and why these wives come to identify with their traumatized husbands. As Figley (1998) relates it, the process starts with the wife's efforts to emotionally support her troubled husband, which lead her to try to understand his feelings and experiences and, from there, to empathize with him. In the process of gathering information about his suffering, she takes on his feelings, experiences, and memories as her own—and hence his symptoms.

Some support for these explanations may be found in two of the studies discussed above. The extensive talk of the women in the Dekel et al. (2005a) study of their efforts to set boundaries, to maintain their individuality, and not to be drawn into fusion with their PTSD husbands provides some qualitative support for these explanations. Indirect quantitative support is provided by Ben-Arzi, Solomon, and Dekel's (2000) findings that wives of PTSD veterans with higher levels of separation individuation reported lower sense of burden and less psychological distress than their counterparts with lower levels of separation-individuation. For more solid support, however, studies would have to be carried out using identification and empathy as independent variables.

The empathy and identification explanation is qualitatively different from the next two explanations, chronic stress and ambiguous loss. While the chronic stress and ambiguous loss explanations attribute the wife's secondary traumatization largely to the hardships of living with the traumatized veteran, the empathy and identification explanation attribute it more to processes within the wife herself.

Chronic Stress Produced by Close and Prolonged Contact with a Malfunctioning Partner. Living with a PTSD husband is a chronic stressor, which, like other chronic stressors, may lead over time to somatic and psychiatric difficulties. This explanation is consistent with findings of heightened distress among spouses of persons suffering from a variety of chronic physical or mental disabilities, such as depression (Krantz & Moos, 1987), schizophrenia (Hatfield & Lefley, 1987), and brain injury (Lezak, 1986), among others. It gains more specific support from subsequent studies showing the contribution of caretaking burden to the wives' secondary traumatization.

Ambiguous Loss. The term "ambiguous loss" was coined by Boss (1987, 1999) to describe situations in which a person is present psychologically but absent physically (e.g., in prison, kidnapped, missing in action) or present physically but absent psychologically, as is the case when the person suffers from a debilitating mental disorder, of which PTSD is one of many examples. According to Boss (1999) there is considerable unclarity regarding the role and responsibility of the person with the mental disorder in the family. The lack of clarity immobilizes other family members: decisions are put on hold, and the boundaries of the relationship are unclear. The ambiguity often becomes as debilitating as the illness itself. The person experiencing the ambiguous loss struggles to reduce this ambiguity and to improve the clarity in the relationship. Due to the persistent nature of the loss, however, the effort becomes physically and psychologically exhausting, with attendant symptoms of depression, anxiety, guilt, and distressing dreams.

This explanation is consistent with qualitative findings on spouses of Alzheimer sufferers (Caron, Boss, & Mortimer, 1999; Kaplan & Boss, 1999) as well as dementia (Boss et al., 1990). Somewhat more specific support regarding the wives of PTSD veterans may be gleaned from the findings of the Dekel et al. (2005a) qualitative study. The women in the focus group spoke extensively of their confusion about the husbands' roles and the great strain that this caused them.

> It's as if I'm living alone. I have to do everything alone. If I want to go out, he tells me "go by yourself" or "go and I'll join you later." What am I? Am I a widow? Divorced? I'm not divorced and I'm not a widow. I have a husband.

Upset in World Assumptions. Gilbert (1998) proposes a more cognitive explanation. This is that just as the basic world assumptions of direct victims of trauma are often upset (Janoff-Bulman, 1992), so too are the assumptions of indirect victims. The partner of a traumatized man learns, just as he had, that the world is unsafe and chaotic and that being a good person does not protect one from harm. Her basic assumptions about the relationship are also upset. Gilbert implies that it is these cognitive upsets, along with the great difficulty of understanding the behavior of the traumatized husband, that lead to the wife's secondary traumatization.

Gilbert (1998) supports her claims with clinical evidence. Some empirical support is provided by studies showing that negative world assumptions contribute to the distress of persons directly exposed to traumas (e.g., Magwaza, 1999); that relatives of crime victims report more negative world assumptions than relatives of nonvictims (Denkers & Winkel, 1995); and that children of PTSD veterans report both more negative world assumptions and greater distress than children of veterans without PTSD (Dinshtein, Dekel, & Polliack, in press). None of these studies, however, has been carried out among wives of PTSD veterans.

Assortative Mating. Finally, various authors claim that the heightened distress of women married to men with PTSD may be attributed to assortative mating (i.e., the tendency to choose a partner similar to oneself). The argument here is that the wife's distress stems from prior vulnerabilities which led her to be attracted to and marry a vulnerable man. This claim is supported by findings of high levels of depression in both spouses in a marriage (Merikanages, Bromet, & Spiker, 1983).

While this explanation cannot be ruled out, there are grounds for questioning it. For one thing, a study of Israeli PTSD veterans and their wives indicates that many were married before the men developed PTSD, meaning that the women had not chosen partners with emotional difficulties (Solomon et al., 1992b). Furthermore, there is reason to believe that most combat veterans, at least in Israel, where army recruits must pass stringent examinations of their physical and mental health, were psychologically healthy prior to their traumatization. Jordan et al. (1992), in their study of secondary traumatization following the Vietnam War, refutes the assortative mating theory on the grounds that they found no significant differences in sociodemographic and other background characteristics among wives of PTSD veterans and wives of non-PTSD veterans.

TREATMENT

Despite the considerable distress of wives of PTSD veterans, there is little discussion in the literature about how to help them to alleviate or cope with it. Most of the accounts of interventions with veterans' wives are embedded in accounts of marital and/or family therapy focused on the needs of the psychologically injured veteran. In a review article, Riggs (2000) divides these therapies into two broad types, systemic treatments and support treatments. In both, the trauma is at the center. Systemic treatment involves the use of marital and family therapy models to reduce relationship distress caused by PTSD. Its aim is to alleviate conflict, improve communication, and otherwise bring about better family functioning. Support treatment aims to enhance familial and other social support for the identified client. It views the wife as an important source of support for the traumatized veteran, and focuses on her role in helping him recover.

Neither of these approaches focuses on the needs of the secondary victims themselves. To our knowledge, there are only a small number of accounts of direct clinical work with veterans' wives. Williams (1980), cited above, and Coughlan and Parkin (1987) give accounts of group therapy conducted with wives of Vietnam veterans. More recently, Remer and Ferguson (1998) offers a detailed, six-stage model for treating veterans' wives suffering from secondary traumatization. The model, however, is based on working with partners of sexual assault victims and seems more suitable for acute crises than to chronic conditions. None of the suggested treatments has been evaluated empirically.

Our review highlights the need for direct therapeutic intervention with the veterans' wives, focused on the women's own needs. At this stage, we have no way of knowing what kinds of intervention would be most efficacious. However, we would like to suggest that the following points be taken into consideration in planning interventions for wives suffering from secondary traumatization.

1. PTSD is a chronic disorder, whose severity and manifestations are affected by both inner and outer events and vary over time, and which is highly resistant to treatment. This means that wives of PTSD veterans have to repeatedly cope with the problems created by the PTSD. With respect to intervention, this means: (a) that one-time intervention, of whatever type, may well not be adequate, especially for women whose husbands are severely impaired by their PTSD; and (b) that different interventions may be appropriate at different stages of the disorder and at different points in the life cycle. Hence 2, 3, and 4:
2. A range of interventions should be available for the wives of veterans, which will address the changes in their husband's situations and their own.

3. Depending on the needs of the women at any particular time, the intervention can be individual or group, and employ any and all of a range of methods: psychoeducational, stress reduction, behavioral-cognitive, psychodynamic, and so forth.
4. A strength-oriented perspective aimed at empowering the wives to seek, recognize, and use their strengths to deal with the long-term and ever-changing consequences of their husband's traumatization may be particularly useful (Cowger, 1994).
5. Intervention need not be limited to sessions with a professional. In addition or instead of such sessions, it may be helpful to offer wives of traumatized veterans a range of cultural and leisure activities, with and without their husbands and children. Such activities may help to recharge their batteries; may help make up for some of the social deficits from which they suffer as a result of their husbands' PTSD; and may increase their social support by bringing them into contact with other women in situations similar to their own.
6. In addition, it should be kept in mind that intervention with wives of PTSD veterans is plagued by low participation rates. Studies show that while veterans' wives may often seek treatment for their husbands and even transport them to the therapy, they themselves often do not engage (Lyons & Root, 2001; Moore & Reger, this volume, chapter 8). Judging from our experience at a clinic that provides mental health services to Israeli Defense Force veterans and their families, wives of traumatized veterans face numerous obstacles to engagement. These include the distance and cost of travel to counseling sessions, difficulties in arranging child care, constraints on their time, and, sometimes, the objections of their husbands. Efforts must thus be made to reach out to wives of veterans and to lower the barriers to their participation.

SUMMARY AND RECOMMENDATIONS FOR FURTHER STUDY

This review points to substantial evidence of secondary traumatization among wives of war veterans, of different wars and in different countries, even decades after the wars' end. Their traumatization is marked both by specific trauma symptoms identified in the *Diagnostic and Statistical Manual of Mental Disorders* and by general psychological distress, including depression, anxiety, heightened hostility, and other symptoms. Various explanations have been offered for the development of secondary traumatization, but only some of them have been examined empirically. The key predictors for which there is empirical evidence include the severity and avoidance symptoms of the veteran's PTSD and his violence against his wife; and the wife's caregiver burden and degree of separation-individuation. In addition, the quality of the marital relationship has been treated both as a manifestation and predictor of the wife's secondary traumatization. With this, data on divorce rates in the United States and Israel suggest that the degree to which these marriages founder varies with the culture. In addition, our own qualitative findings on an admittedly small group of Israeli women suggest that being married to a PTSD veteran may have its compensations, which enable the women to stay in their troubled marriages.

A great deal remains to be learned about the secondary traumatization of veterans' wives, however. Perhaps the most crucial gaps in our knowledge stem from the fact that, with very few exceptions, almost all the studies to date have been cross-sectional. The findings simply do not allow us to speak with much certainty about the directionality of the various associations that have been found. Similar to what Fals-Stewart & Kelly (2005) ask about peacekeepers, we can wonder, does a partner who returns home with PTSD lead to relationship problems, setting the stage for the onset of secondary trauma, or are couples who have preexisting relationship more vulnerable to the development of PTSD and secondary trauma? Or are the associations perhaps bidirectional?

Similar questions can be asked with respect to the wife's personality and disposition. Longitudinal studies, which might help answer some of these questions, are very much needed. Prospective studies, which would enable comparing behaviors "before" and "after" the traumatic experience, would also be in order. While prospective studies might be feasible in peacekeeping missions and planned armed operations, however, they would be very difficult to conduct in unplanned military engagements.

Our understanding of how secondary traumatization develops is also very limited. How much is it a result of the transmission of the traumatic experience, as Maloney (1988) and Figley (1989) imply? How much is it a consequence of the stresses and burdens of living with the traumatized veteran? All the theoretical explanations—empathy and identification, ambiguous loss, and upset to world assumptions—require further examination. Such studies might give us a better idea of how trauma is actually passed on from the direct victim to the indirect one.

In addition, we would like to know considerably more about factors that predict or moderate secondary traumatization. In addition to examining more personal variables and features of the marriage, environmental factors should also be considered. These might include such things as the society's view of the war, the status of women in the society, and the social, psychological, and instrumental support available to the woman.

Further research is also recommended on the marriages of PTSD veterans. In addition to clarifying the direction of the relationship between secondary traumatization and marital problems, we would also like to learn about factors involved in the stability and dissolution of marriages to PTSD veterans. What role is played by the society's views of marriage and of the status of woman? What, if any impact, does the family's economic status have? What personal factors enter into decisions about divorce of women married to veterans with PTSD? In addition, we would like to learn more about how women cope with their difficult marriages to PTSD veterans. How do they handle the various tasks they perform? What gives them the strength to struggle day in and day out with the consequences of their husbands' PTSD? How do they preserve their own emotional stability? Recent literature points to posttraumatic growth among survivors of a range of distressing and traumatic events (e.g., Tedeschi & Calhoun, 1996). One direction for research would be to examine whether the partners of these survivors, among them the wives of traumatized veterans, also experience posttraumatic growth.

Also of scholarly interest are the implications of the wife's secondary traumatization on her functioning in other roles, especially as a mother. A study of children of traumatized veterans suggests that warm and protective mothers are able to temper the negative impact of the fathers' PTSD on their children (Dinshtein, Dekel, & Polliack, 2005). However, there is no research to date on how the mother's secondary traumatization affects her parental functioning.

A great deal also remains to be learned about matters that were not addressed in this chapter. Today, when increasing numbers of women serve in the armed forces, both in United States and elsewhere, the question of how their service-related PTSD impacts on their husbands or partners, as well as on their children, becomes increasingly pressing. The uncertainties and other difficulties experienced by women whose husbands are away in war also merit research attention, and all the more so now that U.S. forces are in Iraq. Moreover, since combat is only one of the many sources of traumatization, there is a need to learn more about how traumatization from other sources may affect those living with the traumatized individual. That is, we would like to know whether different types of traumatic events would affect families in different ways and, if so, how.

Finally, a great deal of work remains to be done to develop, publish, and evaluate interventions for wives of traumatized veterans. To date, no interventions focused on the needs of the secondary victim have been published in the literature. This does not mean that such interventions do not exist. There is probably much more intervention going on with wives of PTSD veterans than is reported. We urge that clinicians report what they are doing in this sphere. We also urge evaluative studies of interventions with wives suffering from secondary traumatization. We hope that the advances in our understanding of the causes and processes of secondary traumatization will help to develop effective interventions to alleviate distress of the secondary victims of traumatic events.

REFERENCES

Azar, S. T. (2000). Preventing burnout in professionals and paraprofessionals who work with child abuse and neglect cases: A cognitive behavioral approach to supervision. *Journal of Clinical Psychology, 56,* 643–663.

Beckham, J. C., Lytle, B. L., & Feldman, M. E. (1996). Caregiver burden in partners of Vietnam War veterans with posttraumatic stress disorder. *Journal of Consulting and Clinical Psychology, 64,* 1068–1072.

Ben-Arzi, N., Solomon, Z., & Dekel, R. (2000). Secondary traumatization among wives of PTSD and post-concussion casualties: Distress, caregiver burden and psychological separation [Hebrew]. *Society and Welfare, 20,* 585–602.

Bleich, A., Solomon, Z., & Dekel, R. (1997). *Evaluating Mental Disability Among Veterans with PTSD.* Rehabilitation Wing, Israel Ministry of Defense.

Blos, P. (1979). *The Adolescent Passage.* New York: International Universities Press.

Boss, P. (1987). Family stress. In M. B. Sussman & S. K. Steinmetz (Eds.), *Handbook of Marriage and the Family* (pp. 695–724). New York: Plenum Press.

Boss, P. (1999). *Ambiguous Loss: Learning to Live with Unresolved Grief.* Cambridge, MA: Harvard University Press.

Boss, P., Caron, W., Horbal, J., & Mortimer, J. (1990). Predictors of depression in caregivers of dementia patients: Boundary ambiguity and mastery. *Family Process, 29,* 245–254.

Bowen, M. (1978). *Family Therapy in Clinical Practice.* New York: Aronson.

Bramsen, I., Van der Ploeg, H. M., & Twisk, J. W. R. (2002). Secondary traumatization in Dutch couples of World War II survivors. *Journal of Consulting and Clinical Psychology, 70,* 241–245.

Byrne, C. A., & Riggs, D. S. (1996). The cycle of trauma: Relationship aggression in male Vietnam veterans with symptoms of posttraumatic stress disorder. *Violence and Victims, 11,* 213–225.

Calhoun, P. S., Beckham, J. C., & Bosworth, H. B. (2002). Caregiver burden and psychological distress in partners of veterans with chronic posttraumatic stress disorder. *Journal of Traumatic Stress, 15,* 205–212.

Caron, W., Boss, P., & Mortimer, J. (1999). Family boundary ambiguity predicts Alzheimer's outcomes. *Psychiatry: Interpersonal and Biological Processes, 62,* 347–356.

Carroll, E. M., Rueger, D. B., Foy, D. W., & Donahoe, C. P. (1985). Vietnam combat veterans with posttraumatic stress disorder: Analysis of marital and cohabitating adjustment. *Journal of Abnormal Psychology, 94,* 329–337.

Caserta, M. S., Lund, D. A., & Wright, S. D. (1996). Exploring the Caregiver Burden Inventory (CBI): Further evidence for a multidimensional view of burden. *International Journal of Aging & Human Development, 43,* 21–34.

Center for Policy Research. (1979). *The Adjustment of Vietnam Veterans to Civilian Life.* New York: Center for Policy Research.

Cohen, O. (2003). Israeli family. In J. Ponzetti (Ed.), *International Encyclopedia of Marriage and Family* (Vol. 2, pp. 960–964). New York: Macmillan.

Cook, J. M., Riggs, D. S., Thompson, R., & Coyne, J. C. (2004). Posttraumatic stress disorder and current relationship functioning among World War 2 ex-prisoners of war. *Journal of Family Psychology, 18,* 36–45.

Coughlan, K., & Parkin, C. (1987). Women partners of Vietnam veterans. *Journal of Psychosocial Nursing and Mental Health Services, 25,* 25–27.

Cowger, C. D. (1994). Assessing client strengths: Clinical assessment for client empowerment. *Social Work, 39,* 262–268.

Danieli, Y. (1986). In the treatment and prevention of long-term effects and intergenerational transmission of victimization: A lesson from Holocaust survivors and their children. In C. R. Figley (Ed.), *Trauma and Its Wake* (pp. 309–313). New York: Basic Books.

Daniels, K. (2005). Intimate partner violence & depression: A deadly comorbidity. *Journal of Psychosocial Nursing Mental Health Services, 43,* 6–7.

Dekel, R., Goldblatt, H., Keidar, M., Solomon, Z., & Polliack, M. (2005a). Being a spouse of a PTSD veteran. *Family Relations, 54,* 24–36.

Dekel, R., & Solomon, Z. (2006). Secondary traumatization among wives of Israeli POWs: The role of POWs' distress. *Social Psychiatry and Psychiatric Epidemiology, 41,* 27–33.

Dekel, R., Solomon, Z., & Bleich, A. (2005b). Emotional distress and marital adjustment of caregivers: Contribution of the care recipient's level of impairment and of the caregiver's appraised burden. *Stress, Anxiety & Coping, 18,* 71–82.

Denkers. A. J. M., & Winkel, F. W. (1995). Reactions to criminal victimization: A field study of the cognitive effects on victims and members of their social network. In D. Graham, S. Bostock, M. McMurran, & C. Wilson (Eds.), *Psychology, Law, and Criminal Justice: International Developments in Research and Practice* (pp. 374–383). Oxford, U.K.: Walter De Gruyter.

Dinshtein, Y., Dekel, R., & Polliack, M. (2005). Secondary traumatization among children of war veterans: The role of the mother [Hebrew]. Bar Ilan University, Ramat-Gan, Israel.

Dirkzwager, A. J., Bramsen, I., Ader, H., & Van Der Ploeg, H. M. (2005). Secondary traumatization in partners and parents of Dutch peacekeeping soldiers. *Journal of Family Psychology, 19,* 217–226.

Evans, L., Mchugh, T., Hopwood, M., & Watt, C. (2003). Chronic posttraumatic stress disorder and family functioning of Vietnam veterans and their partners. *Australian and New Zealand Journal of Psychiatry, 37,* 765–772.

Fals-Stewart, W., & Kelly, M. (2005). When family members go to war—a systemic perspective on harm and healing: Comment on Dirkzwager, Bramsen, Ader, & Van der Ploeg (2005). *Journal of Family Psychology, 19,* 233–236.

Figley, C. R. (1983). Catastrophes: An overview of family reactions. In C. R. Figley & H. I. McCubbin (Eds.), *Stress and the Family: Vol. II. Coping with Catastrophe* (pp. 3–20). Levittown, PA: Brunner/Mazel.

Figley, C. R. (1989). *Helping Traumatized Families.* San Francisco: Jossey-Bass.

Figley, C. R. (1995). Compassion fatigue as secondary traumatic stress disorder: An overview. In C. R. Figley (Ed.) *Compassion Fatigue: Coping with Secondary Traumatic Stress Disorder in Those Who Treat the Traumatized* (pp. 1–20). New York: Brunner/Mazel.

Figley, C. R. (1998). Burnout as systematic traumatic stress: A model for helping traumatized family members. In C. R. Figley (Ed.), *Burnout in Families: The Systematic Costs of Caring* (pp. 15–28). Boca Raton, FL: CRC Press.

Frederikson, L. G, Chamberlain, K., & Long, N. (1996). Unacknowledged casualties of the Vietnam War: Experiences of the partners of New Zealand veterans. *Qualitative Health Research, 6,* 49–70.

Galovski, T., & Lyons, J. (2004). Psychological sequelae of combat violence: A review of the impact of PTSD on the veteran's family and possible interventions. *Aggression and Violent Behavior, 9,* 477–501.

Gilbert, K. (1998). Understanding the secondary traumatic stress of spouses. In C. R. Figley (Ed.), *Burnout in Families: The Systematic Costs of Caring* (pp. 47–74). Boca Raton, FL: CRC Press.

Hankin, C. S., Abueg, F., Gallagher-Thompson, D., & Murphy, R. T. (1992, October). *Caregiver stress: Conceptualizing adaptation of partners of posttraumatic stress disorder combat veterans.* Paper presented at the International Society for Traumatic Stress Studies, San Antonio, TX.

Hartsough, D. M., & Myers, D. G. (1985). *Disaster Work and Mental Health: Prevention and Control of Stress Among Workers.* Rockville, MD: National Institute of Mental Health.

Hatfield, A. H., & Lefley, H. P. (1987). *Families of the Mentally ill.* New York: Guilford Press.

Herman, J. (1992). *Trauma and Recovery.* New York: Basic Books.

Iliffe, G., & Steed, L. (2000). Exploring the counselor's experience of working with perpetrators and survivors of domestic violence. *Journal of Interpersonal Violence, 15,* 393–412.

Janoff-Bulman, R. (1992). *Shattered Assumptions.* New York: The Free Press.

Jordan, B. K., Marmar, C. R., Fairbank, J. A., Schlenger, W. E., Kulka, R. A., Hough, R. L., et al. (1992). Problems in families of male Vietnam veterans with posttraumatic stress disorder. *Journal of Consulting and Clinical Psychology, 60,* 916–926.

Kaplan, L., & Boss, P. (1999). Depressive symptoms among spousal caregivers of institutionalized mates with Alzheimer's: Boundary ambiguity and mastery as predictors. *Family Process, 38,* 85–103.

Kotler, M., Cohen, H., Aizenberg, D., Matar, M., Loewenthal, U., Kaplan, Z., et al. (2000). Sexual dysfunction in male posttraumatic stress disorder patients. *Psychotherapy and Psychosomatics, 69,* 309–315.

Krantz, S. E., & Moos, R. H. (1987). Functioning and life context among spouses of remitted and non remitted depressed patients. *Journal of Consulting and Clinical Psychology, 55,* 353–360.

Kulka, R. A., Schlenger, W. E., Fairbank, J. A., Hough, R. L., Jordan, B. K., Marmar, C. R., et al. (1990). *Trauma and the Vietnam War Generation: Report of Findings from the National Vietnam Veterans Readjustment Study.* New York: Brunner/Mazel.

Letourneau, E. J., Schewe, P. A., & Frueh, C. B. (1997). Preliminary evaluation of sexual problems in combat veterans with PTSD. *Journal of Traumatic Stress, 10,* 125–132.

Lev-Wiesel, R., & Amir, M. (2001). Secondary traumatic stress, psychological distress, sharing of traumatic reminiscences, and marital quality among spouses of Holocaust child survivors. *Journal of Marital Family Therapy, 27,* 433–444.

Lezak, M. (1986). Psychological implications of traumatic brain damage for the patients' family. *Rehabilitation Psychology, 31,* 241–250.

Litz, B. (1992). Emotional numbing in combat-related post-traumatic stress disorder: A critical review and reformulation. *Clinical Psychology Review, 12,* 417–432.

Lyons, J. A. & Root, L. P. (2001). Family members of the PTSD veteran: Treatment needs and barriers. *National Center for Post Traumatic Stress Disorder Clinical Quarterly, 10,* 48–52.

Magwaza, A. S. (1999). Assumptive world of traumatized South African adults. *Journal of Social Psychology, 139,* 622–630.

Mahler, M. S., Pipe, F., & Bergman, A. (1975). *The Psychological Birth of the Human Infant: Symbiosis and Individuation.* New York: Basic Books.

Maloney, L. J. (1988). Posttraumatic stresses of women partners of Vietnam veterans. *Smith College Studies in Social Work, 58,* 122–143.

Mason, P. H. C. (1990). *Recovering from the War: A Guide for All Veterans, Family Members, Friends and Therapists.* High Springs, FL: Patience Press.

Matsakis, A. (1988). *Vietnam Wives.* Washington, D. C: Woodbine House.

McCann, I. L., & Pearlman, L. A. (1989). Vicarious traumatization: A framework for understanding the psychological effects of working with victims. *Journal of Traumatic Stress, 3,* 131–149.

McCann, I. L., & Pearlman, L. A. (1990). *Psychological Trauma and the Adult Survivor: Theory, Therapy and Transformation.* New York: Brunner/Mazel.

Merikanages, K. R., Bromet, E. J., & Spiker, D. G. (1983). Assortative mating, social adjustment and course of illness in primary affective disorder. *Archives of General Psychiatry, 40,* 795–880.

Mikulincer, M., Florian, V., & Solomon, Z. (1995). Marital intimacy, family support, and secondary traumatization: A study of wives of veterans with combat stress reaction. *Anxiety, Stress, and Coping, 8,* 203–213.

North, C. S., Tivis, L., McMillen, C., Pfefferbaum, B., Cox, J., Spitznagel, E., et al. (2002). Coping, functioning and adjustment of rescue workers after the Oklahoma City bombing. *Journal of Traumatic Stress, 15,* 171–175.

Novak, M., & Guest, C. (1989). Application of multidimensional caregiver burden inventory. *Journal of Gerontology, 29,* 798–803.

Remer, R., & Ferguson, R. A. (1998). Treating traumatized partners: Producing secondary survivors of PTSD. In C. R. Figley (Ed)., *Burnout in Families: The Systemic Costs of Caring* (pp. 139–170). Boca Raton, FL: CRC Press.

Riggs, D. S. (2000). Marital and family therapy. In E. Foa & T. Keane (Eds.), *Effective Treatments for PTSD* (pp. 280–301). New York: Guilford Press.

Riggs, D. S., Byrne, C. A., Weathers, F. W., & Litz, B. T. (1998). The quality of the intimate relationships of male Vietnam veterans: Problems associated with posttraumatic stress disorder. *Journal of Traumatic Stress, 11,* 87–101.

Rosenheck, R., & Nathan, P. (1985). Secondary traumatization in the children of Vietnam veterans with post-traumatic stress disorder. *Hospital and Community Psychiatry, 36,* 538–539.

Rosenheck, R., & Thomson, J. (1986). "Detoxification" of Vietnam War trauma: A combined family-individual approach. *Family Process, 25,* 559–570.

Schauben, L. J., & Frazier, P. A. (1995). Vicarious trauma: The effects on female counselors of working with sexual violence survivors. *Psychology of Women Quarterly, 19,* 49–64.

Solomon, Z. (1993). *Combat Stress Reaction: The Enduring Toll of War.* New York: Plenum Press.

Solomon, Z., Waysman, M., Avitzur, E., & Enoch, D. (1991). Psychiatric symptomatology among wives of soldiers following combat stress reaction: The role of the social network and marital relations. *Anxiety Research, 4,* 213–223.

Solomon, Z., Waysman, M., Belkin, R., Levi, G., Mikulincer, M., & Enoch, D. (1992a). Marital relation and combat stress reaction: The wives' perspective. *Journal of Marriage and the Family, 54,* 316–326.

Solomon, Z., Waysman, M., Levy, G., Fried, B., Mikulincer, M., Benbenishty, R., et al. (1992b). From front line to home front: A study of secondary traumatization. *Family Process, 31,* 289–302.

Tedeschi, R. G., & Calhoun, L. G. (1996). The posttraumatic growth inventory: Measuring the positive legacy of trauma. *Journal of Traumatic Stress, 9,* 455–471.

Tubbs, C. Y., & Boss, P. (2000). An essay for practitioners dealing with ambiguous loss. *Family Relations, 49,* 285–286.

Waysman, M., Mikulincer, M., Solomon, Z., & Weisenberg, M. (1993). Secondary traumatization among wives of posttraumatic combat veterans: A family typology. *Journal of Family Psychology, 7,* 104–118.

Williams, C. M. (1980). The "veteran system"—with a focus on woman partners: Theoretical considerations, problems and treatment strategies. In T. Williams (Ed.), *Post Traumatic Stress Disorders of Vietnam Veterans* (pp. 73–124). Cincinnati: Disabled American Veterans.

Wilson, J. P., & Kurtz, R. R. (1997). Assessing posttraumatic stress disorder in couples and families. In J. P. Wilson & R. R. Kurtz (Eds.), *Assessing Psychological Trauma and PTSD* (pp. 349–372). New York: Guilford Press.

Woods, S. J. (2005). Intimate partner violence and post-traumatic stress disorder symptoms in women: What we know and need to know. *Journal of Interpersonal Violence, 20,* 394–402.

Zarit, S. H., Todd, P. A., & Zarit, J. M. (1986). Subjective burden of husbands and wives as caregivers: A longitudinal study. *Gerontologist, 26,* 260–266.

Section III

Combat Stress Management Programs

This final section is, in many ways, a product of the first two. These eight chapters demonstrate that postoperational combat stress injuries can be managed both through prevention and training programs prior to combat, effective stress reduction methods during operations, and especially the desensitization program immediately following to long after combat exposure. The first chapter provides an overview of the history of units responsible for preventing and mitigating combat stress injuries "down range" or in the combat zone itself. It is written by two U.S. Army captains in the off-hours while serving in Iraq doing what they describe in their chapter.

The next two chapters provide a description of state-of-the-arts desensitization methods through virtual reality therapies. The Editors believed that virtual reality approaches represent the best evidence-based treatments for treating combat stress injuries in a way that prevents long-term stress disorders. The challenge is in making these treatments cost-effective and capable of helping the tens of thousands of warfighters with stress injuries.

The next chapter is a review of recent research and current practice in the medication management of combat and operational stress injuries. As this chapter stresses, although modern medications are incapable of repairing the damage dealt by the stress of war, they can sometimes greatly relieve symptoms and disability, and can even promote healing.

The next two chapters are written by current or former military officers in two different countries (the United Kingdom and Canada) who saw the need to provide services that both addressed the problems of combat stress injuries and fit the unique military culture within which it must function. The Editors believe that these are extraordinary innovations and cost-effective methods for helping warfighters cope by using fellow warfighters. They provide important models for American warfighters and their families.

The final two chapters of this section and the book focus on the most challenging of areas of assessment and intervention: spirituality and family relationships. In the chapter focusing on the former, the authors address the complicated, philosophical conundrum faced by returning warfighters: balancing being a human being who values life with being a warfighter charged with taking lives.

The final chapter, providing advice to families and friends of warfighters, draws on the last chapter in the previous section. It is a fitting final chapter to this volume because, in the end, families are the repository and managers of combat stress injuries; families must cope with the warfighter's stress and mental disorders as well as their own; and families must somehow carry on long after military service and the services they and the VA can and will provide.

play and captured in numerous books and movies. Unfortunately, many of these same stories often conclude with the heroic and brave warrior suffering a fateful tragedy. Heroism is not without cost.

Traditionally, the price of war has been viewed primarily in terms of physical injury and death. It is easy to understand why. Consider that in the Civil War alone, over 500,000 Americans perished by the hands of their fellow countrymen. In the 20th century, American families buried another 600,000 of their loved ones as a result of the combined conflicts of this era (Dupuy & Dupuy, 1993). Not to mention that over twice that many suffered nonlethal injuries, many of which were lifelong disabling. It is no wonder why people measure the ill effects of war with a physical yardstick.

Most would certainly agree that death on the battlefield is the most noxious and unsettling outcome of war. However, some would contend that the emotional and psychological scarring that often occurs on the battlefield is a close second. Survey any American family and one would likely find an uncle, grandfather, or distant relative who served in combat. Sometimes, the inquiry would also evoke recollections from family members about their relative coming back from the war "just not quite right." Often times this is a result of the psychological cost of war.

For a moment, conjure up memories of yourself when you were 18 or 19 years old. Many of us would immediately lock on to reminiscences of how emotionally, psychologically, and socially immature we were. Some of us would remember how devastated we became after the loss of our first love or would cringe when we mentally revisited our family conflicts, adolescent rejections, and social mistakes. In itself, early adulthood is a demanding developmental stage with a number of complex milestones to be achieved and psychological demands to be navigated. Young adults must negotiate the challenges of being adult children of their parents and balance increased autonomy with continued degrees of dependence. They take on new responsibilities related to finances, healthcare, and independent living. Interpersonally, they often seek romantic companionships and may consider marriage and families. Occupational decisions loom about long-term career choices, job training, or higher education. Given the numerous demands of this developmental stage and the additional stressors of military service, the reliability and stability of most young soldiers, sailors, airmen, and marines are striking. Yet combat and combat-related operations can be overwhelming for anyone and often interact with or complicate normal developmental stressors. Given service members' dedication to duty, it is incumbent upon behavioral health professionals to minimize service-related stressors, prevent long-term difficulties, and provide treatment when combat and operational stress interferes with the daily life of a service member.

HISTORY OF COMBAT STRESS

The acknowledgment of the negative psychological impact of war on soldiers can be traced back to early cultural mythology. However, it wasn't until the late 17th century that an attempt was made to apply a diagnostic label to a breakdown on the battlefield—"nostalgia." Originally called the "Swiss disease" due to its manifestation in Swiss villagers who were involuntarily placed in rogue armies, it eventually was acknowledged as a universal problem (Jones, 1995). Jones provided an excerpt from Rosen (1975) in his account of Leopold Auenbruger's 18th-century description of this phenomenon. Auenbruger wrote:

> When young men who are still growing are forced to enter military service and thus lose all hope of returning safe and sound to their beloved homeland, they become sad, taciturn, listless, solitary, musing, full of sighs and moans. Finally, they cease to pay attention and become indifferent to everything which the maintenance of life requires of them. (Rosen, 1975, p. 344)

Auenbruger's account of what these young men were going through from a psychological and behavioral standpoint is not far off the mark from what America's troops have dealt with throughout our national history. The cognitive shift that occurs with being faced with one's own mortality and the realization that he or she may never see their family again can have a tremendous impact on the person.

Soldier's Heart

In 1871, a former Army psychiatrist during the American Civil War, J. M. Da Costa, wrote about a cardiac condition known as "Irritable Heart." Also referred to as cardiac neurosis, neurocirulatory asthenia, nervous heart, and eventually soldier's heart, the syndrome was characterized by shortness of breath, sweating, nausea and diarrhea, dull aching of the chest, and a persistent tachycardia during mild levels of exertion. The soldiers also struggled with the reminders of combat (Wooley, 2002). Although Arthur Bowen Richards Myers (1870) originally described the syndrome one year prior in his book, *On the Etiology and Prevalence of Diseases of the Heart Among Soldiers,* Da Costa reported detailed cases of soldiers suffering from this ailment while fighting in the Civil War. An interesting note about Da Costa's accounts is that the symptom presentations he details are very similar to what the psychological and psychiatric communities today call anxiety and, more specifically, panic.

Another interesting point is that Da Costa noticed that many of the soldiers improved just by removing them from the forward lines and allowing them to rest (Da Costa, 1871). Although these accounts are not the first documented cases of combat stress and its treatment in soldiers, they are most likely the first with a sufficient level of description that allows us to make a comparison to our current nomenclature and knowledge about treatment.

Shell Shock

The term "shell shock" was a product of the First World War. It was used to describe the psychological trauma that men suffered as a result of the intense combat prevalent throughout the European theater. Originally, men were believed to be suffering from the direct physical effects of shell blasts or poisoning, due to the odd and unfamiliar symptom presentation. Over time, most cases were found not to have been close to exploding artillery, and were diagnosed as "war neurosis." Some of the more common symptoms of this phenomenon were agitation, fatigue, increased startle response, loss of concentration, and mood lability. Conversion reactions with localized loss of sensory or motor function that resembled neurological damage were also common (J. Stokes, personal communication, June 7, 2005). Many of the fighters would flee the battlesite due to an overwhelming sense of fear and panic or become paralyzed and incapable of movement. Consequently, many of these warriors were labeled as cowards or "malingerers" (Binneveld, 1998; Gilbert, 1994).

W. H. R. Rivers, a British psychiatrist during World War I studying the phenomenon of shell shock, presented a paper at the Royal Society of Medicine in 1917, which was then published a year later in the journal *Lancet*. Rivers described a type of neurosis suffered by soldiers that he attributed to a form of repression. He argued that soldiers under intense stress would attempt to mentally withdraw from the adverse stimuli of war (Rivers, 1918). Rivers' explanation of the psychological mechanisms seen in these soldiers is similar to what most behavioral health professionals today would call dissociation. However, what made Rivers' work so important was that he was able to describe several cases in which there was successful amelioration of symptoms. It was at this point that the area of combat stress casualty intervention really began to gain momentum.

Battle Fatigue

As the World War II successor to shell shock, "battle fatigue" became a popular term in military medicine that is still used in many of the discussions of combat stress today. Battle fatigue is considered to be caused by one (or usually a combination) of four contributing factors: sudden exposure, cumulative exposure, physical stressors, and home front issues (Department of the Army, 1994a,b). Its symptoms are similar to shell shock and soldier's heart in that the individual may experience fatigue, anxiety, loss of concentration and motivation, depression, memory loss, and disturbances in physical functioning.

During WWII, treatment for battle fatigue focused on returning the soldier to the front in order to keep fighting and keep the unit strong. Initially thought to be cruel and counterproductive, service members were found to be able to reintegrate back into their units and continue as productive warriors. This is consistent with what was found in the Russo-Japanese War, WWI, Korea, Vietnam, the Israeli-Lebanon incursion, and many others. Historical accounts suggest that it was the new replacement who was most likely to fail catastrophically and not the experienced soldier (J. Stokes, personal communication, June, 7, 2005).

Posttraumatic Stress Disorder

Posttraumatic stress disorder (PTSD) is one of the most well-known and publicized mental health disorders in the world. PTSD differs from the terms and concepts mentioned above as it is a specific psychiatric diagnosis. However, the term has become synonymous (although incorrectly) with battle fatigue and combat stress. Therefore, a brief discussion at this juncture, if for nothing more than to clarify the disorder, is warranted.

PTSD is characterized by an exposure to a traumatic event with symptoms from three different clusters: intrusive thoughts/recollections, avoidant/numbing symptoms, and hyperarousal (American Psychiatric Association, 2000). The disorder gained its widespread recognition as a result of the Vietnam War. After thousands of returning veterans lined the halls of VA hospitals, scientists began to take a closer look at this complex disorder (Dicks, 1990). Partly in response to the "social epidemic" of PTSD in America, the American Psychiatric Association listed it in the *Diagnostic and Statistical Manual of Mental Disorders* (DSM-III; American Psychiatric Association, 1987) as an official diagnosis. Consequently, numerous research projects were undertaken to learn more about this disorder and how best to treat those affected by it.

To date, most of the major studies related to PTSD are with Vietnam veterans. One of the more influential of these studies was conducted in 1983 by the National Vietnam Veterans Readjustment Study which was mandated by the U.S. Congress. The aim of the study was to determine the prevalence of PTSD in returning veterans as well as identify any readjustment/reintegration problems that they faced. The investigators found that approximately 30% of males and 26% of females who participated in the Vietnam War had PTSD at some point during their lives. They also found that there was a higher incidence in minority populations (Kulka et al., 1990).

Only one large scale study (Hoge et al., 2004) has been conducted on the combat-related mental health difficulties among service members deployed in support of Operation Iraqi Freedom (OIF) and Operation Enduring Freedom. These researchers found that 11% to 20% of 1,709 redeployed soldiers and Marines met broad screening criteria for PTSD 3 to 4 months after returning. This is intriguing, considering that over 90% of service members deployed to Iraq reported being shot at. Based on these studies, it is noteworthy that the vast majority of service members are quite resilient. Approximately 70% of Vietnam veterans and 80% of OIF veterans are not suffering from PTSD. Military personnel are tough, professional, and well trained, and most apparently endure combat stress adequately. However, for the thousands who do suffer significant psychological or functional impairment, there remains an obligation as well as hope.

Probably the most important consequence of the vast research into PTSD is the continued development and refinement of treatment methods. There are literally dozens of scripted treatment programs for PTSD drawing from various theoretical backgrounds. It is safe to assume that as research into this area continues, more effective and targeted approaches to ameliorating the symptoms of PTSD will emerge.

Combat-Operational Stress Reaction

In February of 1999, DOD Directive 6490.5 (Department of Defense, 1999) mandated the use of the term "combat stress reactions." In the next 1 to 2 years, the Navy, Marines, and Air Force wanted it changed to "operational stress reactions," based on the argument that service members were just as vulnerable to stress reactions during peacetime operations as they were during combat. An agreement was reached, and combat-operational stress reaction (COSR) has become the standard term used in all Services. However, it is still not official, pending revision of DOD 6490.5. The authors believe that there is another important difference. COSR is a more inclusive term. The days of support troops with the "gear in the rear" are for the most part over. Linear battlefields are not likely to reemerge, and combat support troops are often co-located with infantry soldiers and participating in tactical missions. Everyone is a potential stress casualty, hence, the appreciated addition of "operational" in the designation. The identifier of operational also makes it clearer that similar symptoms and sequelae can occur in the absence of combat during, for example, peace support operations, humanitarian assistance missions, grueling field exercises, and even under garrison stressors (e.g., preparing for a major inspection).

COSR has been described as what happens when a person experiences a "normal" reaction to what would be considered an "abnormal" experience. COSR encompasses and illuminates the many different types of stress symptoms that a service member may manifest within four specific areas: physical, cognitive, emotional, and behavioral. With regard to physical signs, the service member may experience fatigue and exhaustion, numbness and/or tingling in extremities, nausea and vomiting, insomnia, and psychomotor agitation. Cognitively, it is not uncommon to see difficulties in concentration, memory loss, nightmares, flashbacks, and depersonalization. Emotionally, feelings of fear and hopelessness, mood lability, and anger are often present. And lastly, the service member may exhibit behavioral symptoms that could include misconduct, careless behavior, and impulsivity (Department of the Army, 1994a,b).

Summary

The reactions of troops to stressful events in combat and combat operations, for the most part, have stayed the same over the past several centuries. What has not stayed the same is how they are classified and understood. This is not a trivial difference. In order to develop preventative measures for warding off combat and operational stress reactions and to implement tested and effective treatment strategies, a thorough understanding of the process of combat-operational stress is crucial. As a result of this understanding, the military has been able to develop successful programs in "combating" combat-operational stress.

THE ARMY COMBAT STRESS CONTROL TEAM: A LOOK AT ITS HISTORY, MISSION, CONFIGURATION, AND PROFESSIONALS

History

The primary purpose of any military force is to win war. Simply, in theory, the objective is to overwhelm the enemy with so much stress that they submit and surrender to their adversary. This includes not only the physical stress of injury and death, but the emotional and psychological stress that often plagues soldiers on the battlefield. The loss of personnel through emotional and psychological stress can be a war-stopper. Therefore, any successful army must have procedures and units in place to provide support and care to those in need.

In World War I, the U.S. Army learned from their French and British counterparts that if combat stress cases were evacuated to the rear they seldom returned to their units. Moreover, these soldiers were more likely to become chronic and have difficulty readjusting upon their return home. Contrary to this, when treated close to the front the soldiers were more likely to return to duty. Consequently, the units could remain strong, which increased the chances for successful military operations. As a result, the U.S. Army adopted a three-echelon system for prevention and treatment of combat stress casualties (Salmon & Fenton, 1929).

The first-echelon of care consisted of a psychiatrist positioned within the division. The job of the psychiatrist was to screen for those susceptible to combat stress, consult with command on the prevention of combat stress casualties, triage cases just behind the front so that soldiers with simple exhaustion were rested by their units, and personally treat more symptomatic cases while supervising medical personnel in the division rear. This is where the importance of "treating far forward" can be seen for the first time within U.S. Army medicine. The second-echelon of care consisted of a psychiatrist, psychiatric nurses, occupational therapy volunteers, and trained medics (which included some clinical psychologists and social workers). These professionals formed specialized "neurological hospitals" in old French buildings, which treated exclusively the soldiers suffering from combat stress that the psychiatrist in division was not able to return to duty. After 1 to 3 weeks of rest and replenishment, most of the soldiers at this level were able to return to their units. This represented the emergence of rest and replenishment as concepts central to the fitness teams of the Army's modern combat stress teams. Finally, the third-echelon of care consisted of a rear hospital whose only function was to provide several weeks to months of ongoing treatment to the soldier. Although many of the soldiers were able to return to their units after care at this level, it was considerably less than if they were treated at levels one or two, and most returned to rear area duties (Salmon & Fenton, 1929).

Combat stress teams continued to develop throughout the Korean War, in which a clinical psychologist, a social worker, and some six enlisted specialists were added to the division psychiatrist to form the Division Mental Health Section. Also, Korea saw the first autonomous, mobile psychiatric detachments. This mental health structure continued though Vietnam and the Persian Gulf War. However, in 1994 the Department of the Army published two comprehensive documents which were field manuals that specify doctrinal guidelines for organization and implementation. The first, *Leaders' Manual for Combat Stress Control* (Department of the Army, 1994a) is known within military circles simply as FM 22-51. FM 22-51 provides in-depth explanations of causes, symptoms, and treatment of combat stress casualties. The other is *Combat Stress Control in a Theater of Operations—Tactics, Techniques, Procedures* (Department of the Army, 1994b) or FM 8-51, which specifies the organization and tactical operation of division mental health sections and combat stress control units. Known as the "bible of combat stress," FM 8-51 is responsible for outlining the modern-day combat stress teams mission and purpose and how they are configured and placed on the battlefield. An additional manual, *Combat Stress,* or FM 6-22.5 (Department of the Army, 2000), was written for the Marines and is accepted and used by all Services; it provides the guidelines for small unit leaders to deal with combat stress within their units.

Mission and Purpose

The mission of the Army's combat stress control (CSC) team is straightforward and simple: provide prevention and treatment as close to the soldier's unit as possible for the purpose of keeping the soldier with the unit. The CSC team functions as a "force multiplier," meaning that it focuses on preserving the fighting strength of the soldier. Importantly, this is done in collaboration with the soldier's chain of command as this is always the first line of combat stress prevention. In theory, only when problems become too great for the direct line leaders do CSC teams become involved.

The guidelines ("doctrine") for treating soldiers suffering from COSR follow six basic principles: brevity, immediacy, centrality, expectancy, proximity, and simplicity (B.I.C.E.P.S.) (Department of the Army, 2000).

Brevity refers to the expected length of the CSC intervention. Commanders can expect their service members to return to duty quickly as most interventions take just minutes or hours and soldiers requiring rest or replenishment often need no further treatment after a few days. Those soldiers who do require additional treatment are moved to the next echelon of care.

Immediacy refers to intervening as soon as feasibly possible. If COSR symptoms go untreated, the potential for symptom exacerbation is increased as is the development of new symptoms. This is where a proactive approach by the CSC unit is crucial in that training command and leadership on how to recognize COSR signs allows for immediate action.

Centrality addresses the location of the CSC assets that take care of soldiers who cannot be managed within the unit. By doctrine (Department of the Army, 2000), these assets should be located in a central, nonmedical location in order to maintain the soldier's warfighter identity. Centrality of these assets also fosters good communication with commanders and allows rapid return to the unit once treatment is complete.

Expectancy is important as it does not focus on the soldier as a patient but as someone that is having a normal reaction to an extreme circumstance or condition. This is much more than a "splitting of the cognitive hairs." If the soldier believes that he or she will get better and that the reaction will remit with time, the soldier is able to focus on the tasks required to function as a soldier and be able to perform well on missions.

Proximity is based on the principle of providing services to the soldier within his or her own unit, or as close to the unit as possible. The idea underlying this principle is that in order for the soldier to get better, he or she must negotiate the dichotomy of wanting to seek refuge from war and remaining loyal to his or her fellow soldiers. If the soldier is taken from the unit, the pull to flee from the battlefield grows stronger. If this happens, the soldier may see an exacerbation of symptoms and increase the potential for a long-lasting psychiatric disturbance. For the Army, this means losing a much needed troop that impacts the units' strength and possible future missions.

Simplicity addresses the degree of complexity that most interventions involve. The 4 R's provide a memory aid to guide many of these straightforward CSC interventions. *Reassuring* the soldier of the normality of COSR, providing *rest* from combat or work, *replenishment* of first-order needs (sleep, rest, food, water, hygiene), and *restoring* confidence through purposeful activities are simple CSC actions that are regularly taken to control a COSR.

Configuration of the CSC Unit

The configuration of the CSC unit is based on striking a balance between positioning behavioral health assets as far forward as possible and maintaining assets in the rear to support the forward teams. If needed, in theory, the rear teams are capable of dispatching teams forward in case of a mass casualty situation (Department of the Army, 1994a).

Each CSC unit is either designated as a company or detachment-sized element. The basic differences are the size of the unit, its resources, and whether the unit is active duty or a reserve component (detachments are active duty and companies are reserve). Detachment-sized CSC units can possess anywhere from approximately 25 to 45 personnel. Company-sized units may be twice that size. Even though there is specific doctrine and guidelines on how many soldiers are in a company or detachment, size fluctuates depending on available resources and the most current doctrine at the time.

Prevention. The preventive teams' primary responsibility is the prevention, triage, and short-term treatment of COSRs. The preventive team typically consists of a psychologist, social worker, and two mental health specialists. However, it is not uncommon to find a psychiatrist in the place of a psychologist. As stated earlier, there is specific doctrine which outlines the team configuration; however, necessity often dictates configuration. The team is strategically placed with forward units to prevent stress breakdown and help keep unit manpower strong. This is done through a variety of means. One of the more common approaches utilized is command consultation.

Through educating unit command about COSR, preventive teams empower unit leadership with the ability to recognize the initial signs of COSR in their soldiers. This can be done simply by giving presentations at command meetings or informally through passing out fliers and brochures. Preventive teams are also able to conduct unit climate surveys. This application of psychometric and qualitative techniques is a very useful tool for commanders in that specific issues that may contribute to decreased morale can be identified and subsequent recommendations for improved unit function can be provided.

Preventive teams also provide preventive measures to the soldiers themselves through providing briefings on suicide prevention, stress and anger management, homefront issues, and reintegration tips for returning home. Another important and effective strategy at the soldier level is what is referred to as "walkabouts." A proactive preventive team will send at least one of its members out to different units to talk with soldiers on an informal basis. Typically this is done by an enlisted soldier from the team. Soldiers tend to be reluctant to seek out the formal services of a clinic-based behavioral health program due to the fear of being stigmatized. With "walkabouts" the soldier can talk with the CSC enlisted member where they work, in their living quarters, or even in the dining facility. A major selling point to the soldier on this approach is that detailed records of the encounters are not kept and they do not face the stigma of "mental health." If a higher level of care is needed, the soldier can be referred to the licensed provider for more "in-depth" intervention, which may include two or three counseling sessions.

Another crucial service provided by preventive teams is crisis debriefings. After a traumatic event, CSC professionals can help soldiers normalize feelings and challenge distressing beliefs. Although not group therapy, crisis debriefings often times can become emotionally charged. The debriefing provides a safe environment for the soldiers to process what happened on several levels without the fear of reprimand or stigmatization from their command.

Fitness. Overall, the primary mission of the CSC unit is the prevention of combat and operational stress reactions. However, for those suffering from COSR the fitness team also helps facilitate the restoration of the soldier's confidence in his or her abilities as a soldier. Staffed with a psychiatrist or psychologist, psychiatric nurse, occupational therapist, and mental health and occupational therapy specialists, fitness teams provide basic services to aid in stress recovery.

The fitness concept is based on ensuring rest and replenishment. If the soldier is in need of services greater than what can be provided by the preventive team, the soldier can be sent to the fitness team for as little as one day or up to several days depending on individual needs. While at fitness, the soldier is provided the opportunity for sleep which can be accomplished through providing basic sleep hygiene techniques or in more severe cases through medication. They can receive more intensive help with stress management, relaxation training, and homefront issues. If the soldier is dealing more with depressive or anxiety symptoms, brief psychotherapeutic interventions such as cognitive or solution-focused therapy can be provided.

A primary goal of fitness is to make sure the soldier's basic needs are met. This is often difficult to accomplish within the unit as the soldier may be needed to maintain a high operational tempo if he or she remains with the unit. With proper rest and replenishment, the vast majority of cases seen by fitness teams are returned to their unit and are mission capable.

Command. For preventive and fitness teams to be effective there must be proper assignment and placement of personnel. This is the role of command. As stated earlier, doctrine serves only as a guide for how CSC units are structured and placed on the battlefield. For a CSC unit to be truly effective on the battlefield, command must be flexible and adapt to the needs of whatever unforeseen situation or situations may arise.

One way of ensuring that the CSC unit is being utilized as efficiently and effectively as it possibly can is to keep track of workload numbers. By keeping track of the number of soldier contacts and intensity of services being provided in the different areas of operations, command can make informed decisions on team placement. If one area is suffering more casualties or has an inherently more difficult mission than others, command can strengthen assets in that region by pulling providers from areas with a lower casualty rate or operational tempo.

Another major role of command is to make sure that the morale of CSC unit members remains strong. Behavioral health providers are not immune to the stressors of war. They often share the same environmental, physical, and emotional burdens with line soldiers. Moreover, being required to manage the emotional and psychological problems of others in such an intense and dangerous environment can take its toll. By maintaining strict lines of communication, coordinating midtour leave, staying in contact with family members back home, and providing overall adequate support and resources, command is able to buffer many of the stressors faced by the CSC unit members.

Roles of the CSC Members

All team members, whether officers or enlisted, privates or commanders, participate in the CSC preventive mission. Command consultation, psychoeducational briefs, walkabouts, crisis debriefings, distribution of informational handouts, etc., are activities conducted by all personnel, regardless of specialty. The unique skills and contributions of the various team members are described below. Note that many of these contributions refer to treatment when conducted as part of fitness.

Psychiatrists. Psychiatrists are responsible for diagnostic formulation, treatment, and disposition of soldiers with COSRs and psychiatric disorders. As a prescribing physician, the psychiatrist conducts medication consultations and prescribes psychotropics or other medications when appropriate. In addition, the psychiatrist assists with CSC triage by ruling out medical etiologies that may better explain a soldier's clinical presentation. Though rare, the psychiatrist typically assists in the coordination of air evacuation when necessary. In addition, the psychiatrist assists in the training of both CSC personnel and unit leaders regarding the identification of and appropriate response to COSRs and psychiatric symptoms. If the psychiatrist is the senior clinical provider, he or she may supervise the unit's clinical work. The psychiatrist may also serve as the fitness or prevention Officer in Charge (OIC). When the CSC unit is located near a combat support hospital or medical treatment facility, the psychiatrist may be consulted on injured soldiers with co-occurring psychiatric presentations (Moore, 2005a).

Psychologists. As experts in the assessment, evaluation, and treatment of psychological disorders, clinical psychologists are well suited to distinguish between COSRs and mental health disorders. Regarding triage, psychologists evaluate both soldiers and units using clinical interviewing and psychometric assessment tools. Such assessments assist psychologists in identifying COSRs and neuropsychiatric disorders and help guide recommendations to soldiers and commanders in prevention and treatment (Moore & Reger, 2006). Given the frequency with which homefront concerns result in or exacerbate combat-operational stress, psychologists' familiarity with marriage and family counseling and interpersonal dynamics is also put to good use. When identified, soldiers with COSRs are treated with a variety of individual and group psychological interventions and techniques. In addition, the psychologist supervises subordinate personnel providing clinical services.

Occupational Therapists. The physical and mental demands of the combat zone can result in a variety of changes in behavior, affect, and cognitions. When these changes rise to the level of a COSR, a negative impact on work performance may result. Occupational therapists (OTs) are trained to assess and rehabilitate functional impairments affecting individuals' daily lives. By definition, this includes occupational performance. In CSCs, OTs use their skills and training to assess and improve occupational functioning among soldiers affected by combat-operational stress.

Although OTs participate in certain shared preventive and fitness-related CSC tasks, they also bring unique skills to the mission. With individual soldiers, OTs assess duty task requirements and the soldier's current capabilities in order to structure therapeutic environments to recondition soldiers and return them to their place of duty. OTs may also consult with unit commanders on ways of minimizing the impact of combat-operational stress on an entire unit's work performance. Though nondoctrinal, in practice OTs may also be consulted by combat support hospitals or medical treatment facilities when a client's disease or injury necessitates upper extremity rehabilitation.

Social Workers. Social workers bring their unique psychosocial perspective to the CSC mission by examining COSRs and their prevention through the lenses of systemic factors. They help to identify and resolve systemic risk factors for combat-operational stress and implement organizational preventive factors. Through command consultations and work with individual soldiers, social workers enhance the combat strength of supported units. They provide individual and group counseling and psychological assessment, if it is an area of clinical competence. In addition, CSC social workers are often the consulted professionals in the case of domestic violence or sexual assault in theater.

Psychiatric Nurses. CSC psychiatric nurses possess a variety of clinical skills and expertise that can be drawn upon in various ways depending on the location and needs of a particular team. If the psychiatric nurse has prescribing privileges, he or she may assist the psychiatrist with medication consultations. In addition, the psychiatric nurse assists in the individual and group treatment of COSRs. This may include both preventive interventions as well as treatment. Another important role of the psychiatric nurse is assuming the command role. For example, in 2005–2006, psychiatric nurses were the commanders for all three CSC detachments during Operation Iraqi Freedom III (Moore, 2005b).

Mental Health and Occupational Therapy Specialists. Paraprofessionals have a long and respected history of significant contributions to the discipline of mental health. Army combat stress control is no exception. Mental health specialists and occupational therapy specialists are enlisted personnel who have completed Army basic training as well as several months of specialized training in basic clinical skills and interviewing. In addition to all preventive activities, these specialists are trained to conduct intake interviews, participate in mass casualty interventions, structure and oversee occupational therapy programs, and escort psychiatric casualties during aeromedical evacuations. In addition, as enlisted personnel, they are technical and tactical experts and train the unit in numerous activities as diverse as driving tactical vehicles, responding to dismounted fire, and identifying and responding to nuclear, biological, chemical, radiological, or explosive attacks.

Summary

The history of Army CSC units has its roots in the Army's recognition of the detrimental impact that COSRs can have on soldiers and subsequent mission performance. As a result, the development and refinement of specialized behavioral health teams was undertaken. Through utilizing both professionals and paraprofessionals trained in combating stress reactions in soldiers, the Army has met the challenge of helping soldiers on the "frontlines" deal with harsh and dangerous conditions. Although not a cure, when appropriately applied, CSC units can restore the stress stricken soldier to their prior level of functioning in order to maintain unit and mission capabilities.

It is noteworthy that Operations Enduring and Iraqi Freedom have provided lessons that have continued to develop the Army model of combat stress control (see Reger & Moore, 2006). These conflicts have no "frontlines." As a result, modular preventive teams capable of performing a wide variety of combat-operational stress control activities have emerged as a new aspect to the CSC model. While fitness teams and preventive teams continue to be utilized, organization and placement of these teams is no longer based on the location of the forward line of troops. Instead, geographic areas are parsed and provide appropriate fitness and prevention coverage. Given the diversity of particular geographic locations, the units working within any given area are regarded as having unique characteristics that necessitate formal needs assessment and individualized combat-operational stress control interventions.

CASE STUDIES

Case 1: Preventing Exacerbation of Combat Stress Symptoms

At a small refueling point along a frequently traveled road in Iraq, a CSC preventive team conducted frequent walkabouts with the numerous temporary residents. Convoys typically stopped for approximately 12 to 24 hours for rest, food, and fuel. During one such walkabout, SGT X and SPC Y, mental health specialists with the preventive team, introduced themselves to SPC Z, a truck driver who reported that she had witnessed the death of a unit member during a nighttime military vehicle rollover approximately 1 month prior. She was driving a 5-ton truck 30 meters behind the military vehicle when it suddenly swerved off the road and overturned in a ditch. As the convoy halted, SPC Z responded appropriately according to her convoy training. The reason for the loss of vehicle control was unknown.

Intervention. SGT X informally assessed for signs of combat stress. SPC Z generally denied current difficulties but admitted she was not sleeping as well as she used to and noted that she was less confident in her own abilities. "I'm always looking around now when I drive. I feel like I can't relax. He was a good driver and if it happened to him, it could happen to me!" SGT X drew upon the CSC model and reassured her that her reactions were perfectly normal and even adaptive. He highlighted her strengths (responding according to her training while under stress) and provided advice on sleep hygiene. Furthermore, he covered the different symptoms that often arise after a traumatic event so that she could be an active member in the monitoring of her emotional health. The role of the CSC team was explained and she was encouraged to seek further assistance, should the need arise.

Outcome. The following morning, SPC Y located SPC Z preparing to convoy out of the area. She reported feeling better after sleeping in a decent cot, getting a bit of rest, and having the chance to get her experiences "off her chest." SPC Z thanked the mental health specialist for his help and concern and departed with the convoy.

Case Lessons. This case study demonstrates a typical interaction between service members and forward deployed preventive team members. These teams are often deployed to small, remote locations with limited available medical and psychiatric resources. They assist in appropriate triage of service members with symptoms of combat-operational stress and provide support and help at the duty location. While preventive teams can support restoration and stabilization of service members assigned to their area of operations, it is not uncommon for such teams to capitalize on "one-shot" interventions, such as that described above.

In the combat zone, military bearing and adherence to the rank structure do not represent mere military courtesies. These are core organizational features with life and death consequences. However, these aspects of the organization can create barriers for CSC officers trying to informally assess service members. Although willing to "vent" to fellow enlisted, the presence of an officer may result in professional, censored reports of current functioning.

Given this background, this case study highlights the contributions of the mental health specialist. Highly trained enlisted mental health specialists can provide the ability to make professional, interpersonal connections with fellow service members in order to accurately assess and assist at the place of duty. Such connections represent the "frontline" of combat-operational stress consultation.

Case 2: Treatment in Theater—Preventing Unnecessary Evacuations

The day started out like any other for SPC X, a 27-year-old, married National Guard .50-caliber machine gun operator. As he had done dozens of times before, he began the long and typically boring convoy between his forward operating base and the forward operating base that he ran supplies to 70 miles away. However, today would be a little different.

The improvised explosive device (IED) attack occurred along a strip of highway on which previous contact with insurgents had occurred. SPC X heard a massive explosion before briefly losing consciousness. As his perception cleared, he realized his charred truck had come to a stop against the highway median and other service members were assisting him out of the turret. Physically, he was not wounded. His truck commander, however, took shrapnel to his right leg. No one else in the convoy was injured. In the weeks to follow, the truck commander healed and was returned to duty.

During the CSC team's group debriefing 2 days later, SPC X was quiet, reserved, and sullen. The preventive team assigned to his unit followed up with him the next day at his place of duty. He initially denied difficulties related to the IED and stated he just needed to "suck it up and drive on." However, as the conversation evolved and rapport began to develop, the service member reported making superficial cuts on his wrist following the incident. He was having regular nightmares, crying spells while he reflected on the attack, an exaggerated startle response, and increased irritability with others in his unit. Additionally, the service member suspected his wife of infidelity during the deployment and he was having difficulties concentrating on the job. His sleep was severely disturbed and he had developed dark circles under his eyes. He no longer worked out and he was eating only one meal a day. Tearful and agitated, his thoughts centered on ways to convince his wife that he loved her. Though he reported passive suicidal ideation ("I wish I hadn't lived through the attack"), he denied a plan or intent to end his life. His past history reflected one incident in high school when he had also made superficial cuts on his wrist following an argument with a girlfriend. There was no other history of prior psychiatric treatment or mood difficulties that significantly interfered with social or occupational functioning. The service member was hesitant to meet with a psychologist but was willing to do so in order to focus on saving his marriage.

Intervention. The service member was referred to the CSC psychologist who combined the traditional CSC approach with other individualized psychological interventions. The service member was provided psychoeducation on the effects of sleep deprivation, nutrition, and exercise on mood. Simple, behavioral interventions related to these areas were planned and initiated. His COSRs were normalized and he was referred to a CSC psychiatrist with the fitness team to consider the possible role of medications to increase the likelihood of restorative sleep. A brief course of medications was prescribed. The service member was enrolled in a stress management and coping group and short-term, solution-focused individual therapy was initiated to address depressive and anxious symptoms related to his marital concerns. He was returned to his unit with the recommendation of a temporary suspension of participation in convoys but continued meaningful work within the unit. The service member was given the expectation that his mood and his ability to work on his problems would both improve in a brief period of time, which would allow him to return to his job fully mission capable.

Outcome. Improved sleep resulted in almost immediate increases in concentration, energy, and motivation to work out his problems. His stress reactions progressively decreased in intensity and frequency and were no longer distressing him at 3 weeks. Individual therapy utilized an eclectic approach including cognitive-behavioral and existential interventions. Cognitive restructuring, development of improved problem solving skills, and an increased coping repertoire resulted in a stabilization of mood and full participation in his unit's missions in 3.5 weeks. There was no return of passive suicidal ideation. Improved communication skills and confrontation of self-defeating cognitions allowed the service member to accept reassurance from his wife. Two months later he went home for his scheduled 2 weeks leave and then returned to theater confident in his marriage and his ability to complete the deployment successfully.

Case Lessons. This case study illustrates several significant points related to the treatment of combat and operational stress. First, the interface between preventive and fitness teams is crucial to successful treatment. Preventive teams must appropriately triage identified service members in need of restoration or individual attention and successfully refer to this "higher" level of care. This transfer can be challenging. As discussed above, a soldier may be comfortable with the informal, peer interactions of the mental health specialist. When referral is made to a CSC professional, new challenges and resistance are occasionally encountered.

Second, combat-operational stress reactions do not occur within a vacuum. As with other forms of trauma, reactions to combat stress are shaped by numerous individual and organizational preventive and risk factors. Although preventive teams typically operate in small, briefly organized groups of soldiers that do not focus on the characteristics of individuals, treatment requires that these factors be taken into account. Factors unique to deployed military personnel that might be considered by providers include length of military career and amount of time in theater, past combat experiences in other conflicts, previous contact with the current enemy, access to and quality of communication with homefront social support, and environmental/cultural stressors in the area of operations.

Third, many soldiers have a tendency to "suck it up and drive on," that is, to endure whatever challenges are presented and focus on mission completion. Professionals in the civilian sector might consider such a reaction to trauma to represent a lack of insight, denial, repression, or suppression. In contrast, such an attitude among soldiers is often predictive of a positive CSC treatment outcome. It is true that the soldier must become motivated to recognize their needs, accept help, and adopt new tools. However, if the "soldier-on" attitude is encountered initially it may indicate an individual with a personally meaningful mission, an internal locus of control, a supportive chain of command, or significant internal resources. Regardless, due to the prevalence of this attitude CSC units provide services in a variety of relevant but nonstigmatized contexts to be readily available and familiar resources to service members (e.g., smoking cessation, stress management, long-distance marriage maintenance).

Case 3: Command Contributions—The Role of Leaders in Managing Combat-Operational Stress

In preparation for his CSC unit's deployment to Iraq the Commander of the unit, LTC X, had done his homework. He had 15 years of army leadership experience, not a moment of which was taken for granted as he led his unit into the combat zone. Prior to deployment, his training schedule was effectively and efficiently implemented by the unit's noncommissioned officers. The unit was tactically, technically, and clinically proficient. He had been briefed at length by the leadership of the unit he was replacing and he had a good understanding of the current situation on the ground.

There were 15 forward operating bases (FOBs) within the large geographic area for which the unit had CSC responsibility. These FOBs ranged in size and function from large logistical support areas to small FOBs that were little more than truck stops. Given the unit's personnel strength, the Commander had determined that he could forward deploy 6 to 12 teams depending on their size and composition. Based on intelligence reviewed, feedback from the outgoing unit, and the strengths and challenges within his own unit, the Commander decided to deploy three 4-person teams and four 2-person teams to those seven FOBs with the greatest anticipated needs. COSRs identified in other areas would be brought to one of these seven locations.

Following 1 month in theater, review of the recent CSC workload reports documented a dramatic increase in the number of COSRs. Examining the data closer, the Commander noted that the change was attributed primarily to soldiers from two FOBs, currently unoccupied by the CSC unit. The Commander quickly learned that IEDs had tripled along a particular stretch of highway not far from these two FOBs. As a result, LTC X assessed the distribution of his combat stress assets and reallocated one 4-man team to each of these two FOBs.

These teams initiated preventive and fitness activities within the area, including walkabouts, crisis debriefings, force protection briefs, psychoeducation on combat-operational stress, group treatment on stress/anger management, and individual treatment. In addition, a commander from a large unit at one of the FOBs sensed a dramatic decrease in unit morale and asked the psychologist to conduct a unit survey and make recommendations.

Outcome. Workload reports over the following weeks documented a slow decrease in stress reactions. Although IED attacks in the area continued, the reallocation of personnel reduced stress reactions to slightly above the original baseline by the end of 6 weeks.

Case Lessons. CSC commanders have a variety of responsibilities including planning, directing, and supervising the operations of the unit. When deployed, commanders must conduct ongoing needs assessment within their area of operations and adjust operations as the needs of the mission dictate. The vignette above brings to light this crucial contribution. Without good leadership even world class mental health professionals will lack peak performance. Command decisions about team composition, placement, allocation, and reallocation are among the many key decisions related to successful combat and operational stress management.

This case also underscores the importance of available technologies in assisting commanders in important clinical decision making. The military is cognizant of the important role technology will play in preparing it for the unique warfighting challenges of the 21st century. This is true of the mission of the military behavioral sciences as well. Computerized workload tracking is just one example of many illustrating the mission enhancement available to commanders through technological advances.

CONCLUSION

The conceptual framework of how combat stress is understood has changed over the centuries. The ways in which the military has dealt with its service members suffering from combat stress has also changed. What has not changed is the impact that this inevitable cost of war has on the heroes that serve our country. At a minimum, we owe these brave men and women a return home and a future not plagued by emotional and psychological problems. However, the authors are not so naïve as to believe that these warriors will go completely unaffected by their experiences.

As has been previously noted, the resilience of the men and women serving in our Armed Forces is tremendous. The vast majority will reintegrate back into their civilian lives with relative ease. Unfortunately, history has shown us that thousands will not. Units like the CSC team make it possible for behavioral health providers to mitigate the lasting effects of combat exposure. Let us continue to acknowledge and honor those that have fallen on the battlefield. Let us also provide for those that are still among us—the walking wounded.

ACKNOWLEDGMENTS

The authors would like to thank Colonel (rtd) James Stokes, U.S. Army, and Lt. Colonel Steven Gerardi, U.S. Army, for their thoughtful and helpful suggestions during the development of this chapter.

REFERENCES

American Psychiatric Association. (1987). *Diagnostic and Statistical Manual of Mental Disorders* (3rd ed., rev.). Washington, DC: Author.

American Psychiatric Association. (2000). *Diagnostic and Statistical Manual of Mental Disorders* (4th ed., text rev.). Washington DC: Author.

Binneveld, H. (1998). *From Shell Shock to Combat Stress: A Comparative History of Military Psychiatry*. Amsterdam: Amsterdam University Press.

Da Costa, J. M. (1871). On irritable heart: A clinical study of a form of functional cardiac disorder and its consequences. *American Journal of the Medical Sciences, 61*, 17–52.

Department of Defense. (1999). *Combat Stress Control (CSC) Programs; DOD Directive 6490.5*. Washington, DC: Author.

Department of the Army. (1994a). *Leaders' Manual for Combat Stress Control; Field Manual 22-51*. Washington, DC: Author.

Department of the Army. (1994b). *Combat Stress Control in a Theater of Operations—Tactics, Techniques, Procedures; Field Manual 8-51*. Washington, DC: Author.

Department of the Army. (2000). *Combat Stress; Field Manual 6-22.5*. Washington, DC: Author.
Dicks, S. (1990). *From Vietnam to Hell: Interviews with Victims of Post-Traumatic Stress Disorder.* Jefferson, NC: McFarland & Company.
Dupuy, R. E. & Dupuy, T. N. (1993). *The Harper Encyclopedia of Military History: From 3500 B.C. to Present.* New York: HarperCollins.
Gilbert, M. (1994). *First World War: A Complete History.* New York: Henry Holt.
Hoge, C. W., Castro, C. A., Messer, S. C., McGurk, D., Cotting, D. I., & Koffman, R. L. (2004). Combat duty in Iraq and Afghanistan, mental health problems, and barriers to care. *New England Journal of Medicine, 351,* 13–22.
Jones, F. D. (1995). Psychiatric lessons of war. In F. D. Jones, L.R. Sparacino, V. L. Wilcox, & J. M. Rothberg (Eds.), *Textbook of Military Medicine* (pp. 1–33). Falls Church, VA: Office of The Surgeon General, U.S. Department of the Army.
Kulka, R. A., Schlenger, W. E., Fairbank, J. A., Hough, R. L., Jordan, B. K., Marmar, C. R., et al. (1990). *Trauma and the Vietnam War Generation: Report of Findings from the National Vietnam Veterans Readjustment Study.* New York: Brunner/Mazel.
Moore, B. A. (2005a). How psychiatrists are helping our troops. *Current Psychiatry, 4,* 3.
Moore, B. A. (2005b). The crucial roles of advanced practice psychiatric nurses during wartime. *Journal of the American Psychiatric Nurses Association, 11,* 52.
Moore, B. A. & Reger, G. M. (2006). Clinician to frontline soldier: A look at the roles and challenges of Army clinical psychologists in Iraq. *Journal of Clinical Psychology, 62,* 395–403.
Myers, A. B. R. (1870). *On the Etiology and Prevalence of Diseases of the Heart Among Soldiers.* London: John Churchill and Sons.
Reger, G. M. & Moore, B. A. (in press). Combat stress control in Iraq: Lessons learned during Operation Iraqi Freedom. *Military Psychology.*
Rivers, W. H. R. (1918). The repression of war experience. *Lancet, 194,* 173–177.
Rosen, G. (1975). Nostalgia: A forgotten psychological disorder. *Psychological Medicine, 5,* 340–354.
Salmon T. & Fenton N. (1929). *The Medical Department of the United States Army in the World War, v. 10: Neuropsychiatry in the American Expeditionary Forces.* Washington, DC: U.S. Government Printing Office.
Wooley, C. F. (2002). *The Irritable Heart of Soldiers and the Origins of Anglo-American Cardiology: The U.S. Civil War (1861) to World War I (1918).* Burlington, VT: Ashgate.

9

Virtual Reality Applications for the Treatment of Combat-Related PTSD[1]

ALBERT "SKIP" RIZZO, BARBARA O. ROTHBAUM,[2] AND KEN GRAAP

INTRODUCTION

Posttraumatic stress disorder (PTSD) is a chronic, debilitating, psychological condition that occurs in a subset of persons who experience or witness life-threatening traumatic events. PTSD is characterized by reexperiencing, avoidance, and hyperarousal symptoms that occur over time and lead to significant disruption in one's life (American Psychiatric Association, 1994). Historically, PTSD can be traced to at least the 17th century (Shalev and Rogel-Fuchs, 1993; Grinker and Spiegel, 1945) characterized "war neuroses," including tremors, fearful expressions, marked startle reactions, and "a child like appeal for help" (p. 5) in World War I and II veterans. This definition was originally codified into the *Diagnostic and Statistical Manual of Mental Disorders* in 1980 (DSM-III; American Psychiatric Association, 1980). Today, symptoms of PTSD are recognized in subsets of those persons who survive auto accidents, sexual assaults, terrorist attacks, natural disasters, wars, and in those first responders and medical professionals who care for these survivors during the immediate aftermath of the trauma.

This chapter deals with the use of virtual reality (VR) specifically in combat veterans and will look toward the day when VR is used in both evaluative and therapeutic modalities. In what follows, we briefly review our operational definition of VR and the theoretical basis for using VR in a cognitive behavioral PTSD treatment. This is followed by a review of the literature on VR treatment with Vietnam veterans and an explication of current research in Iraq or Afghanistan veterans.

VIRTUAL REALITY

Virtual reality offers a new human–computer interaction paradigm in which users are active participants within a computer-generated, three-dimensional virtual world. VR environments discussed here differ from traditionally displayed programs in that computer graphics are displayed in a head mounted display (HMD), and are augmented with motion tracking, vibration platforms, localizable 3D sounds within the VR space, and, in some scenarios, scent delivery technology to facilitate an immersive experience for participants. The immersive nature of the VR environments typically leads to a strong sense of *presence* reported by those immersed in the virtual environment.

VR exposure is intended to be a component of a treatment program administered by a qualified professional. VR exposure therapy (VRET) is employed at the point in therapy when exposure therapy would normally be introduced and has the advantages of extending the range of options available to a clinician and introducing a shared experience with the participant. Such a shared experience is for practical purposes impossible without VR. For example, it would be impossible to get clinicians on the battlefield with combat PTSD clients and it is currently impossible to share the clients' imagined scenes. VR offers both a variety of stimuli and a shared realistic experience without leaving the office. More control over exposure stimuli, an ability to repeat needed exposures, opportunities to monitor clients' responses in multiple domains, and less exposure of the client to possible harm or embarrassment are other advantages of using VR for exposure therapy.

RATIONALE FOR TREATMENT OF PTSD IN VR

PTSD is a severe, and often chronic, disabling anxiety disorder, which develops in some persons following exposure to a traumatic event that involves actual or threatened injury to themselves or to others. Prospective studies indicate that most traumatized individuals experience symptoms of PTSD in the immediate aftermath of the trauma. In a prospective study of rape victims, 94% met symptom criteria for PTSD in the first week following the assault (Rothbaum et al., 1992). Therefore, the symptoms of PTSD are part of the *normal reaction* to trauma. The majority of trauma victims naturally recover as indicated by a gradual decrease in PTSD symptoms severity over time. However, subsets of persons continue to exhibit severe PTSD symptoms long after a traumatic experience. Therefore, PTSD can be viewed as a failure of natural recovery that reflects in part a failure of fear extinction following trauma.

Consequently, several theorists have proposed that conditioning processes are involved in the etiology and maintenance of PTSD. These theorists invoke Mowrer's (1960) two-factor theory, which posits that both Pavlovian and instrumental conditioning are involved in the acquisition of fear and avoidance behavior. Through a generalization process many stimuli may elicit fear and avoidance. Consistent with this hypothesis, emotional and physiological reactivity to stimuli resembling the original traumatic event, even years after the event's occurrence, is a prominent characteristic of PTSD and has been reliably replicated in the laboratory (e.g., Blanchard et al., 1986; Pitman et al., 1987). Further, cognitive and behavioral avoidance strategies are hypothesized to develop in an attempt to avoid or escape these distressing conditioned emotional reactions. The presence of extensive avoidance responses can also interfere with extinction by limiting the amount of exposure to the conditioned stimulus in the absence of the unconditioned stimulus.

Conceptualizing PTSD within the framework of emotional processing theory, Foa, Steketee, and Rothbaum (1989) suggested that the traumatic memory could be conceived as a mental fear structure comprising a network of information about the feared stimuli; information about verbal, physiological, and overt behavioral responses; and interpretative information about the meaning of the various stimuli and responses contained in the network. Foa and Kozak (1986) suggested that two conditions are required for the reduction of fear. First, the fear memory must be activated. That is, as suggested by Lang (1977), if the fear structure remains in storage and unaccessed, it will not be available for modification. Second, information must be provided which includes elements "incompatible with some of those that exist in the fear structure, so that a new memory can be formed. This new information, which is at once cognitive and affective, has to be integrated into the evoked information structure for an emotional change to occur" (p. 22). Cognitive and behavioral therapy utilizing VR aims to reduce fear presumably by first activating the fear structure and then through the therapeutic process modifying it. VR exposure therapy has been shown to be effective at accessing the fear structure as evidenced by emotional responses in participants across a wide range of studies.

VRET refers to several behavioral and cognitive behavioral treatment techniques that involve exposure to feared stimuli (e.g., thoughts, images, objects, situations, or activities) in order to reduce pathological (unrealistic) fear, anxiety, and anxiety disorder symptoms. In the treatment of PTSD, exposure therapy usually involves prolonged, imaginal exposure to the client's memory of the trauma and in vivo exposure to various reminders of the trauma. This approach is believed to provide a context in which one can begin to therapeutically process the emotions that are relevant to the trauma as well as to provide extinction training, allowing the fearful symptoms to decrease even in association with the feared stimuli. While the efficacy of imaginal exposure has been established in multiple studies with diverse trauma populations (Rothbaum & Schwartz, 2002; Rothbaum et al., 2000), many clients are unwilling or unable to effectively visualize the traumatic event. In fact, avoidance of reminders of the trauma is inherent in PTSD and is one of the defining symptoms of the disorder. It is often reported that "some patients refuse to engage in the treatment, and others, though they express willingness, are unable to engage their emotions or senses" (Difede & Hoffman, 2002, p. 529). Research on this aspect of PTSD treatment suggests that the inability to emotionally engage (in imagination) is a predictor for negative treatment outcomes (Jaycox, Foa, & Morral, 1998). This idea was supported by three studies in which clients with PTSD were unresponsive to previous *imaginal* exposure treatments but went on to respond to VR exposure therapy (Difede & Hoffman, 2002; Rothbaum et al., 2001). As well, VR provides an objective and consistent format for documenting the sensory stimuli that the client is exposed to that is not possible when operating within the unseen world of the client's imagination.

During a course of treatment using prolonged exposure, typically four treatment components are administered over 9 to 12 sessions lasting 90 to 120 minutes each: (1) psychoeducation about the symptoms of PTSD and factors that maintain PTSD and the rationale for exposure therapy; (2) training in controlled breathing or other stress reduction techniques that clients may use as a stress management skill, though it should be noted that clients are discouraged from using it during exposure exercises; (3) prolonged imaginal exposure to the trauma memory conducted in therapy sessions and repeated as homework; and (4) prolonged in vivo exposure implemented as homework. There is substantial evidence that exposure programs are highly effective in the treatment of PTSD. There is no compelling evidence that any cognitive behavioral therapy program is more effective than exposure therapy, and no evidence for the usefulness of adding other components to exposure therapy (Foa, Rothbaum, & Furr, 2003).

In summary, there is evidence that symptoms of PTSD are present within a very short period after exposure for trauma and may in fact be part of the normal coping process. However, a minority of individuals develops a chronic disorder that interferes with functioning. There is strong evidence for exposure therapy in the treatment of PTSD and a coherent theoretical model that has been tested clinically and has suggested efficacious approaches to the treatment of PTSD. Such approaches have been applied using VR environments controlled by therapists to accompany the imaginal exposure in clients with PTSD.

VIRTUAL REALITY EXPOSURE THERAPY FOR COMBAT-RELATED PTSD

The application and value of virtual reality for the treatment of cognitive, emotional, psychological, and physical disorders has been well specified (Glantz, Rizzo, & Graap, 2003; Rizzo et al., 2004), and a number of controlled studies over the last 10 years have documented its clinical efficacy as an exposure therapy treatment for anxiety disorders (Wiederhold & Wiederhold, 2004). The first use of VR for a Vietnam veteran with PTSD was reported in a case study of a 50-year-old, Caucasian male veteran meeting DSM-IV criteria for PTSD (Rothbaum et al., 1999). Results indicated posttreatment improvement on all measures of PTSD and maintenance of these gains at 6-month follow-up. Examples of stimuli from these environments are included in Figures 9.1 to 9.4.

This case study was followed by an open clinical trial of VR for Vietnam veterans (Rothbaum et al., 2001). In this study, 16 male PTSD clients were exposed to two HMD-delivered virtual environments, a virtual clearing surrounded by jungle scenery, and a virtual Huey helicopter, in which the therapist controlled various visual and auditory effects (e.g., rockets, explosions, day/night, yelling). After an average of 13 exposure therapy sessions over 5 to 7 weeks, there was a significant reduction in PTSD and related symptoms (see Table 9.1).

FIGURES 9.1 and 9.2 The landing zone clearing in the Virtual Vietnam Scenario.

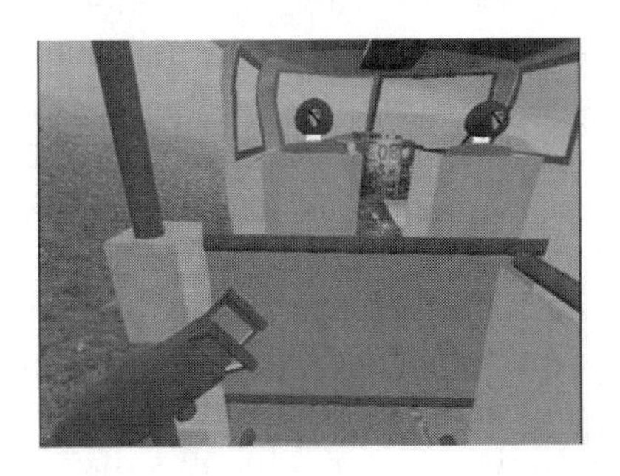

FIGURES 9.3 and 9.4 The view inside the virtual helicopter in Virtual Vietnam.

TABLE 9.1
Pre- and Posttreatment, and 3- and 6-month Follow-up Means (SD)

Measure	Baseline (N = 9)	Post-Tx (n = 9)	3-Mo FU (n = 5)	6-Mo FU (n = 8)
CAPS Total Score(%) Decrease Range	68.00 (15.26)	57.78 (20.61) p = 0.0727 –15% +41% to –38%	54.6 (17.47) p = 0.0256* –27% –13% to –48%	47.12 (17.04) p = 0.0021* –31% –15% to –67%
CAPS Cluster B Reexperiencing	16.33 (6.06)	13.89 (6.33) p = 0.2812	9.40 (6.99) p = 0.0231*	11.12 (4.45) p = 0.0103
CAPS Cluster C Avoidance	28.22 (8.18)	24.78 (10.74) p = 0.2814	23.20 (7.33) p = 0.0507	17.25 (9.35) p = 0.0116*
CAPS Cluster D Arousal	23.44 (4.47)	19.11 (8.91) p = 0.1163	22.00 (4.69) p = 0.0777	18.75 (5.31) p = 0.0021*
IES Total Score	42.89 (10.20)	36.11 (21.64) p = 0.3988	19.4 (14.71) p = 0.0327*	29.88 (19.39) p = 0.0912
IES Intrusion	20.33 (6.10)	16.11 (8.56) p = 0.2126	8.00 (9.07) p = 0.0135*	13.88 (10.48) p = 0.0949
IES Avoidance	22.55 (7.88)	20.00 (15.43) p = 0.6259	11.40 (5.86) p = 0.1585	16.00 (10.61) p = 0.1412
Beck Depression Inventory	26.11 (11.36)	21.77 (10.12) p = 0.09	25.6 (12.28) p = 0.38	17.85 (11.01) p = 0.01*

* = significant < .05.
CAPS = Clinician Administered PTSD Scale; IES = Impact of Events Scale; Mo FU = months follow-up; SD = standard deviation; Tx = treatment.

After VRET, the majority of clients' ratings of their global improvement indicated improvement. At 6 months, 6 of 8 reported improvement. Clinician's ratings of clients' global improvement as measured by the Clinical Global Improvement Scale indicated that 5 of 6 showed improvement immediately after the study while one appeared unchanged. At 6 months, 7 of 8 were rated as demonstrating some improvement. Clinician-rated PTSD symptoms as measured by the Clinician Administered PTSD Scale, the primary outcome measure, at 6-month follow-up indicated an overall statistically significant reduction from baseline in symptoms associated with specific reported traumatic experiences. Eight of 8 participants at the 6-month follow-up reported reductions in PTSD symptoms ranging from 15% to 67%. Significant decreases were seen in all three symptom clusters. Client self-reported intrusion and avoidance symptoms as measured by the Impact of Events Scale were significantly lower at 3 months than at baseline but not at 6 months, although there was a clear trend toward fewer intrusive thoughts and somewhat less avoidance.

The authors concluded that VRET led to significant reductions in PTSD and related symptoms and was well tolerated. No person decompensated due to exposure to the VREs. No participant was hospitalized during the study for complications related to the treatment. Most of those who dropped out of the study were provided opportunities for other treatment within the PTSD Clinical Team clinic at the Atlanta VA Medical Center and did not appear to suffer any long-term problems attributable to their participation. This preliminary evidence suggested that VRET was a promising component of a comprehensive treatment approach for veterans with combat-related PTSD.

Positive findings in the study of Vietnam veterans has led other groups to propose VR environments to facilitate PTSD treatment in civilians. For example, subsequent to the September 11, 2001 terrorist attacks on the World Trade Center in New York, Difede and Hoffman (2002) constructed a scenario in which civilians, firefighters, and police officers with PTSD could be exposed to relevant events in VR. In their first report, a case study was presented using VR to provide exposure to the trauma memory with a client who had failed to improve with traditional exposure therapy. The authors reported significant reduction of PTSD symptoms after repeatedly exposing the client to explosions, sound effects, virtual people jumping from the burning buildings, towers collapsing, and dust clouds, and attributed this success partly because of the increased realism of the VR images as compared with the mental images the client could generate in imagination. Positive VR treatment results from a wait-list controlled study with clients who were not successful in previous imaginal therapy were submitted for publication by this group (Joanne Difede, personal communication, December 13, 2005). These early results suggest that VR is a useful technology to apply for the treatment of PTSD and that it may be a promising component of a comprehensive treatment approach for persons with combat-related PTSD.

Two other groups have begun the development and initial user-centered testing of VR scenarios to treat PTSD in survivors of war and terrorist attacks (Gamito et al., 2005; Josman et al., 2005). In Portugal, there are an estimated 25,000 survivors with PTSD from their 1961–1974 wars in Mozambique, Angola, and Guiné (Gamito et al., 2005). This research group has constructed a single VR "ambush" scenario by modifying a common PC-based combat game. They reported having recently conducted an initial user-centered test with one PTSD client who has provided feedback, suggesting the need for the construction of a system that provides more graduated delivery of anxiety-provoking trigger stimuli. In Israel, Josman et al. (2005) implemented a terrorist "bus bombing" PTSD treatment scenario in which the client is positioned in an urban cafe across the street from the site where a civilian bus may explode. The system controls allow the client to sit in the outdoor cafe and be exposed to a range of progressive conditions—from the street being empty with no bus or sound effects, to the bus passing in an uneventful manner with or without sound, to the bus arriving and exploding with full sound effects. Clinical tests have recently commenced but findings are not available at the time of this writing.

DESIGN AND DEVELOPMENT OF THE FULL-SPECTRUM VIRTUAL IRAQ OR AFGHANISTAN PTSD THERAPY APPLICATION

In 1997, researchers at Georgia Tech released the first version of the Virtual Vietnam VR scenario for use as an exposure therapy tool for treating PTSD in Vietnam veterans. This occurred over 20 years following the end of the Vietnam War. During those intervening 20 years, despite valiant efforts to develop and apply traditional psychotherapeutic approaches to PTSD, the progression of the disorder in some veterans significantly impacted their psychological well-being, functional abilities, and quality of life, as well as that of their family members and friends. The tragic nature of this disorder also had significant ramifications for the U.S. Department of Veteran Affairs healthcare delivery system, often leading to designations of lifelong service-connected disability status among those diagnosed with PTSD. The Virtual Vietnam scenario landmarked the first time that VR was applied to the treatment of PTSD and this initial effort produced encouraging results.

In the early 21st century the conflicts in Iraq and Afghanistan again drew U.S. military personnel into combat. In the first systematic study of mental health problems due to these conflicts, Hoge et al. (2004) reported that "[t]he percentage of study subjects whose responses met the screening criteria for major depression, generalized anxiety, or PTSD was significantly higher after duty in Iraq (15.6 to 17.1 percent) than after duty in Afghanistan (11.2 percent) or before deployment to Iraq (9.3 percent)" (p. 13). With this history in mind, the University of Southern California Institute for Creative Technologies (ICT) and Virtually Better, Inc. (VB) have initiated a project that is creating an immersive virtual environment system for the treatment of veterans from the Iraq and Afghanistan Wars diagnosed with combat-related PTSD. The U.S. Office of Naval Research has now funded this project as part of a larger multiyear effort. The VR treatment environment is being created from a cost-effective approach to recycling virtual graphic assets that were initially built for the U.S. Army–funded combat tactical simulation scenario entitled *Full Spectrum Command*, which later inspired the creation of the commercially successful Xbox game, *Full Spectrum Warrior*. Increasingly, the military has been able to take advantage of simulation technology, primarily for training soldiers. Such software is often referred to as mission rehearsal simulations, and the USC ICT has been at the forefront of constructing such software since the late 1990s. The presence of expertise in designing combat simulations, the graphics technology adapted from the Xbox game, and the collaboration with VB has led to an opportunity to once again apply

VR to combat-related PTSD, albeit this time within a tighter timeframe than the technology allowed for Vietnam-era PTSD.

TECHNICAL BACKGROUND AND DEVELOPMENT HISTORY

One primary aim of the current project is to use the already existing ICT Full Spectrum Warrior graphic assets as the basis for creating a clinical VR application for the treatment of PTSD in returning Iraq or Afghanistan War military personnel (see Web site ftp://ftp.ict.usc.edu/arizzo/PTSD%20Materials/ for video demos of the content). The ICT Games Project has created two training tools for the U.S. Army to teach leadership and decision-making skills. Full Spectrum Command is a PC application that simulates the experience of commanding a light infantry company. It teaches resource management, adaptive thinking, and tactical decision making. Full Spectrum Warrior, developed for the Xbox game console, puts the trainee in command of a nine-person squad. Trainees learn small unit tactics as they direct fire teams through a variety of immersive urban combat scenarios. These tools were developed through collaboration between ICT, entertainment software companies, the U.S. Army Training and Doctrine Command, and the Research, Development, and Engineering Command, Simulation Technology Center. In addition, subject matter experts from the Army's Infantry School contributed to the design of these training tools.

The current VR PTSD application is designed to run on two Pentium 4 notebook computers each with 1 GB RAM, and a 128 MB DirectX 9 compatible graphics card. The two computers are linked using a null Ethernet cable. One notebook runs the therapist's control application while the second notebook drives the user's head mounted display and orientation tracker. We are exploring the usability of three different HMDs for use in this application, aiming to find the best instrument available to deliver this treatment at the lowest cost. This goal is important in order to promote maximum accessibility to this system in the future. The three HMDs being tested for this purpose are: (1) the 5DT HMD 800, capable of 800x600 (SVGA) resolution (for specs see http://www.5dt.com/products/phmd.html); (2) the Icuiti v920 HMD, capable of 640x480 (VGA) resolution (for specs see http://www.icuiti.com); and (3) the eMagin OLED z800 HMD, capable of 800x600 (SVGA) resolution (for specs see http://www.emagin.com). The Intersense InertiaCube2 tracker is being used for three degrees of freedom (pitch, roll, and yaw) head orientation tracking, and the user navigates through the scenario using a USB gamepad device. It should also be noted that while we believe that the HMD display approach will provide the optimal level of immersion and interaction characteristics for this application, the system is fully configurable to be delivered on a standard PC monitor or within a large screen projection display format. The application is built on ICT's FlatWorld Simulation Control Architecture (FSCA).

The FSCA enables a network-centric system of client displays driven by a single controller application. The controller application broadcasts user-triggered or scripted event data to the display client. The client's real-time 3D scenes are presented using Numerical Design Limited's Gamebryo graphics engine. The content originally used in Full Spectrum Warrior was edited and exported to the engine using Alias' Maya software.

Olfactory and tactile stimuli are also being added into the experience of the virtual Iraq environment. The olfactory stimuli are delivered via the ES-1 Scent Machine®, developed by Envirodine Studios, Inc. (http://www.envirodine.com). The ES-1 Scent Machine® is a computer peripheral, USB device that uses up to eight scent cartridges, a series of fans, and a small air compressor to deliver scents to participants. The scents can be computer controlled by placing triggers into the VR programs (e.g., participant walks by a fire and smells smoke), delivered via key press by the clinician, or simply turned off to decrease sensory impact within the virtual environment. Scent is activated in this application by triggers programmed into the environment via the ICT FSCA (Pair et al., 2006). Scents may be used as direct stimuli (e.g., scent of burning rubber) or as general cues to help immerse persons in the world (e.g., ethnic food cooking). This allows for the simultaneous delivery of these stimuli with visual and audio events to create a more realistic multimodal experience for the client in order to enhance the sense of presence in the environment. The amount of scent to be released is specified in seconds. For example, one could deliver a 1 second burst of a concentrated but subtle scent of a flower garden when in passing. Conversely, the machine could be programmed to deliver a longer burst of scent such as might be experienced when approaching someone wearing cologne. The scents are concentrated and gelled much like an air freshener cartridge and enclosed within the Scent Palette in an airtight chamber that fills with compressed air. When activated, the scent is released into an airstream from four electric fans so that it moves past the user and dissipates into the volume of the room. The scents that have been selected for this application thus far include burning rubber, cordite, garbage, body odor, smoke, diesel fuel, Iraqi spices, and gunpowder. Scent has been shown to be related to emotional responding, and it is believed that the addition of scent will allow clinicians a greater range of options for manipulating the realism within the virtual environment.

The addition of tactile input in the form of vibration is designed to add another sensory modality to the virtual environment, again to enhance presence. Vibration is obtained through the use of a Logitech Force-Feedback Gamepad and from sound transducers (Aura Bass Shakers, Aura Sound, Inc., http://www.aurasystems.com) located beneath the client's floor platform, driven by an audio amplifier. The sound files embedded in the software are customized to provide vibration consistent with relevant visual and audio stimuli in the scenario. For example, explosions and gunfire can be accompanied by this additive sensation and the vibration can also be varied as when a virtual vehicle moves across seemingly uneven ground.

CLINICAL APPLICATION CONTROL OPTIONS: SCENARIO SETTINGS, USER PERSPECTIVES, TRIGGER STIMULI, AND THE CLINICAL INTERFACE

Prior to acquiring the funding required to create a comprehensive VR application to address a wide range of possible combat-related PTSD experiences, the USC ICT created a prototype virtual environment designed to resemble a Middle Eastern city (see Figure 9.6). This virtual environment was designed as a proof of concept demonstrator and as a tool for initial user testing to gather feedback from both Iraq War military personnel and clinical professionals. This feedback has been used to refine the city scenario and to drive development of other relevant scenario settings. Current Office of Naval Research funding has now allowed us to evolve this existing prototype into a full-featured version 1.3 application that is currently undergoing user-centered design feedback trials at the Naval Medical Center, San Diego, with non-PTSD soldiers who have returned from an Iraq tour of duty. As well, user-centered feedback is also being collected on this version within an Army Combat Stress Control Team in Iraq (see Figure 9.7). The vision for the project includes not only the design of a series of diverse *scenario settings* (e.g., city, outlying village, and desert convoy scenes), but, as well, the creation of options for providing the user with different first-person *user perspective options*. These choice options when combined with real-time clinician input via the "Wizard of Oz" *clinical interface* are envisioned to allow for the creation of a user experience that is specifically customized to the varied needs of clients who participate in treatment. This is an essential component for giving the therapist the capacity to modulate client anxiety as is required for an exposure therapy approach. Such experience customization and real-time stimulus delivery flexibility are key elements for these types of VR exposure applications.

Scenario Settings

The software has been designed such that clients can be teleported to specific scenario settings based on a determination as to which environments most closely match the client's needs, relevant to their individual combat-related experiences. All scenario settings are adjustable for time of day or night, weather conditions, and lighting illumination. The following are the scenario settings being created for the application:

1. **City Scenarios**—In this setting, we create two variations. The first city setting (currently developed in our prototype version 1.2) has the appearance of a desolate set of low populated streets with old buildings, ramshackle apartments, warehouses, a mosque, factories, and junkyards (Figures 9.5, 9.6, 9.8, 9.9, 9.10). The second city setting has similar street characteristics and buildings, but is more highly populated and has more traffic activity, marketplace scenes, and monuments.

FIGURE 9.5 City scenario.

FIGURE 9.6 City scenario.

FIGURE 9.7 User-centered feedback session in Iraq.

FIGURE 9.8 "Flocking" companion.

FIGURE 9.9 "Flocking" patrol.

2. **Checkpoint**—This area of the City Scenario is constructed to resemble a traffic checkpoint with a variety of moving vehicles arriving, stopping, and then moving onward into the city.
3. **City Building Interiors**—Some of the City Scenario buildings have interiors modeled allowing the client to navigate through them. These interiors will have the option of being vacant (Figure 9.10) or inhabited by various numbers and types of virtual human characters.

FIGURE 9.10 Interior view.

4. **Small Rural Village**—This setting consists of a more spread-out rural area containing ramshackle structures, a village center, and much decay in the form of garbage, junk, and wrecked or battle-damaged vehicles. It also contains more vegetation and has a view of a desert landscape in the distance that is visible as the user passes by gaps between structures near the periphery of the village.
5. **Desert Base**—This scenario is designed to appear as a desert military base of operations consisting of tents, soldiers, and an array of military hardware.
6. **Desert Road Convoy**—This scenario consists of a paved roadway that will eventually connect the City, Village, and Desert Base scenarios. The view from the road currently consists of desert scenery and sand dunes (Figures 9.11, 9.13, 9.14) with occasional areas of vegetation, ramshackle structures, battle wreckage, debris and an occasional virtual human figure standing by the side of the road.

FIGURE 9.11 Desert road scenario.

FIGURE 9.12 City Humvee view.

User Perspective Options

The VR system is designed such that once the scenario is chosen, it is possible to select from a variety of user perspective and navigation options. These designed to provide flexibility in how the interaction in the scenario can be customized to suit the client's needs.

User perspective options in the final system will include:

1. Client walking alone on patrol from a first-person perspective (Figures 9.5, 9.6, 9.10).
2. Client walking with one soldier companion on patrol. The accompanying soldier will be animated with a "flocking" algorithm that will place him always within a 5-meter radius of the client and will adjust position based on collision detection with objects and structures to support a perception of realistic movement (Figure 9.8).
3. Client walking with a patrol consisting of a number of companion soldiers using a similar "flocking" approach as in #2 above (Figure 9.9). These flocking options are under development and will be integrated during year 2 of this project.
4. Client view from the perspective of being either inside of the cab of a Humvee or other moving vehicle or from a more exposed position in a gun turret above the roof of the vehicle. Options are provided for automated travel as a passenger through the various setting scenarios (Figures 9.11–9.14), or at the driving column that allow for user control of the vehicle via the gamepad controls or with an actual steering wheel and pedal controls. The interior view also has options for other occupant passengers that has ambient movement. This view is also adjustable to support the perception of travel within a convoy or as a lone vehicle.
5. Client view from the perspective of being in a helicopter hovering above or moving over any of the scenario settings (Figure 9.15).

In each of these user perspective options, we are considering the wisdom of having the client possess a weapon. This will necessitate decisions

FIGURE 9.13 Desert Humvee view.

FIGURE 9.14 Desert Humvee view.

FIGURE 9.15 Helicopter view.

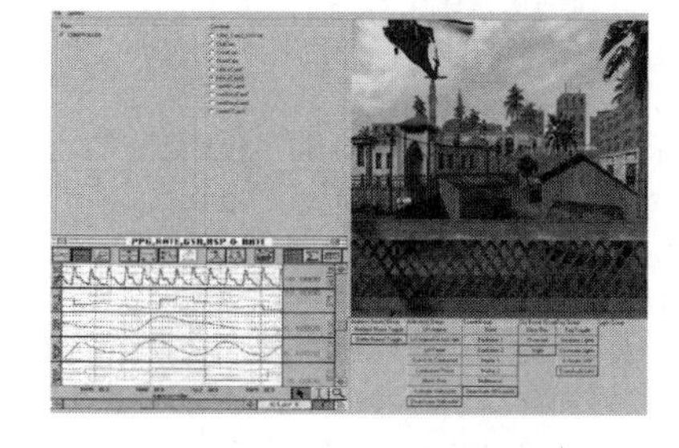

FIGURE 9.16 Clinical interface.

as to whether the weapon will be usable to return fire when it is determined by the clinician that this would be a relevant component for the therapeutic process. Those decisions will be made based on the initial user and clinician feedback from the version 1.2 application.

Trigger Stimuli

The specification, creation, and addition of trigger stimuli will likely be an evolving process throughout the life of the application, based on relevant client and clinician feedback. We began this part of the design process by including options that have been reported as relevant by returning soldiers and military subject matter experts. For example, Hoge et al. (2004), in their study of self-reported anxiety, depression, and PTSD-related symptomatology in returning Iraq War veterans, present a useful listing of combat-related events that were commonly experienced in their sample. These events provided a useful starting point for conceptualizing how relevant trigger stimuli could be presented in a VR environment. Such commonly reported events included: "Being attacked or ambushed, Receiving incoming artillery, rocket, or mortar fire, Being shot at or receiving small-arms fire, Shooting or directing fire at the enemy, Being responsible for the death of an enemy combatant, Being responsible for the death of a noncombatant, Seeing dead bodies or human remains, Handling or uncovering human remains, Seeing dead or seriously injured Americans, Knowing someone seriously injured or killed, Participating in demining operations, Seeing ill or injured women or children whom you were unable to help, Being wounded or injured, Had a close call, was shot or hit, but protective gear saved you, Had a buddy shot or hit who was near you, Clearing or searching homes or buildings, Engaging in hand-to-hand combat, Saved the life of a soldier or civilian" (p. 18). From this and other sources, we have begun our initial effort to conceptualize what is both functionally relevant and pragmatically possible to include as trigger stimuli in the virtual environment.

The current design envisions four classes of triggers: (1) visual (e.g., animate and inanimate views of vehicles, persons, bodies, markets, and other soldiers); (2) auditory (e.g., yelling, weapons sounds, explosions, ambient city sounds, ambient weather sounds); (3) vibrotactile (e.g., riding over rough terrain, explosions, weapons fire); (4) olfactory (e.g., body odor, Middle Eastern spices, and cordite). These stimuli can be provided alone or in combinations which allow the clinician to control the level of exposure presented to the client via a clinical interface panel. The creation of more complex events that can be intuitively delivered from a clinician's interface while providing a client with options to interact or respond in a meaningful manner is one of the ongoing focuses in this project. Thus far in the version 1.3 prototype, a variety of auditory trigger stimuli have been created (e.g., incoming mortars, weapons fire, voices, wind) that can be actuated by the clinician via mouse clicks on a clinical interface. We can also similarly trigger dynamic audiovisual events such as helicopter flyovers above the user's position and verbal orders from a commanding officer gesturing in an excited manner. Clinicians remain in contact with clients via a microphone into the environment. The creation of complex, multimodal stimulus cues, in combination with the actual experiences of the participants, allows immersion to be facilitated by a skilled clinician. Perhaps it may be of value to actually immerse the client in varying degrees of combat in which they may see members of their patrol (or themselves) get wounded or in fact have the capability to fire a weapon back at enemy combatants. One can imagine varying degrees of exposure culminating in the most traumatic event occurring again, but with a dramatically different outcome. However, such trigger options will require not only interface design expertise, but also clinical wisdom as to how much and what type of exposure is needed to produce a positive clinical effect. These issues will be keenly attended to in our initial clinical trials.

The Clinical Interface

To deliver and control all the above features in the system, a "Wizard of Oz" type clinical interface was created (Figure 9.16). This interface is a key element in the application, as it needs to provide a clinician with a usable tool for selecting and placing the client in VR scenario locations that resemble the contexts that are clinically relevant for a graduated exposure approach. As important, the clinical interface must also allow the clinician to further customize the therapy experience to the client's individual needs via the systematic real-time delivery and control of "trigger" stimuli in the environment. This is essential for fostering the anxiety modulation needed for therapeutic habituation.

In our initial configuration, the clinician uses a separate computer monitor/mouse or tablet laptop to display and actuate the clinical interface controls. While the results from our initial user feedback trials are currently guiding the interface design, our initial candidate setup provides four quadrants in which the clinician can monitor ongoing user status information, while simultaneously directing trigger stimulus delivery. The upper left quadrant contains basic interface menu buttons used for placement of the client (and immediate removal if needed) in the appropriate scenario setting and user perspective. This quadrant also contains menu keys for the control of time of day or night, atmospheric illumination, weather conditions, and initial ambient sound characteristics. The lower left quadrant will provide space for real-time display of the clients' heart rate and galvanic skin response readings for monitoring of physiological status when that feature is integrated. The upper right quadrant contains a window that displays the imagery present in the user's field of view in real time. And the lower right quadrant contains the control panel for the real-time delivery of specific trigger stimuli that are actuated by the clinician in an effort to modulate appropriate levels of anxiety as required by the theory and methodology of exposure-based therapy. The overall design of the system is such that once the scenario setting is selected, the clinician can then adjust the time of day, weather options, ambient sounds, scent and vibration configurations, and user perspective. Once these options are selected, the client can experience this customized environment setting while the clinician then may focus on the judicious delivery of trigger stimuli. These interface options have been designed, with the aid of feedback from clinicians, with the goal of providing a usable and flexible control system for conducting thoughtfully administered exposure therapy that can be readily customized to suit the needs of the client.

CONCLUSIONS

War is perhaps one of the most challenging situations that a human being can experience. The physical, emotional, cognitive, and psychological demands of a combat environment place enormous stress on even the best-prepared military personnel. Such stressful experiences that commonly occur in warfighting environments have a considerable likelihood for producing significant numbers of returning soldiers at risk for developing PTSD. The initial data coming from both survey studies and anecdotal observations suggest that a large population of returning soldiers from the Iraq and Afghanistan conflicts are in fact reporting symptoms congruent with the diagnosis of PTSD. It is our view that this situation requires our best efforts to find ways to maximize treatment access and efficacy, and VR is a logical and attractive medium to use to address these aims.

Continuing advances in VR technology, along with concomitant system cost reductions, have supported the development of more usable, useful, and accessible VR systems that can uniquely target a wide range of psychological disorders (Rizzo & Kim, 2005). The unique match between VR technology assets and the needs of various clinical treatment approaches has been recognized by a number of authors, and an encouraging body of research has emerged, particularly in the area of exposure therapy for anxiety disorders (Glantz, Rizzo, & Graap, 2003; Rizzo et al., 2004; Zimand et al., 2003). As well, a growing body of research has suggested that VR is a powerful medium through which a professional may extend the clinical options available to treat clients with PTSD. Technological advances have made the presentation of compelling multisensory experiences in VR a reality, and the use of such tools is being investigated in several ongoing trials for survivors of the Vietnam, Iraq, and Afghanistan conflicts. VR tools are also improving with advances in video game technology in the areas of graphics processing, development software, interface tools, and some emerging developments in display technology. Essentially many high-end digital game scenarios are, in fact, well-done virtual environments and this emerging reality has helped to accelerate the development and quality of our VR PTSD system. Although cost factors limit the creation of custom virtual environments specific to the unique experiences of every person, it is possible to construct flexible archetypic VR worlds for groups of clients who have survived traumatic situations that lend themselves to abstraction and some degree of commonality. Examples considered in this chapter include Vietnam (helicopter and landing zone scenarios), the World Trade Center (street view of the terrorist attack), terrorist bus bombings (street view in Israel), and a range of scenes in VR Iraq or Afghanistan with six general scenario settings.

One of the more foreboding findings in the recent Hoge et al. (2004) report was the observation that, among Iraq and Afghanistan veterans, "those whose responses were positive for a mental disorder, only 23 to 40 percent sought mental health care. Those whose responses were positive for a mental disorder were twice as likely as those whose responses were negative to report concern about possible stigmatization and other barriers to seeking mental health care" (p. 13). While military training methodology has better prepared soldiers for combat in recent years, such hesitancy to seek treatment for difficulties that emerge upon return from combat, especially by those who may need it most, suggests an area of military mental healthcare in need of attention. To address this concern, perhaps a VR system for PTSD treatment could serve as a component within a reconceptualized approach to how treatment is accessed by veterans returning from combat.

One option would be to integrate VR-delivered combat exposure as part of a comprehensive "assessment" program administered upon return from a tour of duty. Since past research is suggestive of differential patterns of physiological reactivity in soldiers with PTSD when exposed to combat-related stimuli (Keane et al., 1998; Laor et al., 1998), an initial procedure that integrates a VR PTSD application with psychophysiological monitoring could be of value. If indicators of such physiological reactivity are present during an initial VR exposure, a referral for continued assessment and/or care could be negotiated and/or suggested. This could be provided in a format whereby the perceived stigma of independently seeking treatment could be lessened as the soldier would be simply involved in some form of "noncombat reintegration training" in a similar fashion to other designated duties to which they would participate.

VR PTSD therapy may also offer an additional attraction and promote treatment seeking by certain demographic groups in need of care. The current generation of young military personnel, having grown up with digital gaming technology, may actually be more attracted to and comfortable with participation in a VR application approach as an alternative to what is viewed as traditional "talk therapy" (even though such talk therapy would obviously occur in the course of a recommended multicomponent approach for this disorder). The potential for a reduction in the perceived stigma surrounding treatment has been anecdotally reported by practitioners who use VR to treat civilians with aerophobia (fear of flying) (Wiederhold & Wiederhold, 2004). These observations indicate that some clients have reported that prior to treatment, they had "just lived with the problem" and never considered seeking professional treatment. Upon hearing of VR therapy for fear of flying, often via popular media reports, they then sought out VR exposure treatment, typically with resulting positive outcomes.

In addition to the ethical factors that make an unequivocal case for the importance of exploring new options for assessment and treatment of combat-related PTSD, economic drivers for the Department of Veterans Affairs healthcare system and the military also provide incentives for investigating novel approaches in this area. As of September 2004, there were 13,524 Gulf War veterans receiving compensation for PTSD from the Department of Veterans Affairs (VA Fact Sheet, 12/2004). In addition to the direct costs for benefit compensation, medical care usage by persons with PTSD is estimated to be 60% higher than average (Marshall et al., 2000), and lost income-based tax revenues raise the "hidden" costs even higher. These figures make the initial development and continuing infrastructure costs for running PC-based VR systems pale by comparison. The military could also benefit economically by way of reduced turnover of soldiers with mild PTSD. These personnel might be more likely to reenlist if their mental health needs were addressed soon after combat in a progressive manner via earlier VR assessment and treatment. As well, such a VR tool initially developed for exposure therapy purposes offers the potential to be "recycled" for use both in the areas of combat readiness assessment and for stress inoculation. Both of these approaches could provide measures of who might be better prepared for the emotional stress of combat. For example, novice soldiers could be pre-exposed to challenging VR combat stress scenarios delivered via hybrid VR/real world stress inoculation training protocols as has been reported by Wiederhold and Wiederhold (2005) with combat medics.

Finally, one of the guiding principles in our development work concerns how VR can *extend* the skills of a well-trained clinician. This VR approach is not intended to be an automated treatment protocol that could be administered in a "self-help" format. The presentation of such emotionally evocative VR combat-related scenarios, while providing treatment options not possible until recently, will most likely produce therapeutic benefits when administered within the context of appropriate care via a thoughtful professional appreciation of the complexity and impact of this disorder.

ACKNOWLEDGMENT

Some of the projects described here have been sponsored by the U.S. Army Research, Development, and Engineering Command. Statements and opinions expressed do not necessarily reflect the position or the policy of the United States Government, and no official endorsement should be inferred.

REFERENCES

American Psychiatric Association. (1980). *Diagnostic and Statistical Manual of Mental Disorders* (3rd ed). Washington, DC: Author.

American Psychiatric Association. (1994). *Diagnostic and Statistical Manual of Mental Disorders* (4th ed). Washington, DC: Author.

Blanchard, E. B., Kolb, L. C., Gerardi, R. J., Ryan, D., & Pallmeyer, T. P. (1986). Cardiac response to relevant stimuli as an adjunctive tool for diagnosing post-traumatic stress disorder in Vietnam veterans. *Behavior Therapy, 17,* 592–606.

Difede, J., & Hoffman, H. (2002). Virtual reality exposure therapy for World Trade Center post traumatic stress disorder. *Cyberpsychology and Behavior, 5,* 529–535.

Foa, E. B., & Kozak, M. J. (1986). Emotional processing of fear: Exposure to corrective information. *Psychological Bulletin, 99,* 20–35.

Foa, E. B., Rothbaum, B. O., & Furr, J. M. (2003). Is the efficacy of exposure therapy for posttraumatic stress disorder augmented with the addition of other cognitive behavior therapy procedures? *Psychiatric Annals, 33,* 47–53.

Foa, E. B., Steketee, G., & Rothbaum, B. O. (1989). Behavioral/cognitive conceptualizations of post-traumatic stress disorder. *Behavior Therapy, 20,* 155–176.

Gamito, P., Ribeiro, C., Gamito, L., Pacheco, J., Pablo, C., & Saraiva, T. (2005). *Virtual war PTSD: A methodological thread.* Paper presented at the 10th Annual Cybertherapy Conference, Basel, Switzerland, June 13–17.

Glantz, K., Rizzo, A. A., & Graap, K. (2003). Virtual reality for psychotherapy: Current reality and future possibilities. *Psychotherapy: Theory, Research, Practice, Training, 40,* 55–67.

Grinker, R. R. & Spiegel, J. P. (1945). *War neuroses.* Philadelphia,: The Blakiston Company.

Hoge, C. W., Castro, C. A., Messer, S. C., McGurk, D., Cotting, D. I., & Koffman, R. L. (2004). Combat duty in Iraq and Afghanistan, mental health problems, and barriers to care. *New England Journal of Medicine, 351,* 13–22.

Jaycox, L. H., Foa, E. B., & Morral, A. R. (1998). Influence of emotional engagement and habituation on exposure therapy for PTSD. *Journal of Consulting and Clinical Psychology, 66,* 186–192.

Josman, N., Somer, E., Reisberg, A., Garcia-Palacios, A., Hoffman, H., & Weiss, P. L. (2005). *Virtual reality: Innovative technology for the treatment for victims of terrorist bus bombing with post-traumatic stress disorder.* Paper presented at the 10th Annual Cybertherapy Conference, Basel, Switzerland, June 13–17.

Keane, T. M., Kaloupek, D. G., Blanchard, E. B., Hsieh, F. Y., Kolb, L. C., Orr, S. P., et al. (1998). Utility of psychophysiological measurement in the diagnosis of posttraumatic stress disorder: Results from a Department of Veterans Affairs Cooperative Study. *Journal of Consulting and Clinical Psychology, 66,* 914–923.

Lang, P. J. (1977). Imagery in therapy: An information processing analysis of fear. *Behavior Therapy, 8,* 862–866.

Laor, N., Wolmer, L., Wiener, Z., Reiss, A., Muller, U., Weizman, R., et al. (1998). The function of image control in the psychophysiology of posttraumatic stress disorder. *Journal of Traumatic Stress, 11,* 679–696.

Marshall R. P., Jorm, A. F., Grayson D. A., & O'Toole, B. I. (2000). Medical-care costs associated with posttraumatic stress disorder in Vietnam veterans. *Australian and New Zealand Journal of Psychiatry, 34,* 954–962.

Mowrer, O. A. (1960). *Learning and Behavior.* New York: Wiley.

Pair, J., Allen, B., Dautricourt, M., Treskunov, A., Liewer, M. C., Graap, K., et al. (2006). A virtual reality exposure therapy application for Iraq War post traumatic stress disorder. In *Proceedings of the IEEE VR2006 Conference* (64–71). Los Alamitos, CA: IEEE.

Pitman, R. K., Orr, S. P., Forgue, D. F., de Jong, J. B., & Claiborn, J. M. (1987). Psychophysiologic assessment of post-traumatic stress disorder imagery in Vietnam combat veterans. *Archives of General Psychiatry, 44,* 970–975.

Rizzo, A. A., & Kim, G. (2005). A SWOT analysis of the field of virtual rehabilitation and therapy. *Presence: Teleoperators and Virtual Environments, 14,* 1–28.

Rizzo, A. A., Schultheis, M. T., Kerns, K., & Mateer, C. (2004). Analysis of assets for virtual reality applications in neuropsychology. *Neuropsychological Rehabilitation, 14,* 207–239.

Rothbaum, B. O., Foa, E. B., Riggs, D., Murdock, T., & Walsh, W. (1992). A prospective examination of post-traumatic stress disorder in rape victims. *Journal of Traumatic Stress, 5,* 455–475.

Rothbaum, B. O., Hodges, L., Alarcon, R., Ready, D., Shahar, F., Graap, K., et al. (1999). Virtual reality exposure therapy for PTSD Vietnam veterans: A case study. *Journal of Traumatic Stress, 12,* 263–271.

Rothbaum, B. O., Hodges, L., Ready, D., Graap, K., & Alarcon, R. (2001). Virtual Reality exposure therapy for Vietnam veterans with posttraumatic stress disorder. *Journal of Clinical Psychiatry, 62,* 617–622.

Rothbaum, B. O., Meadows, E. A., Resick, P., & Foy, D. W. (2000a). Cognitive-behavioral therapy. In E. B. Foa, T. M. Keane, & M. J. Friedman (Eds.), *Effective Treatments for PTSD: Practice Guidelines from the International Society for Traumatic Stress Studies* (pp. 60–83). New York: Guilford Press.

Rothbaum, B. O., & Schwartz, A. (2002). Exposure therapy for posttraumatic stress disorder. *American Journal of Psychotherapy, 56,* 59–75.

Shalev, A. Y., & Rogel-Fuchs, Y. (1993). Psychophysiology of posttraumatic stress disorder: From sulfur fumes to behavioral genetics. *Psychosomatic Medicine, 55,* 413–423.

VA Fact Sheet, U.S. Department of Veteran Affairs. 12/2004, Retrieved on September 2, 2006, at http://www1.va.gov/opa/fact/ptsd.asp.

Weiss, D. S., Marmar, C. R., Schlenger, W. E., Fairbank, J. A., Jordan, B. K., Hough, R. L., et al. (1992). The prevalence of lifetime and partial post-traumatic stress disorder in Vietnam theater veterans. *Journal of Traumatic Stress, 5,* 365–376.

Wiederhold, B. K., & Wiederhold, M. D. (2004). *Virtual-Reality Therapy for Anxiety Disorders: Advances in Education and Treatment.* Washington, DC: American Psychological Association Press.

Wiederhold, M. D., & Wiederhold, B. K. (2005). *Military mental health applications.* The 13th Annual Medicine Meets Virtual Reality Conference, Long Beach, CA, January 29.

Zimand, E., Anderson, P., Gershon, G., Graap, K., Hodges, L., & Rothbaum, B. (2003). Virtual reality therapy: Innovative treatment for anxiety disorders. *Primary Psychiatry, 9,* 51–54.

ENDNOTES

1. This research was supported by NIMH grant No. 5 R21 MH55555-02 awarded to Dr. Barbara Rothbaum and by an ONR grant ("Use of Virtual Reality War Scenarios with Returning Gulf War and Afghanistan Combatants Experiencing Symptoms of Acute Stress Response [ASR] and Post Traumatic Stress Disorder") awarded to Dr. Albert Rizzo and Mr. Ken Graap.
2. Disclosure Statement: Dr. Barbara Rothbaum receives research funding and is entitled to sales royalty from and owns equity in Virtually Better, Inc., Decatur, Georgia, which is developing products related to the research described in this paper. The terms of this arrangement have been reviewed and approved by Emory University in accordance with its conflict of interest policies.

10

Experiential Methods in the Treatment of Combat PTSD

JAMES L. SPIRA, JEFFREY M. PYNE, AND
BRENDA K. WIEDERHOLD

There are many different experiences that can lead to acute stress disorder or posttraumatic stress disorder (PTSD) in military personnel. Unfortunately, PTSD is a relatively common outcome of combat exposure, especially in the type experienced in recent deployments. The primary focus of this chapter will be the role of experiential psychotherapy treatments that teach cognitive, affective, and behavioral control to better cope with combat-related PTSD. In particular, we will focus on self-help skills during exposure therapy, particularly those utilizing virtual reality (VR) systems, to assist returning troops in gaining control over PTSD symptoms.

INTRODUCTION

There are many different experiences that can lead to acute stress disorder or posttraumatic stress disorder in military personnel. Combat-related experiences that can lead to PTSD include witnessing another service member being killed or wounded, feeling responsible for the death of a military member, being ambushed, a near-death experience, and witnessing the death or wounding of civilians including children. Each of these experiences is outside the range of what is considered normal human experience. In addition, PTSD appears to be more severe and longer lasting when the event is caused by human means and design, e.g., warfare (American Psychiatric Association, 1994).

Combat-related PTSD may be a condition that existed from the start of humankind. The written history of PTSD dates back to the account of Achilles in the *Iliad* by Homer (c. 800 BCE). More recently in the United States, symptoms of PTSD have been described as "soldier's heart" during the Civil War, "shell shock" during WWI, "combat fatigue" or "war neurosis" during WWII, and PTSD after the Vietnam War. The third edition of the *Diagnostic and Statistical Manual of Mental Disorders* published in 1980 was the first to publish diagnostic criteria for PTSD (American Psychiatric Association, 1980).

In a study of Vietnam veterans, 31% of men and 27% of women suffered from PTSD at some point since their return from the war (Schlenger et al., 1992). A few months after returning from Operation Iraqi Freedom (OIF), 12.2% to 19.9% of Marine Corps personnel and 12.9% to 18.0% of Army personnel met diagnostic criteria for PTSD (Hoge et al., 2004). The numbers of OIF/OEF service members with PTSD is expected to increase over time, because the delayed PTSD symptom onset has been shown in connection with other recent military conflict (Gray, Bolton, & Litz, 2004), and because of the duration of the conflict, the repeated and longer deployments, the duration of deployment not being known or changing during deployment, the nonconventional type of warfare, the increased use of reservists, and all military occupations being at increased risk.

According to the Veterans Health Administration, the British National Health Service National Institute for Clinical Excellence, the American Psychiatric Association, and recent expert panels (Anonymous, 1999; Foa, Keane, & Friedman, 2000; Ursano et al., 2004; Veterans Health Administration & Department of Defense, 2000), the current treatment recommendations for PTSD include the use of medication and psychotherapy. Each of these sources recommends selective serotonin reuptake inhibitors (SSRIs) as the firstline medication treatment for PTSD. However, the remission rates for combat-related PTSD (20%–30%) using SSRIs remain low (Stein, Kline, & Matloff, 2002). The success of traditional psychotherapeutic methods, either alone or in combination with SSRI treatment, is better (Van Etten & Taylor, 1998), but still is far from the level of success this combination has with treating other common disorders, such as depression or anxiety. Other methods are therefore needed for this escalating problem.

Following a brief review of the PTSD symptom clusters, medication treatments, and traditional psychotherapies for treating PTSD, this paper will primarily focus on experiential approaches, specifically skill-based approaches that can be used in combination with exposure therapy, such as virtual reality–assisted exposure therapy for the treatment of combat-related PTSD.

PTSD SYMPTOM CLUSTERS

After experiencing a traumatic event, the three core PTSD symptom clusters include re-experiencing, avoidance/numbing, and hyperarousal. The reexperiencing symptoms include recurrent and intrusive distressing recollections of the event; recurrent distressing dreams of the event; flashbacks; intense psychological distress when reminded of the event; and physiological reactivity when reminded of the event. The avoidance/numbing symptoms include efforts to avoid thoughts, feelings, or conversations associated with the event; efforts to avoid activities, places, or people that are reminders of the event; inability to recall important aspects of the event; decreased interest in significant activities; feeling detached from others; decreased range of affect; and sense of foreshortened future. The hyperarousal symptoms include problems sleeping; increased irritability; difficulty concentrating; hypervigilance; and exaggerated startle response.

MEDICATION TREATMENTS FOR PTSD BY SYMPTOM CLUSTER

The current APA treatment recommendations for PTSD recommend SSRIs as the firstline medication treatment for all three PTSD symptom clusters (Ursano et al., 2004). The evidence supporting the use of other classes of medications by PTSD symptom cluster was also summarized in a recent review (Schoenfeld, Marmar, & Neylan, 2004). Tricyclic antidepressants were found generally effective except in relieving symptoms in the avoidance/numbing cluster. Monoamine oxidase inhibitors (MAOIs) were found to be generally effective but there is limited evidence about the effectiveness of the more tolerable reversible MAOIs. Benzodiazepines were generally not effective for core PTSD symptom clusters but may improve sleep in the short run, although long term use of benzodiazepines for sleep is contraindicated. Anticonvulsants appeared to be more helpful for reexperiencing symptoms. Second-generation antipsychotic medications appeared to be helpful for all core PTSD symptom clusters. Adrenergic inhibitors may be useful as an early intervention to prevent the development of PTSD following a traumatic event or to decrease reexperiencing symptoms or as an adjunctive treatment. However, in a recent meta-analysis the overall effectiveness of medication management was found to be half as effective as psychotherapy and had twice the dropout rate (Van Etten & Taylor, 1998).

TRADITIONAL PSYCHOTHERAPY TREATMENTS FOR PTSD

The psychotherapy options include preventive and treatment strategies. The psychotherapeutic preventive strategies following traumatic exposure include one-session critical incident stress debriefing (CISD) and cognitive behavioral therapy (CBT). Two recent meta-analyses found no evidence to support the use of CISD to decrease psychological distress or prevent the onset of PTSD (Rose et al., 2002; van Emmerik et al., 2002). A limited number of well-designed studies demonstrate some success in preventing PTSD using a few sessions of CBT starting 2 to 3 weeks after the traumatic event (Bryant, Moulds, & Nixon, 2003; Bryant et al., 1999).

Psychotherapeutic treatment strategies extend along the continuum from therapies where the focus is more cognitively reflective to therapies where the focus is more directly sensorily experiential and skill based. Each of the therapies outlined below include reflective and experiential elements, yet to varying degrees. More reflective therapies include interpersonal and psychodynamic therapies. Combination reflective and experiential therapies include cognitive behavioral and dialectical behavioral therapies. More experiential and skill-based therapies include somatic (relaxation training and biofeedback), attentional (meditation), and exposure (flooding, graded exposure, eye movement desensitization reprocessing [EMDR], and hypnosis) approaches.

REFLECTIVE PSYCHOTHERAPIES

Interpersonal psychotherapy (IPT) is a structured psychotherapy that was developed to address the interpersonal and social problems stemming from the development of a client's personality and influenced by social interactions (Klerman et al., 1984). Interpersonal and social problems are often responsible for a client with PTSD seeking treatment and often influence the symptom course. A pilot study of group-based IPT demonstrated improvement in social functioning but had limited effect on more PTSD-specific symptoms (Robertson et al., 2004). Therefore, there is minimal evidence to support the use of IPT for the treatment of PSTD (Robertson, Humphreys, & Ray, 2004).

Psychodynamic psychotherapy is a less structured psychotherapy with a long tradition in mental health treatment that broadly explores a client's underlying personality structure that gives rise to the way a client responds to life events including traumatic events. Psychodynamic formulations provide richness to our understanding of the traumatic stress associated with PTSD and are often incorporated into other treatment strategies used for PTSD. There is one controlled trial comparing brief psychodynamic psychotherapy, hypnotherapy, desensitization, and a wait-list control where all active treatment groups resulted in significant symptom improvement (Brom, Kleber, & Defares, 1989). Therefore, there is minimal evidence to support the use of psychodynamic psychotherapy in the treatment of PTSD (Robertson, Humphreys, & Ray, 2004).

Combination Reflective and Experiential Psychotherapies

Cognitive behavioral therapy targets the client's distorted threat appraisal assumptions in order to reverse dysfunctional thinking patterns that are associated with and perpetuate PTSD symptom clusters. Through reflective dialogue, therapists help clients identify distorted automatic cognitive, affective, physiological, and behavioral responses to current events and focus instead on more rational responses appropriate to the situation. Experiential homework is given to assist the client in putting into practice what was discussed in therapy sessions. Proponents of CBT sometimes incorporate other therapeutic modalities, including aspects of the more experientially oriented therapies discussed below. Similar to psychodynamic and interpersonal approaches, therapists using a CBT approach often examine underlying factors that may influence current responses to traumatic events, such as core beliefs about oneself or the world. In the case of PTSD, many CBT therapists also have the client describe the traumatizing event while utilizing relaxation techniques in what can be considered a mild form of exposure therapy. CBT can be conducted in group or individual formats, yet there are fewer group CBT studies than individual and no studies comparing group and individual CBT (Ruzek, Young, & Walser, 2003). The evidence base for PTSD focused CBT is supported by a number of controlled studies (see in Robertson, Humphreys, & Ray, 2004).

Dialectical behavioral therapy (DBT) is a structured psychotherapy that was specifically developed for the treatment of borderline personality disorder (BPD). The combination of BPD and PTSD is often referred to as complex PTSD. BPD often develops within the context of childhood abuse or neglect and is a known risk factor for developing PTSD related to a traumatizing event later in life. DBT combines reflective cognitive and experientially based skill development by focusing on affect regulation, distress tolerance, and principles of mindfulness meditation to address distressing symptoms and behavior. While DBT has been used clinically with PTSD clients, the evidence base is limited to only case report at this time (Robertson, Humphreys, & Ray, 2004).

Experiential Psychotherapies

The term "experiential psychotherapies" refers to the use of sensory-based, nonreflective methods within a reflective psychotherapeutic context (usually CBT based) that focus on developing attentional control and autonomic regulation in an attempt to gain mastery over troublesome symptoms. The more experiential therapies can be divided into *somatic* (autonomic regulation through relaxation training and biofeedback), *attentional* (meditation, developing control over cognitive processing), and *exposure* (flooding, graded exposure, EMDR, and hypnosis). As is true of all the psychotherapies, there is considerable overlap across the reflective to experiential continuum as well as considerable overlap among the components of the more experientially oriented therapies. While evidence for the value of these approaches when used alone for the treatment of PTSD is lacking, almost all the most effective treatments for PTSD use them in one form or another as an integral part of their treatment.

Many popular treatments for PTSD include, as part of their treatment, a skill-based somatic component. Relaxation exercises have an autonomic emphasis. Typically, clients are trained to reduce the sympathetic arousal associated with PTSD symptoms and enhance parasympathetic recuperation using progressive muscle relaxation and slow abdominal breathing with or without biofeedback.

Biofeedback can be useful as a method of self-regulation. Utilized for more than 40 years, there are a variety of approaches to biofeedback. The oldest forms have the client watch a monitor or listen to a tone that reflects autonomic arousal as measured by skin temperature, skin conductance, muscle tension, respiratory rate, and/or heart rate. Clients are told to manipulate the monitor (sound or graphic) by any means they can (e.g., recalling or imagining a pleasant scene or slow abdominal breathing). Rather than simply instructing the client to "make the tone go up" any way they can, modern-day biofeedback practitioners are part of the feedback loop in that the provider continually monitors the physiological data and suggests attentional and somatic exercises for the client to use. In this way, the specific approach to self-regulation is tailored to each client. Such physiological monitoring and feedback is a useful tool for use in conjunction with other interventions (including virtual reality–graded exposure therapy, described below) to continually monitor objective arousal in clients with PTSD and assist clients in regaining a sense of mastery over their symptoms (Wiederhold et al., 2002).

Attentional therapies employ various meditative traditions, which emphasize different aspects of attention. Eastern meditative traditions enhance attention to the moment and less intrusive thoughts by reducing background "noise" and enhancing foregrounded signal. For example, Zen meditation emphasizes signal enhancement—attending to and becoming absorbed in what one sees, hears, feels, and smells at each moment. When thoughts arise, practitioners note that intrusive "noise," let it go, and return to the sensations ("signal") at hand. Vipassana (mindfulness) meditation emphasizes noise reduction—notice what thoughts and feelings arise, but do not react to it or judge it. The client is instructed to notice what arises, passively attending to it until it dissipates, and then return to the moment at hand (such as feeling the breath flowing in and out, or continuing with one's work activity). These practices are complementary, merely emphasizing different aspects of the same principle (Table 10.1).

When a combat veteran returns from the most intense experience of his life, those intense experiences continue to repeatedly "play" in his mind. It is natural to pay attention to life-threatening experiences, even if those are now incorporated into one's memory. However, the more he attends to

TABLE 10.1
Attentional Retraining

1. Whatever one attends to, one enhances
   - attending to worry or pain will enhance those experiences
   - attending to comfort or work will enhance those experiences
2. One's nervous system gears up to support what one attends to
   - attending to worry or pain will activate stress responses
   - attending to comfort or neutral sensations will activate recuperative responses
   - attending to work one is engaged in rallies the ability to focus in the moment
3. If one can act on what one is worrying about, then one should do so. Otherwise, one should redirect attention back to the situation at hand, or a positive comforting sensation.

those thought intrusions, the more intrusive they become. Learning to shift attention away from those intrusive thoughts and into the moment at hand allows those intrusions to lessen, and daily functioning to increase. Experiential methods have been developed to facilitate this attentional control.

Meditative traditions share with CBT the belief that cognitive processes drive affective, physiological, and behavioral reactivity. However, while CBT focuses on underlying belief systems as the cause of current dysfunction, meditative traditions emphasize mastery over fundamental cognitive processes or attentional retraining (i.e., whatever one attends to, one enhances, and the body supports that processing). Indeed, when clients attend to a distressing thought or feeling, their sympathetic arousal is significantly increased. By contrast, when engaged in Zen meditation (attending to what they see, hear, and feel), clients significantly reduce their sympathetic arousal, even without controlling their breathing or otherwise consciously manipulating their physiology (Spira, 2004; Spira & Kotay, 2004).

Exposure-based therapy helps clients decrease their fear response to internal and external cues that otherwise cause symptom intensification. Exposure therapy is based on emotional processing theory (EMT). Applying EMT to PTSD, fear memories are stored as a "fear structure" and include psychological and physiological information about stimuli, meaning, and responses (Foa & Kozak, 1986). Once accessed and emotionally engaged, the fear structure is then open to modification and, if treated appropriately, over time will result in habituation and extinction of the fear response. Common approaches to exposure therapy include flooding, graded exposure, EMDR, tolerating narrative report of the traumatic event, and hypnosis.

Flooding exposure therapies attempt to present the client with as much stimulation as possible, and have the client sustain attention to that stimulation until it begins to extinguish, usually in about 20 minutes. Several theories support the use of flooding-type exposure. Classical conditioning is the original theoretical basis of this approach, where the conditioned stimulus (loud sound, internal memory) no longer is paired with a conditioned response (fear arousal), and therefore this conditioned response extinguishes over time. Case studies using flooding exposure therapy have reported mixed results (Keane & Kaloupek, 1982; Keane et al., 1989; Pitman et al., 1996), therefore the evidence base for flooding therapy is not strong at this time.

Graded exposure therapy attempts to elicit arousal at the level the client can tolerate and then increase exposure gradually over time as the client learns skills to modulate arousal. This approach is most often coupled with a skill-based de-arousal method, such as relaxation training (progressive muscle relaxation, biofeedback), distancing (hypnosis, visual imagery), and/or attentional retraining (Zen or Vipassana meditation). Graded exposure can include imaginal, in vivo, or virtual reality exposure techniques. To date, the most commonly used graded exposure technique in PTSD treatment is imaginal exposure. Virtual reality graded exposure will be discussed in detail below.

Eye movement desensitization reprocessing most typically involves the client focusing on a disturbing memory while the therapist initiates saccadic eye movements by asking the client to track the horizontal motion of the therapist's finger moving rapidly in front of them. Following the therapist's finger movement is thought to disassociate memories from associated emotions. In studies with and without the saccadic eye movements, it is not clear that the eye movements are necessary for treatment efficacy (Cahill, Carrigan, & Frueh, 1999). A meta-analysis of EMDR and other exposure techniques found no significant differences in outcomes (Davidson & Parker, 2001). At this time, the evidence base for EMDR treatment of PTSD shows it to be at least equivalent to CBT and in some cases to other exposure therapies (Robertson, Humphreys, & Ray, 2004).

Hypnotherapy using light or deep trance techniques has been used clinically for decades to treat combat-related stress disorders (Watkins, 2000). Typically, the client is induced into a comfortable, relaxed mental and physical state while simultaneously reviewing and distancing from the traumatic episode, thus learning to dissociate the traumatic event from arousing sequelae. However, results from controlled studies are not available at this time (Cardena, 2000). Therefore, there is minimal evidence to support the use of hypnotherapy for the treatment of PTSD.

In summary, exposure-based therapies (including CBT with exposure) have been found to be the most effective form of treating PTSD (Foa, 2000). Van Etten and Taylor (1998) analyzed 61 treatment trials that included pharmacotherapy and modalities such as behavior therapy (particularly exposure therapy), EMDR, relaxation training, hypnotherapy, and dynamic psychotherapy. Specifically, the effect size for all types of psychotherapy interventions was 1.17 compared with 0.69 for medication, and the mean dropout rate in medication trials was 32% compared with

14% in psychotherapy trials. In addition, this meta-analysis found that exposure therapy was more efficacious than any other type of treatment for PTSD when measured by clinician-rated measures.

VIRTUAL REALITY–ASSISTED EXPOSURE THERAPY

Virtual reality can be used to deliver graded exposure or flooding exposure therapies. Graded exposure therapy has been used clinically for a variety of anxiety disorders (Moore et al., 2002; Wiederhold, Gevirtz, & Spira, 2001). The advantages of VR exposure over imaginal exposure include the control the therapist has over the exposure presented and VR exposure's nonreliance on individual imagery ability or even the ability of the client to verbalize his or her experiences (although the ability to talk about the traumatic event[s] can be utilized within a VR environment to increase personal relevancy and increase arousal). Many clients are unwilling or unable to effectively visualize the traumatic event. In fact, avoidance of reminders of the trauma is inherent in PTSD, and is one of the defining symptoms of the disorder (Difede & Hoffman, 2002). One disadvantage of the use of VR exposure for PTSD is that the VR environment is content specific and must be developed for a particular context. The evidence base for use in combat-related PTSD treatment at this time is limited to case studies (Rothbaum & Foa, 1999; Rothbaum et al., 2001), but will be expanding with treatment trials described below that are currently underway and supported by the Department of Defense.

A VR environment can be used to present both general and specific stimuli to VR environment (e.g., Iraqi village) and is often sufficient to elicit a general reminder of the arousal one experienced during deployment. In addition, if the VR environment allows for operator control over a repertoire of various optional stimuli, then a graded exposure of relevant arousing stimuli can be individually tailored to allow for an arousal hierarchy to be developed and presented to each client. For example, a Marine who conducted night operations may not get sufficiently aroused in a daytime environment. Similarly, a Navy Construction Battalion (Seabee) driver may require a convoy scenario to elicit arousal. Since the goal is to teach mastery over cognitive, affective, and physiological arousal, the ability to generate arousal is critical for successful treatment. An optimal VR environment would therefore contain a general reminder of the deployment and have a range of options that the therapist can employ to bring out the arousal that is more specific to each client's unique experience.

Other aspects of VR environments that are important to treatment include realism, immersion, and interaction. Although technology has been steadily improving with regard to video graphics and VR in particular, it is not necessary that the environment be completely "realistic." In fact, it has been observed in various VR studies that exact reproductions are not necessary to elicit anxiety (Hodges et al., 1999). If the VR environments are similar enough to the index traumatic events, then it should be possible to "trigger" emotional responses similar to those which may have occurred originally, thereby providing access to the memories of the trauma. In the future, degree of realism (e.g., the addition of vibration, scent, and other stimuli to the VR) will likely enhance the options available to clinicians and provide greater coverage of traumatic situations.

Immersion appears to be related to the degree of arousal that can be achieved with a given exposure. Using a head mount with the greatest clarity, viewing range, and comfort, along with the client's ability to see the environment move as they move their head or body, allows the client greater immersion and perhaps greater arousal. Sounds presented through headphones are also a critical element for improved immersion. It is also possible to enhance immersion by placing a vibration platform underneath the client (to vibrate with helicopters going overhead, rockets exploding, etc.), matching climate (dry heat blowing on the client), or even using a machine to present smells to the client (burning rubber, gun powder, etc). The more sensory modalities that are stimulated, in theory, the greater the immersion.

Another factor that affects immersion is the degree to which a client can interact with the VR environment. Usually the client will use a joystick or computer mouse to navigate through the environment and move their head to change the visual field. The level of client interaction with the VR environment is another aspect of exposure that the therapist can utilize to influence arousal.

Two Department of Defense–funded studies are underway to study the use of VR therapies for combat-related PTSD. One of the studies will use a graded exposure approach in a randomized controlled design and another will use a flooding approach in a case series design. In both studies, the primary outcomes will be symptom severity, physiologic reactivity to a test VR environment, and health-related quality of life.

To develop the general and specific content for the VR environments we interviewed 18 Marine and Navy personnel recently diagnosed with combat-related PTSD and receiving outpatient mental health treatment. Specifically, clients were asked about the precise sights, sounds, smells, and feelings associated with the recurring intrusive thoughts they experienced upon returning from their combat tours. Some of the most salient memories include voices of Iraqi civilians, Arabic prayer, sounds of gunfire and rocket fire and explosions, helicopters flying overhead or landing, terrorists running and firing guns, comrades being wounded by gunfire, buildings and vehicles burning, driving through dangerous areas, etc. This information has been used to create VR environments for use with medical and Marine Corps personnel, and can be presented within the VR environment as needed for treatment.

In the flooding VR exposure study, the therapist will ask the client to relate his or her narrative of the sentinel traumatic event or sequence of events and then presents the client with VR stimuli sufficient to maintain a high level of arousal for at least 20 minutes. All clients will also be treated with an SSRI prescribed by their mental health provider. It is critical to not overarouse the client to the extent that they begin to dissociate cognitively, shut down emotionally, or become overwhelmed during or after the session. The therapist will also record the client narrative and the sounds of the VR environment during the VR session so that that the client can continue to listen to this recording daily, in between sessions, in order to facilitate the extinguishing of arousal. In a small, single group design study of Vietnam veterans with chronic combat-related PTSD, the use of a similar protocol twice a week for 6 weeks was found to be beneficial in reducing their PTSD symptoms (Rothbaum et al., 2001).

In the graded exposure VR study, we will determine the relative value of 10 weekly sessions of VR graded exposure plus SSRI treatment compared with 10 weekly group CBT sessions plus SSRI treatment. The VR graded exposure therapy will incorporate Zen absorption techniques to focus comfortably into the moment (attentional retraining) and Vipassana internal noise reduction techniques to distance arousing thoughts and feelings. The graded exposure VR intervention will also incorporate biofeedback to monitor physiologic response so that the therapist can better (a) determine when a client is becoming aroused and (b) train the client to modulate these responses. Over the past 5 years, heart rate variability (HRV) has become the indicator of choice for many biofeedback therapists and those who wish to monitor physiological reactivity in their clients or research subjects (Spira & Kotay, 2004). In particular, the very low frequency (VLF)/low frequency (LF) ratio (part of the HRV spectral analysis) is the best single indicator of when a client is focused comfortably in the moment without significant cognitive/affective/physiological arousal. Simply, when VLF is more than 50% greater than LF, the therapist should instruct the client to relax and focus in the moment or, if this is not possible for the client, to reduce the intensity of the VR stimulus presented to the client. When VLF is less than 50% LF, then the client is more calm and relaxed and the VR stimuli can be increased so that the client has more opportunity to practice experiential methods of self-regulation. As the client becomes more skilled at modulating his physiologic response to the VR environment, the client will gain a sense of mastery over arousal, develop confidence to be able to handle even more arousal, and reestablish the calm and relaxed state as their natural baseline. As with other exposure therapies, the goal is to generalize these skills into everyday activities. At this time, it is not known which clients will be more likely to benefit from VR-assisted exposure therapies or how best to integrate VR therapies with other existing treatments for PTSD.

CONCLUSION

Existing medications and psychotherapies have limited success in treating PTSD, including PTSD developed in combat. However, experiential psychotherapies utilized within a therapeutic framework are promising additions to existing approaches. Ongoing studies testing VR-assisted interventions will help define the role of novel VR interventions in the treatment of combat-related PTSD.

ACKNOWLEDGMENTS

Dr. Jeffrey M. Pyne is supported by a VA HSR&D Advance Research Career Development Award.

REFERENCES

American Psychiatric Association. (1980). *Diagnostic and Statistical Manual of Mental Disorders* (3rd ed.). Washington, DC: Author.

American Psychiatric Association. (1994). *Diagnostic and Statistical Manual of Mental Disorders* (4th ed.). Washington, DC: Author.

Anonymous. (1999). The expert consensus guideline series. Treatment of Posttraumatic Stress Disorder. The Expert Consensus Panels for PTSD. *Journal of Clinical Psychiatry, 60*(Suppl 16), 3–76.

Brom, D., Kleber, R. J., & Defares, P. B. (1989). Brief psychotherapy for posttraumatic stress disorders. *Journal of Consulting and Clinical Psychology, 57,* 607–612.

Bryant, R. A., Moulds, M. L., & Nixon, R. V. (2003). Cognitive behaviour therapy of acute stress disorder: A four-year follow-up. *Behaviour Research and Therapy, 41,* 489–494.

Bryant, R. A., Sackville, T., Dang, S. T., Moulds, M., & Guthrie, R. (1999). Treating acute stress disorder: An evaluation of cognitive behavior therapy and supportive counseling techniques [see comment]. *American Journal of Psychiatry, 156,* 1780–1786.

Cahill, S. P., Carrigan, M. H., & Frueh, B. C. (1999). Does EMDR work? And if so, why?: A critical review of controlled outcome and dismantling research. *Journal of Anxiety Disorders, 13,* 5–33.

Cardena, E. (2000). Hypnosis in the treatment of trauma: A promising, but not fully supported, efficacious intervention. *International Journal of Clinical and Experimental Hypnosis, 48,* 225–238.

Davidson, P. R., & Parker, K. C. (2001). Eye movement desensitization and reprocessing (EMDR): A meta-analysis [see comment]. *Journal of Consulting and Clinical Psychology, 69,* 305–316.

Difede, J., & Hoffman, H. G. (2002). Virtual reality exposure therapy for World Trade Center post-traumatic stress disorder: A case report. *Cyberpsychology and Behavior, 5,* 529–535.

Foa, E. B. (2000). Psychosocial treatment of posttraumatic stress disorder. *Journal of Clinical Psychiatry, 61*(Suppl 5), 43–48; discussion 49–51.

Foa, E. B., Keane, T. M., & Friedman, M. (Eds.). (2000). *Effective Treatments for PTSD: Practice Guidelines from the International Society for Traumatic Stress Studies.* New York: Guilford Press.

Foa, E. B., & Kozak, M. J. (1986). Emotional processing of fear: Exposure to corrective information. *Psychological Bulletin, 99,* 20–35.

Gray, M. J., Bolton, E. E., & Litz, B. T. (2004). A longitudinal analysis of PTSD symptom course: Delayed-onset PTSD in Somalia peacekeepers. *Journal of Consulting and Clinical Psychology, 72,* 909–913.

Hodges, L. F., Rothbaum, B. O., Alarcon, R. D., Ready, R., Shahar, D., Graap, F. et al. (1999). A virtual environment for the treatment of chronic combat-related post-traumatic stress disorder. *Cyberpsychology and Behavior, 2,* 7–14.

Hoge, C. W., Castro, C. A., Messer, S. C., McGurk, D., Cotting, D. I., & Koffman, R. L. (2004). Combat duty in Iraq and Afghanistan, mental health problems, and barriers to care. *New England Journal of Medicine, 351,* 13–22.

Keane, T. M., Fairbank, J. A., Caddell, J. M., & Zimering, R. T. (1989). Implosive (flooding) therapy reduced symptoms of PTSD in Vietnam combat veterans. *Behavior Therapy, 20,* 245–260.

Keane, T. M., & Kaloupek, D. G. (1982). Imaginal flooding in the treatment of a posttraumatic stress disorder. *Journal of Consulting and Clinical Psychology, 50,* 138–140.

Klerman, G. L., Weissman, M. M., Rounsaville, B. J., & Chevron, E. S. (1984). The interpersonal approach to understanding depression. In G. L. Klerman, M. M. Weissman, B. J. Rounsaville, E. S. Chevron (Eds.). *Interpersonal Psychotherapy of Depression* (pp. 51–69). New York: Basic Books.

Moore, K., Wiederhold, B. K., Wiederhold, M. D., & Riva, G. (2002). Panic and agoraphobia in a virtual world. *Cyberpsychology and Behavior, 5,* 197–202.

Pitman, R. K., Orr, S. P., Altman, B. (1996). Emotional processing and outcome of imaginal flooding therapy in Vietnam veterans with chronic posttraumatic stress disorder [see comment]. *Comprehensive Psychiatry, 37,* 409–418.

Robertson, M., Humphreys, L., & Ray, R. (2004). Psychological treatments for posttraumatic stress disorder: Recommendations for the clinician based on a review of the literature. *Journal of Psychiatric Practice, 10,* 106–118.

Robertson, M., Rushton, P. J., Bartrum, D., & Ray, R. (2004). Group-based interpersonal psychotherapy for posttraumatic stress disorder: Theoretical and clinical aspects. *International Journal of Group Psychotherapy, 54,* 145–175.

Rose, S., Bisson, J., Churchill, R., & Wessely, S. (2002). Psychological debriefing for preventing post traumatic stress disorder (PTSD). *Cochrane Database of Systematic Reviews,* (2), CD000560.

Rothbaum, B. O., & Foa, E. B. (1999). Exposure therapy for PTSD. *National Center for PTSD Research Quarterly, 10*(2).

Rothbaum, B. O., Hodges, L. F., Ready, D., Graap, K., & Alarcon, R. D. (2001). Virtual reality exposure therapy for Vietnam veterans with posttraumatic stress disorder. *Journal of Clinical Psychiatry, 62,* 617–622.

Ruzek, J. I., Young, B. H., & Walser, R. D. (2003). Group treatment of posttraumatic stress disorder and other trauma-related problems. *Primary Care Companion to the Journal of Clinical Psychiatry, 10,* 53–57.

Schlenger, W. E., Kulka, R. A., Fairbank, J. A., Hough, R. L., Jordan, B. K., & Marmar, C. R. (1992). The prevalence of post-traumatic stress disorder in the Vietnam generation: A multimethod, multisource assessment of psychiatric disorder. *Journal of Traumatic Stress, 5,* 333–363.

Schoenfeld, F. B., Marmar, C. R., & Neylan, T. C. (2004). Current concepts in pharmacotherapy for posttraumatic stress disorder. *Psychiatric Services, 55,* 519–531.

Spira, J. (2004). *Using meditation and hypnosis to modify EEG and heart rate variability.* Paper presented at the Annual Meeting of the American Association of Biofeedback and Psychophysiology, Colorado Springs, CO.

Spira, J., & Kotay, A. (2004). *Influence of relaxation and stress on very low frequency heart rate variability.* Paper presented at the Annual Meeting of the American Association of Biofeedback and Psychophysiology, Colorado Springs, CO.

Stein, M. B., Kline, N. A., & Matloff, J. L. (2002). Adjunctive olanzapine for SSRI-resistant combat-related PTSD: A double-blind, placebo-controlled study [see comment]. *American Journal of Psychiatry, 159*, 1777–1779.

Ursano, R. J., Bell, C., Eth, S., Friedman, M., Norwood, A., Pfefferbaum, B., et al. (2004). Practice guideline for the treatment of patients with acute stress disorder and posttraumatic stress disorder. *American Journal of Psychiatry, 161*(11 Suppl), 3–31.

van Emmerik, A. A., Kamphuis, J. H., Hulsbosch, A. M., & Emmelkamp, P. M. (2002). Single session debriefing after psychological trauma: A meta-analysis. *Lancet, 360*, 766–771.

Van Etten, M. L., & Taylor, S. (1998). Comparative efficacy of treatments for posttraumatic stress disorder: A meta-analysis. *Clinical Psychology & Psychotherapy, 5*, 144–154.

Veterans Health Adminstration & Department of Defense. (2000). *VHA/DoD clinical practice guideline for management of major depressive disorder in adults.* Retrieved August 2005, from www.oqp.med.va.gov/cpg/MDD/MDD_GOL.htm

Watkins, J. G. (2000). The psychodynamic treatment of combat neuroses (PTSD) with hypnosis during World War II. *International Journal of Clinical and Experimental Hypnosis, 48*, 324–335; discussion 336–341.

Wiederhold, B. K., Gevirtz, R., & Spira, J. (2001). Virtual reality exposure therapy vs. imagery desensitization therapy in the treatment of flying phobia. In: G. Riva, C. Galimberti (Eds.). *Towards CyberPsychology: Mind, Cognition, and Society in the Internet Age* (pp. 254–272). Amsterdam: IOS Press.

Wiederhold, B. K., Jang, D. P., Kim, S. I., & Wiederhold, M. D. (2002). Physiological monitoring as an objective tool in virtual reality therapy. *Cyberpsychology and Behavior, 5*, 77–82.

11

Medication Management of Combat and Operational Stress Injuries in Active Duty Service Members

NANCY M. CLAYTON AND WILLIAM P. NASH

> There is no medicine like hope, no incentive so great, and no tonic so powerful as expectation of something better tomorrow.
>
> *Orison Swett Marden*

Clearly there is no pharmacologic panacea for the spectrum of psychological, biological, spiritual, and existential injuries that can be sustained from combat or operational stress. No magic pill can erase the image of a best friend's shattered body, or assuage the guilt from having traded duty with him that day. Medication cannot reestablish a person's trust in her mastery of her world, or restore her lost innocence, or revive one's fractured faith in God, or create wholeness for someone who feels as though he's left pieces of himself on the battlefield, which he desperately wants to retrieve.

Medication can, however, alleviate some debilitating and nearly intolerable symptoms of combat and operational stress injuries. By decreasing physiologic arousal levels, medication can decrease hypervigilance, allowing people to feel more at ease in a nonthreatening environment, to fall asleep, and stay asleep without being awakened by horrifying nightmares.

Medication can restore individuals' tolerance for day-to-day frustrations so they don't fly into a rage when their children play too loudly, or so they can remain seated when the church door slams shut on Sunday morning. Medication can help reduce the anxiety associated with traumatic memories so that clients can do the work they need to do to begin to integrate their combat experiences into the fabric of their nonoperational lives. Moreover, medication may help to halt further injury to the mind and brain caused by chronically elevated levels of toxic stress hormones and excitatory neurotransmitters, which, when unchecked, can lead to permanent damage. Recent studies suggest that certain medications may even *prevent* the development of traumatic stress injuries when prescribed in the hours or days following exposure to traumatic events.

For an active duty population, where the maintenance of physical and psychological readiness is a requirement, medication can help restore personnel to full functioning capacity. This includes far more than combat readiness. It includes the preservation of relationships with spouses, children, peers, superiors, and the community. It includes the avoidance of substance abuse and other self-destructive behaviors. It includes the restoration and preservation of self-respect—the knowledge that they can sustain psychological wounds of battle, that their wounds can heal, and that they may continue to serve and be valued by their commanders.

HISTORY OF PHARMACOTHERAPY FOR COMBAT/OPERATIONAL STRESS

Medication did not feature prominently in the treatment of posttraumatic stress disorder (PTSD) in the first decade after it was officially coded in the *Diagnostic and Statistical Manual of Mental Disorders*, 3rd ed. (DSM-III) in 1980 (American Psychiatric Association, 1980). Although the symptoms of autonomic arousal characteristic of posttraumatic stress were clearly biological in nature, the syndrome of posttraumatic stress was still perceived and classified as more of a "psychoneurotic" disorder than one rooted in neurobiological changes (Nicholi, 1988). Service members who suffered functional impairment from combat stress did not remain on active duty, and many suffered from untreated, chronic symptoms for many years before ultimately being diagnosed and treated for PTSD. Treatment focused on behavioral desensitization, psychological support, and occupational and social rehabilitation. Though sedative-hypnotic and anti-psychotic medications were frequently used to subdue acute agitation for stress injured patients on inpatient psychiatric wards, these drugs caused a host of adverse side effects, sometimes creating more problems than they alleviated. Psychiatric textbooks throughout the 1980s, therefore, discouraged the use of pharmacotherapy in clients with posttraumatic stress, in favor of nonpharmacologic interventions (Nicholi, 1988; Kaplan & Sadock, 1985).

In the 1990s, as new classes of psychiatric medication were developed, and as nonpsychiatric medications were proven to benefit other psychiatric disorders, several classes of medication were studied, formally and informally, for their effectiveness in reducing specific, targeted *symptoms* of posttraumatic stress. Antidepressants were studied for their effectiveness in treating symptoms of anxiety, depression, and hyperarousal. Anticonvulsant medications, long used in psychiatry as "mood stabilizers," were studied for their effects on anger, rage, and mood lability. Antiadrenergic antihypertensive medications were studied for their effectiveness in reining in autonomic hyperarousal. Antipsychotics were studied for their effectiveness against aggression, agitation, and dissociation. Little was understood, however, about the mechanism of action of most of these drugs, why some classes of medication were more effective than others, or why there was so much variability among results of any particular medication from one study to the next. There was no professional consensus, and no treatment guidelines existed.

Advances in the field of neurobiology over the past 10 years, particularly on the neurobiological effects of stress, have illuminated some of the biological derangements that occur when the human stress response is challenged beyond its adaptive capacity (for a brief review, see Nash & Baker, this volume, chapter 4). The elucidation of the neurobiology of the effects of stress on the brain has also contributed enormously to our ability to understand better the mechanisms of action of several existing medications, and has provided a blueprint for the development of entirely new classes of medication with the potential to treat, and perhaps even to prevent, neurological damage incurred from combat and operational stress.

A vast (and rapidly growing) literature emerged from all of this research, which can be overwhelming to sort through for clinicians in the trenches seeking effective treatment strategies for their soldiers and Marines. Fortunately, a group of subject matter experts from the Departments of Veterans Affairs and Defense recognized "the need to diagnose and treat PTSD among the military population" (Department of Veterans Affairs & Department of Defense, 2004, p. iii), and assembled a working group to address the matter.

The VA/DoD guideline working group, aided by a Cochrane Review and other specialized staff work, spent a year reviewing all the literature addressing the prevention, diagnosis, and treatment of the *entire spectrum* of traumatic stress disorders, from short-lived, spontaneously remitting "acute stress reactions," all the way to complicated, severe, chronic PTSD. For each preventive, diagnostic, and treatment modality they considered, an algorithm was created of "best practices" based on the strength of scientific evidence available. Where research was lacking, they arrived at a consensus of expert opinion, and made recommendations for appropriate future research. The final product, a 175-page document titled *VA/DoD Clinical Practice Guideline for the Management of Post-Traumatic Stress* was released in January 2004 (available online at http://www.oqp.med.va.gov/cpg/PTSD/PTSD_Base.htm).

The module on Pharmacotherapy Interventions addresses pharmacotherapy for two specific categorical DSM-IV-TR (American Psychiatric Association, 2000) diagnoses—acute stress disorder (ASD) and PTSD—even though, as the *Guideline* working group acknowledged, ASD and PTSD are only parts of "a spectrum of traumatic stress disorders" that do not always fit so neatly into one or another narrow category (Department of Veterans Affairs & Department of Defense, 2004, p. i). Some clients whose combat/operational stress injuries do not meet full criteria for either PTSD or ASD may yet obtain significant symptomatic relief by one or another medication. In clinical practice, prescribing clinicians must always critically assess to what extent drug research findings, in general, apply to their patients, in particular.

In the sections that follow, the major classes of medication that have been studied in the treatment of posttraumatic stress will be introduced, with emphases on the strength of evidence of their efficacy, the mechanisms of their action, their therapeutic benefits, and special considerations related to their use. This brief survey is intended to familiarize the nonclinical reader with useful, demystified information about these medications. It is by no means comprehensive, as new medications and data are constantly emerging. Practitioners are encouraged to consult the most recent available clinical practice guidelines for prescription information.

SELECTIVE SEROTONIN REUPTAKE INHIBITORS

Unlike many of the earlier classes of psychiatric medications whose psychotropic benefits were discovered accidentally, the *selective serotonin reuptake inhibitor* (SSRI) medications were the first class rationally designed as antidepressants, molecularly engineered in the laboratory to target the presynaptic reuptake pump in serotonin neurons in the brain (Domino, 1999). In 1988, Eli Lilly released the first SSRI, fluoxetine (Prozac), and since then, five more have found their way into the American formulary: sertraline (Zoloft; Pfizer), paroxetine (Paxil; GlaxoSmithKline), fluvoxamine (Luvox; Solvay), citalopram (Celexa; Forest Laboratories), and escitalopram (Lexapro; Forest Laboratories). Compared with earlier generations of antidepressant medications, the SSRIs are safer, and they have a far more tolerable side-effect profile. They have turned out to be effective not only against depression, but against anxiety as well, earning FDA approval for the treatment of panic disorder, obsessive compulsive disorder, social anxiety disorder, and generalized anxiety disorder. In 1998, sertraline became the first medication to receive FDA approval for the treatment of posttraumatic stress disorder, followed shortly thereafter by paroxetine (e.g., Brady et al., 2000; Marshall et al., 2001). Although the other SSRIs do not have FDA approval for the treatment of PTSD, their mechanism of action is the same, and they are likely to be equally effective clinically. This is the only class of medication considered by the

VA/DoD guideline, on the basis of controlled clinical trials, to have "significant benefit" in the treatment of PTSD (Department of Veterans Affairs & Department of Defense, 2004).

How SSRI Antidepressants Work

Although SSRIs were born out of "rational drug design," it is not entirely clear how they exert their numerous therapeutic benefits. On one hand, the mechanism of action of the SSRIs is remarkably simple: the drugs bind to the presynaptic reuptake pumps on serotonergic axons and dendrites, thereby preventing those pumps from sweeping serotonin back into the cells from the extracellular space. The lower rate of serotonin reuptake by neurons affected by SSRI antidepressants causes an increase in the numbers of serotonin molecules within synapses surrounding these neurons. But exactly how and why this mechanical disruption, and the consequent increase in availability of serotonin in synaptic spaces, result in therapeutic benefits is extremely complex and only partially understood. Our comprehension of the process will continue to evolve as our tools for deciphering molecular and genetic biology become more refined. Nevertheless, the following is our best current understanding of how SSRI antidepressants work.

In chapter 4 of this volume, Nash and Baker described the "neurotransmitter receptor hypothesis" of depression and anxiety. This theory posits that sustained or overwhelming stress can deplete certain neurons of their neurotransmitter chemical stores, including serotonin. The connections between serotonin neurons and their target postsynaptic neurons require a certain minimum amount of serotonin in synaptic gaps to transmit electrochemical signals properly. Hence, once serotonin has been sufficiently depleted due to stress, serotonergic circuits in the brain fail to function as they should. To compensate for this deficiency in neurotransmission, neurons involved in serotonergic circuits undergo an allostatic change in order to make better use of what serotonin is available to them (Friedman, 2001). Specifically, these postsynaptic cells synthesize more of—or "up-regulate"—their serotonin-specific receptors. With more receptors present on postsynaptic membranes to bind to serotonin released from presynaptic neurons, more signal gets through and the function of serotonergic circuits is normalized, to some extent.

However, one problem with receptor up-regulation as an allostatic compensation for depletion of serotonin is that this process affects not only the serotonin receptors that are necessary to carry serotonin signals from one neuron to the next (the postsynaptic receptors), but also the serotonin receptors that serve as negative feedback brakes on presynaptic serotonergic neurons. These inhibitory presynaptic receptors, also called "autoreceptors," function to slow down the flow of neuronal impulses in the presynaptic cell. So, while up-regulation of postsynaptic receptors serves to increase the flow of serotonergic signals in the brain, the up-regulation of auto-receptors serves to do the opposite, and to make serotonin neurons susceptible to turning themselves off at times of high serotonergic activity.

When an SSRI antidepressant is introduced into this environment of depleted neurotransmitter and up-regulated receptors, a sequence of changes occurs over a period of days to weeks. In the days immediately following blockade of the serotonin reuptake pump, there is an accumulation of serotonin in the somato-dendritic area (near the body of the neuron rather than near the terminus of its long axon) of the presynaptic neuron, saturating the auto-receptors in that area. This causes a decrease in the rate of serotonergic neuronal impulse flow in that neuron, and therefore a decrease in the rate of serotonin release at its axonal synapses. Thus, in the first days after beginning treatment with an SSRI antidepressant, clients may experience a *decrease* rather than an increase in the effectiveness of their serotonin system; initially, their symptoms of anxiety, depression, or anger may worsen instead of improve. But then, in response to the persistent saturation of auto-receptors, presynaptic serotonin neurons begin to down-regulate the numbers of their auto-receptors, and the initial inhibition of serotonin release at their axons is diminished. Serotonergic activity then increases, causing an accumulation of serotonin downstream in the axonal (rather than the dendritic) synapses, which saturates the serotonin receptors on the surface of the postsynaptic cell. Like the presynaptic neuron, the postsynaptic neuron responds by eventually down-regulating its serotonin receptors over the ensuing days or weeks. According to the "neurotransmitter receptor hypothesis," it is the eventual down-regulation of postsynaptic serotonin receptors that causes the primary therapeutic benefits of antidepressants. The time it takes for these various changes to occur corresponds with the initiation and resolution of sometimes uncomfortable side effects, and the delayed onset of the desirable therapeutic effects of SSRI antidepressant medications (Stahl, 2000).

It was once thought that this down-regulation of postsynaptic receptor sites accounted for all the therapeutic benefits of the SSRIs. But if that were the case, why was there such a high rate of relapse in clients who discontinued the SSRI after 4 to 8 weeks, once the receptor down-regulation process was complete? Why was it necessary for clients to continue the SSRI for several months in order to maintain the therapeutic benefits, and reduce the risk of relapse upon discontinuation? And why do some clients, especially those suffering from anxiety disorders, require more than 4 to 8 weeks to respond to SSRI antidepressants?

Our emerging understanding of the role of serotonin, and the adverse effects of stress on the brain, provides more of the story. As reviewed by Malberg and Schechter (2005), another effect of overwhelming or prolonged stress on the brain is the possible destruction of neurons in the *hippocampus*, a center in the temporal lobe of the brain essential for memory, cognition, and possibly emotional regulation. Although stress-induced neuronal damage in the hippocampus of humans has yet to be definitively documented, preclinical animal studies have convincingly demonstrated that stress can injure the hippocampus by three different mechanisms: (1) a decrease in the normal rate of proliferation and regrowth of hippocampal neurons, (2) atrophy of the dendritic projections of those neurons, and (3) hippocampal cell death (see Malberg & Schechter, 2005, for a review). Furthermore, once those destructive changes have occurred, the longer the subjects go untreated, even after the stressor is removed, the greater the potential damage to the hippocampus (Malberg & Schechter, 2005). *The injury continues to progress, even in the absence of the stressor.*

One of the most remarkable findings to emerge from animal studies is the role of antidepressant medication in the treatment of stress-induced hippocampal damage. Preclinical studies have demonstrated repeatedly that antidepressant medications can help repair all three types of hippocampal injury: stimulation of new cell growth, normalization of the rate of cell proliferation, and repair of dendritic projections. Antidepressants that restore the normal effects of serotonin and norepinephrine on intracellular signaling and gene expression ultimately activate trophic factors in the brain, including hippocampal *brain-derived neurotrophic factor* (BDNF), which stimulate neurogenesis (Warner-Schmidt & Duman, 2006). Finally, antidepressants (as well as other classes of medication discussed below) have also been shown to increase levels of *neuropeptide-Y* (NPY) in the brain (Husum et al., 2000; Nikisch et al., 2005). As an antagonist of corticotropin-releasing factor (CRF), which contributes to stress-induced neuronal damage, NPY also may contribute to the repair of hippocampal damage, in addition to exerting its own antianxiety effects (Friedman, 2001).

Symptoms Relieved by SSRI Antidepressants

More than for any other class of medications, empirical evidence supports the use of SSRI antidepressants for the treatment of posttraumatic stress, and they are listed as "firstline" agents in all the current treatment guidelines (Cooper, Carty, & Creamer, 2005; Foa, Keane, & Friedman, 2000; Friedman, 2001; Schoenfeld, Marmar, & Neylan, 2004; Department of Veterans Affairs & Department of Defense, 2004). In addition to the long-term benefits of these medications in countering the effects of stress on the hippocampus, as described above, SSRIs have proven effective in reducing all three symptom clusters of posttraumatic stress, including reexperiencing, hyperarousal, and emotional numbing/avoidance. Specifically, SSRIs help reduce irritability, angry outbursts, and frustration intolerance. Sleep improves, and nightmares decrease. Hypervigilance, panic symptoms, and exaggerated startle responses improve or resolve. Clients describe feeling like they have become more "themselves" again, that they have regained their "shock absorbers," their "resilience."

Of course, not all individuals experience complete or even major relief of PTSD symptoms after taking an SSRI antidepressant. But outcome studies have reported *response* rates (defined as moderate or significant improvement) of approximately 60% (e.g., Marshall et al., 2001) and *remission* rates (defined as no longer meeting criteria for the disorder) of approximately 25% (e.g., Davidson, 2004) for an SSRI antidepressant as a sole agent. Response rates have been reported to be poorer for Vietnam-era combat veterans with chronic, severe PTSD, but it is not known to what degree their chronicity by the time treatment was initiated, or secondary gain issues such as disability compensation, may have limited their responses to SSRI treatments (Schoenfeld, Marmar, & Neylan, 2004).

Practical Considerations in the Use of SSRI Antidepressants

Unlike some medications that work on an as-needed basis, SSRI antidepressants only work if they are taken every day for a significant period of time. Animal studies show that chronic dosing regimens are required to repair hippocampal damage and restore normal rates of cell proliferation (Malberg & Schechter, 2005). An exact duration of antidepressant treatment for traumatic stress injuries has not yet been established, but based on treatment paradigms for depression, where the risk of relapse is inversely proportional to the duration of successful treatment, the VA/DoD (Department of Veterans Affairs & Department of Defense, 2004) guideline recommends continuing SSRI treatment for at least a year from the time of symptom remission.

A whole year of taking a psychiatric medication can seem like a huge commitment for young warfighters who may be loath to admitting to themselves or anyone else that they might need that kind of help. To some, taking an SSRI antidepressant, whose names many service members already know (and often deride as crutches for the weak), may seem like adding insult to injury. Little can be done to make this pill easier to swallow. But both authors have had significant clinical success by thoroughly educating their active duty clients about the biological underpinnings of their stress injuries, and how SSRI antidepressants help promote healing. Still, compliance may be problematic, both in frequency of dosing and duration of treatment.

One of the disadvantages of SSRIs is the latency of their therapeutic effect. Although some individuals report feeling better within the first week or two, it frequently takes 3 to 6 weeks before people begin to notice even the slightest bit of improvement. The VA/DoD (Department of Veterans Affairs & Department of Defense, 2004) guideline recommends a therapeutic trial of at least 12 weeks before changing to another medication regimen. While waiting for the therapeutic benefits to begin, clients often feel *worse* for a period of time, due to the initial adverse effects common in these medications during the process of receptor down-regulation, as described above. Some people report no initial adverse effects, but others experience headache, nausea, restlessness, yawning, teeth clenching, or other odd physical sensations. These initial side effects almost always resolve completely within the first few weeks, but they can make already skeptical military clients question whether it is worth trying to wait for the therapeutic action to kick in. Sometimes a second medication is prescribed during this period to alleviate stress symptoms and to mitigate SSRI side effects. Some people experience decreased sexual function on SSRIs, a reversible side effect that often resolves over time, but which sometimes persists for the duration of treatment, and can be a cause for discontinuation of the medication.

The good news is that once past the initial side effects, SSRIs are very well tolerated by most people, and they are safe. They are not toxic to any tissues or organ systems in the body, and do not interfere with cognitive or motor functioning (Devane, 1995; Wadsworth et al., 2005). Each military service has its own rules about the deployability of service members who are taking SSRI antidepressants, but contrary to rumors that frequently circulate in military units, most active duty military members taking SSRI medication remain fit for full duty, including deployment, throughout their treatment. Only service members who engage in special high-risk duties, such as aviation and the handling of nuclear weapons, are prohibited from taking SSRI antidepressants while performing their duties.

FIRST GENERATION ANTIDEPRESSANTS

The VA/DoD (Department of Veterans Affairs & Department of Defense, 2004) guideline classifies the older tricyclic (TCA) and monoamine oxidase inhibitor (MAOI) antidepressants as having "some benefit" in the treatment of PTSD. Both MAOIs and TCAs have been shown to reduce symptoms of reexperiencing and insomnia, with MAOIs outperforming TCAs when the two were compared, but neither was effective in reducing symptoms of emotional numbing and avoidance (Cooper, Carty, & Creamer, 2005; Friedman et al., 2000). But the significant side effects associated with these classes of antidepressants at therapeutic doses, and life-threatening toxicities at higher doses, render them less than optimal for active duty service members suffering from stress injuries. With the development of novel classes of medication, it is doubtful that these older generation drugs will be the subject of further study in the treatment of combat and operational stress.

NEWER ANTIDEPRESSANTS

There are new classes of antidepressants, including dual-action *serotonin norepinephrine reuptake inhibitors* (SNRI), including venlafaxine (Effexor;

Wyeth), and *noradrenergic and specific serotonergic antidepressant* (NaSSA) classes, such as mirtazapine (Remeron; Organon), which show promise for the treatment of posttraumatic stress (Cooper, Carty, & Creamer, 2005; Friedman, 2001; Malberg & Schechter, 2005), and are considered by the VA/DoD (Department of Veterans Affairs & Department of Defense, 2004) guideline as having "some benefit." These medications have a safety profile similar to the SSRIs, and are generally as well tolerated. Whether or not they prove to have any advantages over SSRIs in the treatment of traumatic stress injuries will be determined by future research.

ANTIADRENERGIC ANTIHYPERTENSIVES

Antiadrenergic medications, better known as "alpha-blockers" or "beta-blockers," have been on the market for decades, and are commonly prescribed to control high blood pressure. In psychiatry, these medications have long been used to treat the symptoms associated with performance and social anxiety, and panic, as well as to reduce unwanted side effects, such as tremor and restlessness, caused by other psychiatric drugs (Kelly, 1985). In the past few years, as the neurobiology of the human stress response has become better understood, interest has grown in this class of medication for the treatment of traumatic stress. As described by Nash and Baker (this volume, chapter 4), norepinephrine (NE), epinephrine, and the sympathetic nervous system are central to the human stress response, and are highly implicated in the symptoms of stress injuries. Individuals who have persistently high levels of sympathetic nervous system (adrenergic) activity following a traumatic event are at greater risk to develop PTSD than those whose adrenergic activity returns to baseline. Individuals with chronic traumatic stress injuries continue to have persistently elevated levels of NE and epinephrine, which contribute to the sometimes disabling symptoms of hyperarousal, and interfere with the memory processing that is vital to integrating and healing from traumatic memories (Geracioti et al., 2001; Southwick et al., 1999).

It should be no surprise, therefore, that medications that block the effects of the sympathetic nervous system in the body or the effects of NE in the brain have been found to be helpful in alleviating some of the symptoms of traumatic stress (Friedman, 2001; Department of Veterans Affairs & Department of Defense, 2004). In addition, there is now a growing body of evidence to support the use of antiadrenergic medications to prevent the development of stress injuries in the first place (Pitman et al., 2002; Vaiva et al., 2003).

How Antiadrenergic Antihypertensives Work, and How They Help

Norepinephrine and epinephrine bind to receptors of two broad types, called alpha and beta receptors. Both receptor types are found on target organs throughout the body as well as on neurons in the central nervous system, and activation of these two adrenergic receptor types produces many of the physiological changes characteristic of the mammalian stress response. But alpha and beta receptors, including their many subtypes, initiate different changes in different parts of the body.

Adrenergic receptors in the body (outside the central nervous system) include three subtypes of beta receptor (beta-1, beta-2, and beta-3), and two subtypes of alpha receptor (alpha-1 and alpha-2). Beta-1 receptors are located in the heart, causing an increase in both the rate and force of cardiac contractions, as well as an increase in the conduction velocity of the heart. Beta-2 receptors are located in the lungs, mediating bronchial dilation; in skeletal muscle, mediating potassium uptake; and in the liver, causing an increase in the breakdown of glycogen, leading to the release of circulating glucose into the circulation as fuel for action. The third subtype of beta receptor, beta-3, causes the breakdown of triglycerides, leading to the release of fatty acids into the circulation, though currently there are no medications that act at that receptor. Alpha-1 receptors in the body cause the constriction of smooth muscles surrounding blood vessels, the constriction of bladder sphincter muscles, and the dilation of the iris. Alpha-2 receptors are located on a number of tissues throughout the body, including platelets, where they facilitate platelet aggregation in blood clotting, and in cardiac tissue, where they serve to *decrease* heart rate, modulating the effects of beta-1 stimulation (van Zwieten et al., 1982).

Both alpha-1 and all the beta receptors are located on postsynaptic neurons throughout the central nervous system, where they bind with and are activated by NE (but not epinephrine, which has no activity in the brain). Alpha-2 receptors in the brain, on the other hand, are found only on *presynaptic* sympathetic neurons, where they serve to *decrease* the release of NE through *feedback inhibition* (Brenner & Stevens, 2006). In individuals with stress injuries, these alpha-2 receptors become *down-regulated* (decreased in numbers), thereby decreasing the body's normal mechanism for "turning off" sympathetic arousal (Friedman, 2001).

Antiadrenergic alpha-blocker and beta-blocker drugs inhibit the excitatory effects of NE and epinephrine by preventing NE and epinephrine from binding to those receptor sites. Different drugs bind to different receptor subtypes, resulting in different therapeutic effects. Of the myriad alpha-blockers and beta-blockers on the market, the two that have been most studied to treat the symptoms of traumatic stress are propanolol (Inderal; Wyeth) and prazosin (Minipress; Pfizer). Another class of medications recently studied for its possible usefulness in the treatment of posttraumatic stress is the *alpha-2 agonists*—medications that activate (rather than block) alpha-2 receptors in the brain, potentially enhancing the ability of the brain to put the brakes on its own stress-induced adrenergic activity (Bremner et al., 1996a,b).

Propranolol. The prototype nonselective beta-blocking medication is propranolol, which has activity at both the beta-1 and beta-2 receptors, but neither of the alpha receptors. Because it is highly lipid-soluble, it easily penetrates the blood–brain barrier to exert its antiadrenergic effects both centrally and peripherally. Propranolol has been investigated recently for its efficacy in preventing the development of posttraumatic stress disorder in individuals exposed to a potentially traumatic event. In a randomized, placebo-controlled, double-blind pilot study, trauma-exposed individuals who received propranolol on a scheduled dosing regimen in the days following a potentially traumatic event exhibited fewer symptoms of posttraumatic stress when tested at 1 and 3 months afterward, compared with individuals who received placebo (Pitman et al., 2002). Vaiva et al. (2003) report similar results from their nonrandomized study. On the other hand, a recently reported randomized, placebo-controlled trial of propranolol or the anticonvulsant gabapentin for PTSD prophylaxis in acutely injured civilian trauma victims demonstrated no significant or even trend level differences in subsequent development of PTSD among treatments (Stein, 2006). Additional data are needed to make firm conclusions about the utility of propranolol for the preventive treatment of active duty military personnel in operational environments.

Although there have been no randomized, controlled studies of propranolol in the treatment of established posttraumatic stress, it has been shown in small open trials to be effective in reducing hypervigilance, exaggerated startle, angry outbursts, and intrusive recollections (Friedman et al., 2000; Schoenfeld, Marmar, & Neylan, 2004). Its efficacy in the management of traumatic nightmares, sleep disturbances, and avoidance/emotional numbing has been mixed; in fact, sleep disturbances and vivid dreams have been widely reported in the nonpsychiatric literature as common adverse effects of beta-blocking medications (Kostis & Rosen, 1987; Stoschitzky et al., 1999).

Prazosin. Raskind and his colleagues (2003) at the University of Washington–affiliated Veterans Affairs Puget Sound Health Care System reviewed the preclinical literature on CNS adrenergic hyperactivity, and found data to support the hypothesis that overstimulation of the alpha-1 receptor (compared with beta receptors) may play a prominent role in disrupted sleep physiology, emergence of trauma nightmares, and increased release of the anxiety-producing neurohormone, CRF. They investigated the highly lipid-soluble alpha-1-blocker, prazosin, in a small randomized, placebo-controlled, double-blind study on combat veterans. Not only did prazosin substantially reduce trauma-related nightmares, it also significantly reduced physiologic arousal, distressing intrusive memories, emotional numbing, and avoidance behavior (Raskind et al., 2003). These researchers have been able to replicate these findings in a second, larger placebo-controlled study (Raskind et al., 2006).

Alpha-2 Agonists. Clonidine and guanfacine are centrally acting medications that *activate* the presynaptic alpha-2 receptors, which become downregulated in chronic stress injuries, thereby enhancing the brain's own mechanism of "turning off" the sympathetic outflow of NE (Friedman et al., 2000; Raskind et al., 2006). There have been no randomized, placebo-controlled studies on clonidine in the treatment of traumatic stress, but case reports and open trials have described its usefulness in reducing some of the symptoms of PTSD, including hyperarousal and sleep disturbances (Friedman, 2001; Department of Veterans Affairs & Department of Defense, 2004). Unfortunately, a recent large placebo-controlled guanfacine trial for PTSD in veterans was clearly negative (Neylan et al., in press).

Practical Considerations in the Use of Antiadrenergic Antihypertensives. Antihypertensive medications are generally well tolerated by clients, though some properties are important to note. Because these drugs are designed to control high blood pressure, they need to be started in low doses, and gradually titrated to therapeutic doses in order to avoid hypotension and consequent dizziness after the first few doses. Due to the beta-1 activity of propranolol, there is a decrease in the rate and force of cardiac contractions, as well as a decrease in the velocity of conduction. This beta-1 blockade can result in poor exercise performance and early fatigue if clients try to exercise in the hours following a dose of the medication. The alpha-2 agonists can also cause exercise fatigue due to increased alpha-2 activation, which causes bradycardia (slow heart rate). Athletes taking propranolol, clonidine, or guanfacine should be warned that they will be unable to get their heart rate up to normal levels while taking these drugs, and that their power and endurance will be slightly diminished. Although prazosin does not directly affect cardiac tissue, clients have reported decreased exercise tolerance while taking that medication as well. Exercise fatigue can usually be circumvented easily if clients time their workouts prior to taking the medication, or after the medication has been eliminated. Since propranolol is a nonselective beta-blocker, binding not only to the beta-1 receptors on the heart, but also to the beta-2 receptors in bronchial tissue causing mild broncho-constriction, it may cause bronchospasm in clients who have asthma or reactive airway disease, and must be avoided in those clients. In those cases, prazosin or an alpha-2 agonist may be used. Alternatively, a selective beta-1 blocker may be prescribed, with the caveat that the less lipid-soluble the drug, the less effect that drug will have on decreasing CNS adrenergic activity due to its inability to cross the blood–brain barrier.

One feature of the alpha-blocker, prazosin, in the treatment of sleep disturbances in deployed warfighters deserves particular note. Sleep loss in stress-injured warriors can be a very pernicious problem. Operations often leave little time for sleep, and few if any days off to ever catch up on sleep. With continued sleep debt, healing from stress injuries is impeded, and symptoms of all types are magnified. So it becomes imperative to ensure maximum quality of sleep whenever sleep is permitted. That includes being able to fall asleep, but also staying asleep for as many hours as are available. Stress-injured warfighters who repeatedly bolt upright in a sweat during nightmares are unable to get adequate restorative sleep. But because of operational contingencies, warfighters are often strongly averse to taking any medication that may make them groggy or hard to awaken. Prazosin has the ability to promote restorative sleep in some individuals without impairing alertness. For this reason, alone, it is an invaluable addition to the clinical armamentarium in theater. However, antiadrenergic antihypertensives are not commonly used to treat high blood pressure in active duty service members deployed to a war zone, so these drugs are not as readily available as SSRI antidepressants in theater. One of the authors (WPN) carried a large bottle of prazosin in his backpack throughout his deployment to Iraq, so it could be dispensed to troops in the field when needed.

ANTICONVULSANTS/MOOD STABILIZERS

The rationale for using antiepileptic medications in the treatment of certain psychiatric disorders is based on the neurological theory of *sensitization* or *kindling*. Although other neurobiological factors are involved, seizures are thought to be caused in large part by excessive firing in the neural pathways mediated by the brain's primary excitatory neurotransmitter, glutamate (Loscher, 1998). The kindling phenomenon occurs when repetitive, focal, neuronal stimulation results in a gradual lowering of the seizure threshold, so that it takes less and less stimulation to trigger seizure activity (Friedman, 2001). A hyperexcitable state is created in the brain, leaving delicate tissues vulnerable to the effects of *excitotoxicity* (Sattler & Tymianski, 2000). Many clients with refractory temporal lobe or complex partial epilepsy suffer neuronal loss and sclerosis in their hippocampi due to the sensitization of *N*-methyl-D-aspartate (NMDA) receptors. As Nash and Baker review in chapter 4, a similar sensitization and vulnerability to excitotoxicity occurs in response to persistently high levels of cortisol, due to prolonged or excessive stress. The kindling mechanism has been hypothesized to contribute particularly to the symptoms of increased physiologic reactivity, spontaneous intrusive memories, and flashbacks seen in individuals with posttraumatic stress (Schoenfeld, Marmar, & Neylan, 2004).

How Anticonvulsants and Mood Stabilizers Work, and How They Help

Anticonvulsant medications work by various mechanisms that are not well understood, but which all involve reducing the level of excitatory activity in the brain. Most of the anticonvulsants block sodium channels, which decrease the spread of excitatory neuronal activity, thereby inhibiting the release of glutamate. Lamotrigine (Lamictal; GlaxoSmith-Kline) is an anticonvulsant that works by inhibiting the release of glutamate from presynaptic cells, and attenuating the entry of calcium at the NMDA receptor (Brenner & Stevens, 2006). In one controlled randomized trial, lamotrigine proved effective in reducing symptoms of avoidance/numbing and intrusive memories in individuals with posttraumatic stress (Department of Veterans Affairs & Department of Defense, 2004). Carbamazepine (Tegretol; Novartis) has strong antikindling properties, and in small, open clinical trials was associated with reduction in reexperiencing symptoms as well as irritability, impulsivity, and violent behavior in individuals with PTSD (Cooper, Carty, & Creamer, 2005). Valproate (Depakote; Abbott), an anticonvulsant that increases the activity of the brain's primary inhibitory neurotransmitter, GABA, has been reported to decrease symptoms of hyperarousal in PTSD (Department of Veterans Affairs & Department of Defense, 2004). Topiramate (Topamax; Ortho McNeil), another drug that enhances GABA activity, has been reported to decrease nightmares (Department of Veterans Affairs & Department of Defense, 2004).

Schoenfeld, Marmar, and Neylan (2004) point out that due to their antikindling effects, the anticonvulsant class of medications, like the antiadrenergic antihypertensives, hold the potential to *prevent* the development of symptoms of traumatic stress, if administered to individuals in the first hours to days after experiencing a traumatic event. The VA/DoD guideline classifies this group of medications as having "unknown benefit" in the treatment of posttraumatic stress (Department of Veterans Affairs & Department of Defense, 2004). Much more clinical research is needed to determine whether or not there will be a major role for anticonvulsants in the treatment and/or prevention of traumatic stress injuries.

Although not an anticonvulsant medication, lithium has long proved a powerful mood stabilizer in the treatment and prevention of manic episodes, though the mechanism of action of this simple salt remains unclear (Brenner & Stevens, 2006). It has been reported to decrease symptoms of hyperarousal, anger, and irritability in clients with posttraumatic stress (Schoenfeld, Marmar, & Neylan, 2004). In addition, like the antidepressant medications, chronic lithium treatment has been shown to increase hippocampal neurogenesis in animal studies (Malberg & Schechter, 2005).

Practical Considerations in the Use of Anticonvulsants and Mood Stabilizers

Lithium and the older anticonvulsants, such as carbamazepine and valproate, are associated with numerous side effects, have narrow therapeutic windows, and require regular monitoring of blood levels to guard against toxicity—all of which deters their use in an active duty military population. Some of the newer anticonvulsants, such as lamotrigine, topiramate, and gabapentin, are safer and have more tolerable side effect profiles, rendering them better candidates for use among military personnel. But use of any of these medications by deploying personnel is rare because of their potential side effects and toxicities.

ANTIPSYCHOTICS

Antipsychotic medications are traditionally prescribed to treat "psychoses," a variety of symptoms where an individual's thinking, behavior, perceptions, and ability to relate to others become severely impaired. The most severe form of psychotic disorder is *schizophrenia*, a chronic, progressive mental illness characterized by a gradual loss of range of emotional and verbal expression, and loss of normal behavioral and cognitive function. There is a sort of "fading away" of the full person. These partial losses of self are referred to as the "negative symptoms" of schizophrenia. Schizophrenics also usually develop "positive symptoms," which are distorted perceptions and behaviors that they did not have before. Positive symptoms can include hallucinations, delusions, and exaggerated, aggressive, or disorganized behavior, thinking, and/or speech.

It is important to distinguish between *psychotic symptoms* and *mental illness*. It is possible for people who do *not* have chronic, progressive mental illnesses to experience transient symptoms of psychosis, which, unlike for schizophrenics, are reversible. Sleep deprivation, drug intoxication, drug withdrawal, dehydration, hyperthermia, a number of metabolic disturbances, and stress can induce psychotic symptoms—including hallucinations, paranoia, aggressiveness, and cognitive/behavioral disorganization—in normal people. These symptoms remit once the underlying cause is corrected, though in some cases, there is a delay in remission.

Disruptions in the activity of the neurotransmitter *dopamine* (DA) in two of its four major pathways in the brain have long been hypothesized as integral to the development and expression of psychotic symptoms, and all antipsychotic medications block DA activity to some degree. DA is involved in the regulation of a multitude of complex neurophysiologic functions, however, including motivation, mood, cognition, perception, behavior, voluntary and involuntary motor control, addiction, lactation, and diurnal rhythms, which has made the pharmacologic treatment of psychosis very challenging (Stahl, 2002).

The traditional antipsychotic medications developed in the 1950s and 1960s, now referred to as "typical" antipsychotics, are potent DA blocking agents effective in treating the *positive* symptoms of schizophrenia (e.g., hallucinations, delusions, disorganized behavior) and other psychotic disorders. Because of their nonselective DA blockade, however, they simultaneously exacerbate the *negative* symptoms, or the cognitive and emotional deficits seen in schizophrenia. Moreover, by blocking DA in areas of the brain that have nothing to do with cognition, emotion, or behavior, they can cause unusual side effects, such as lactation, as well as severe (and sometimes permanent) movement disorders.

The first typical antipsychotic, chlorpromazine (Thorazine; GlaxoSmithKline) was used widely during the Vietnam War as a short-term sedative for highly agitated stress-injured troops. Because of their powerful sedating properties, typical antipsychotics remain today in common use in hospital emergency services everywhere. However, typical antipsychotic medications have not been studied extensively in PTSD (Department of Veterans Affairs & Department of Defense, 2004), and due to their adverse side effect profile, it is doubtful that this group of medication will receive any further attention from researchers in the treatment of combat and operational stress injuries. The VA/DoD guideline classifies the typical antipsychotics as providing "no benefit/harm" for the treatment of posttraumatic stress (Department of Veterans Affairs & Department of Defense, 2004).

The second generation of antipsychotic medications, known as "atypical" antipsychotics, is far more tolerable and less toxic than its predecessors. These medications show more promise for use in operational stress injuries.

How Atypical Antipsychotics Work, and How They Help

Like the older, typical antipsychotics, the newer, atypical antipsychotics also block dopamine throughout the brain. But unlike typical antipsychotics, atypicals also block certain serotonin receptors, including the ones that help to regulate the release of DA in the brain. This complex interaction of serotonin receptor inhibition throughout the DA pathways permits the atypical antipsychotic medications to exert their DA blocking effects in the pathways affecting emotion and behavior, while sparing those pathways that regulate cognition, movement, and prolactin inhibition (see Stahl, 2002, for a detailed description of this complex interaction). In addition to DA and serotonin blocking properties, the drugs in this class have activity in other neurotransmitter systems to varying degrees as well, giving each of them unique properties (Stahl, 2002). But all of the drugs in this class tend to be far less sedating and cause far fewer side effects of other types than the older typical antipsychotics.

There have been few studies so far investigating the effectiveness of the atypical antipsychotics as monotherapy for traumatic stress (Cooper, Carty, & Creamer, 2005; Department of Veterans Affairs & Department of Defense, 2004), but there have been some promising case reports of their use as adjunctive therapy in combination with SSRIs, particularly to target explosive, aggressive, or violent behavior (Schoenfeld, Marmar, & Neylan, 2004). Risperidone (Risperdal; Janssen) and olanzapine (Zyprexa; Eli Lilly) have been reported to reduce some of the core symptoms of traumatic stress (Schoenfeld, Marmar, & Neylan, 2004). Some clinicians report success using atypical antipsychotics as augmentation to SSRI therapy for individuals with trauma-related hallucinations, violent behavior, or intense hypervigilance and paranoia (Friedman, 2001). Quetiapine (Seroquel; AstraZeneca), which has alpha-1 blocking and antihistamine properties, has now been shown to improve the quality of sleep, and reduce nightmares in combat veterans with posttraumatic stress, and is frequently prescribed in low doses as a hypnotic (Robert et al., 2005).

Practical Considerations for the Use of Atypical Antipsychotics

While far safer than the first generation medications in this class, some of the atypical antipsychotics are associated with a spectrum of side effects, such as weight gain and sedation, which may preclude their widespread use among military personnel. Some of them have also been associated with inducing diabetes. Unlike the antidepressant medications and lithium, antipsychotic medications have *not* been shown to stimulate hippocampal neurogenesis or the rate of neuronal cell proliferation (Malberg & Schechter, 2005). Larger randomized controlled trials are needed to assess the effectiveness of this class of medication, both as monotherapy and augmentation therapy, and to determine optimal duration of treatment and dosing strategies.

HYPNOTIC AND SEDATIVE MEDICATIONS

Sleep disturbance is frequently the chief symptom that causes active duty personnel with stress injuries to seek medical attention. Many soldiers and Marines are willing to put up with persistent intrusive memories, anxiety, irritability, nightmares, emotional numbing, avoidance, and flashbacks. But a year or two of getting by on 2 to 4 hours of interrupted sleep each night disables the hardiest of warriors. In addition to the subjective discomfort resulting from inadequate sleep, sleep deprivation and disruption of sleep architecture have been associated with an increase in physical health problems (Friedman & Schnurr, 1995); impaired motor and cognitive performance, and impaired learning (Durmer & Dinges, 2005); and weight gain (Gangwisch et al. 2005).

What exactly goes on neurochemically in *normal sleep* is not well understood, let alone how and why sleep becomes disrupted in traumatic stress, or how those characteristic sleep disturbances contribute to other symptoms and sequelae of traumatic stress injuries. An entire volume could be devoted to the complex subject of sleep, which is only skeletally reviewed here.

Normal sleep architecture is divided into two phases: nonrapid eye movement (NREM) and rapid eye movement (REM) sleep. NREM sleep is further divided into four stages, 1 through 4, which are distinguished from one another on the basis of brain wave activity measured by electroencephalography. In stages 1 and 2, brain waves are fast, but disorganized compared with wakefulness. Stages 3 and 4, also known as "deep sleep," are characterized by very slow brain waves, called "delta" waves. In a single *sleep cycle,* an individual progresses sequentially from stages 1 through 4 sleep, then brain wave activity picks up again, going through NREM stages in reverse order before the onset of REM sleep. In REM sleep, when dreaming occurs, brain wave activity is very active, the body's major antigravity muscles are immobilized, and there are bursts of rapid eye movements. In normal night's sleep, an individual will go through four to five sleep cycles, but the time spent in some phases and stages changes as the night progresses: time spent in REM sleep increases with each cycle throughout the night, and time spent in deep sleep decreases (Harvey, Jones, & Schmidt, 2003). Although the purposes of each stage and phase of sleep are not well understood, Harvey, Jones, and Schmidt (2003) summarize the best current hypotheses—that stages 1 to 2 of NREM serve as transitional stages, that stages 3 to 4 of NREM are involved in tissue restoration, and that REM sleep is needed for such things as learning, memory consolidation, and emotional processing.

Sleep architecture becomes disrupted in the stress-injured individual (Lavie, 2001), though no consistent pattern has been identified from the numerous conflicting data that have emerged from investigations, and the conclusions drawn from the conflicting data can be confusing. It is clear that individuals with stress injuries maintain increased arousal levels at night (Mellman et al., 1995), and that sleep deprivation, particularly REM sleep deprivation, leads to disruptions in hippocampal function and plasticity (McDermott et al., 2003). From these data, it is fair to conclude that restoring normal sleep is a crucial component in the process of recovering from traumatic stress injuries.

Sleep can be treated by nonpharmacological methods (Harvey, Jones, & Schmidt, 2003), but if those fail, pharmacotherapy may be helpful. On the other hand, some medications that are commonly used to induce sleep have effects on sleep architecture that may have a negative effect on the healing process.

Since agitation and anxiety are also common symptoms of all types of stress injuries, and are common presenting symptoms for stress-injured warriors, hypnotic and sedative drugs are frequently used to help manage these disorders, even though the empirical support for their use is not strong. Four classes of commonly used sedatives and hypnotics will be briefly considered: (1) benzodiazepines, (2) nonbenzodiazepine hypnotics, (3) miscellaneous medications which are sedating as a side effect, and (4) alcohol.

Benzodiazepines

Benzodiazepines are second-generation sedative medications, introduced to the market in the 1960s to replace the highly addictive, highly lethal *barbiturates.* The nonlethal benzodiazepines were initially believed to possess all of the benefits of the barbiturates—for anxiety, insomnia, muscle spasm, spasticity, certain types of epilepsy, and the induction of surgical anesthesia—with none of the risks. But just 3 years after the first benzodiazepine, diazepam (Valium; Roche), was released, Mick Jagger's satirical "Mother's Little Helper" was playing on radios around the world, exposing what a powerfully seductive—and mainstream—drug of abuse this had become.

How Benzodiazepines Work, How They Help, and How They Harm. Commonly prescribed benzodiazepines used to treat anxiety or agitation include alprazolam (Xanax; Pharmacia & Upjohn), lorazepam (Ativan; Baxter), diazepam (Valium), and chlordiazepoxide (Librium; Valeant). Benzodiazepines commonly prescribed to promote sleep include triazolam (Halcion; Pharmacia & Upjohn), flurazepam (Dalmane; Valeant), and oxazepam (Serax; Wyeth-Ayerst). All of them work the same way—by enhancing the activity of gamma-aminobutyric acid (GABA), the primary inhibitory neurotransmitter in the brain. This accounts for their notorious ability to decrease arousal and cause sedation. However, they cause a number of adverse effects, similar to the effects of alcohol, which are problematic, particularly for active duty military members. These include loss of motor coordination, dizziness, impaired cognitive processing, impaired judgment, and impaired planning. These medications, like alcohol, also cause emotional and behavioral disinhibition. In addition, benzodiazepines can produce *anterograde amnesia*—memory loss for the period of time that the medication is active in the CNS. These medications can also be associated with a significant amount of rebound anxiety, insomnia, irritability, and headache once the dose is eliminated from the body (Brenner & Stevens, 2006).

In the few published studies assessing the effectiveness of benzodiazepines in traumatic stress, benzodiazepines have not been shown to alleviate the core symptoms of PTSD, and may actually interfere with clients' ability to desensitize their fear response to triggers for traumatic memories (Schoenfeld, Marmar, & Neylan, 2004). Though benzodiazepines, particularly those with shorter half-lives, can induce sleep rapidly, they also *decrease the quality* of sleep by increasing the time spent in stages 1 and 2, decreasing the time spent in stages 3 and 4, and decreasing the time spent in REM (Brenner & Stevens, 2006). Therefore, the use of benzodiazepines as hypnotics may actually interfere with the process of healing from stress injuries. All the clinical practice guidelines recommend against prescribing benzodiazepines for posttraumatic stress, although Cooper, Carter, and Creamer (2005) suggest that they may have limited use as augmentation therapy. The VA/DoD (Department of Veterans Affairs & Department of Defense, 2004) guideline does not support the use of benzodiazepines as augmentation therapy, and places them in the "no benefit/harm" category for the treatment of posttraumatic stress disorder, but in the "unknown benefit" category for acute stress disorder.

Practical Considerations in the Use of Benzodiazepines. Due to the risks for dependency, impairment of cognitive and motor function, emotional and behavioral disinhibition, anterograde amnesia, rebound anxiety, and negative effects on sleep architecture—as well as the availability of many safer alternatives—benzodiazepines should be avoided in active duty clients suffering from combat/operational stress injuries. Short-term use may be appropriate in circumstances where their use would be limited—for example, to manage severe agitation or during medical evacuation—but if prescribed for such purposes, benzodiazepines should be discontinued and/or replaced with a more appropriate medication at the earliest possible opportunity.

Nonbenzodiazepine Hypnotics

Nonbenzodiazepine hypnotics include the four most popular sleep medications in current use: zolpidem (Ambien; Sanofi Aventis), zaleplon (Sonata; Jones), eszopiclone (Lunesta; Sepracor), and zopiclone (Imovane; Aventis; not available in the United States). Although these medications are structurally unrelated to benzodiazepines, they are believed to induce their hypnotic effects by potentiating the activity of GABA, much like benzodiazepines do.

GABA channels are complicated membrane gates on neuronal surfaces, with different binding sites for different neurochemicals. There is a receptor site for GABA, the brain's own inhibitory neurotransmitter. But curiously, there are receptor sites for exogenous chemicals as well. There is a receptor site for alcohol, and one for barbiturates, and there are two receptor sites for benzodiazepines called "omega" receptors. Benzodiazepines bind to both omega-1 and omega-2 receptors in order to enhance the brain's GABA activity. The *nonbenzodiazepine* hypnotics, however, bind only to the omega-1 receptor sites. The nonbenzodiazepine hypnotics induce sleep rapidly, but unlike benzodiazepines, they do *not* increase stages 1 and 2 sleep, nor do they decrease stages 3 and 4, or REM sleep—effects that are believed to result from omega-2 activation (Brenner & Stevens, 2006). These medications vary in their half-lives, with zaleplon having the shortest (1 hour) and eszopiclone the longest (5–7 hours).

Nonbenzodiazepine hypnotics are schedule IV controlled substances in the United States, meaning that they have low potential for abuse, but that there may be some limited physical or psychological dependency if they *are* abused. In their review of case reports on zolpidem and zopiclone, Hajak et al. (2003) concluded that, compared with benzodiazepines, there is a far lower incidence of dependency, but that there is an increased risk for abuse of this class of medication among clients with a history of substance abuse or dependence.

Miscellaneous Medications Which Are Sedating as a Side Effect

Many medications commonly prescribed for insomnia are not actually hypnotics at all. It is common practice for clinicians to prescribe low-dose tricyclic antidepressants such as amitriptyline (Elavil; AstraZeneca), atypical antidepressants such as trazodone (Desyrel; Apothecon), antihistamines such as diphenhydramine (Benadryl; Parke Davis), antipsychotics, and muscle relaxants as off-label hypnotics. Recent research on the atypical antipsychotic, quetiapine (Seroquel), has shown this medication to be of benefit in the treatment of dysregulated sleep for those with posttraumatic stress (Robert et al., 2005). Olanzapine (Zyprexa) has been shown to improve disordered sleep in PTSD clients when prescribed as an adjunct to SSRIs (Stein, Kline, & Matloff, 2002). Much more research is needed, however, to weigh the benefits against the risks of medications to treat sleep disturbances in these individuals.

As a group, medications that are sedating merely as a side effect have the advantage of having a very low potential for habituation, addiction, or abuse. These may be the primary reasons for their use in stress-injured warriors and veterans, especially since posttraumatic stress disorder is known to frequently be associated with substance abuse and dependence. When prescribing these medications to active duty warfighters, however, other important considerations include their potential to cause persistent grogginess and sedation after sleep has ended. These "hangover" side effects of such medications are often a result of their usually long half-lives. While deployed to an operational environment, service members will often simply refuse to take any medication that renders them unable to spring into action when necessary. They would rather get only a couple of hours of sleep per 24-hour day than to get 6 to 8 hours but suffer from grogginess afterward. Obviously, the operational tempo must be taken into account before such medications are offered to active duty service members.

Alcohol

The effects of alcohol on brain chemistry are identical to those of benzodiazepines: loss of motor coordination, dizziness, impaired cognitive processing, impaired judgment, impaired planning, emotional and behavioral disinhibition, and anterograde amnesia. Alcohol also has negative effects on sleep, causing fragmentation of sleep architecture, shorter duration of sleep, reduction in stages 3 and 4 sleep, and disturbances of REM sleep (Landolt & Gillin, 2001). Although not well understood, alcohol can continue to exert its deleterious effects on sleep for months to years, even after the chronic drinker has stopped drinking, which Landolt and Gillen (2001) conclude may have something to do with the effects of alcohol on the GABA and serotonin neurotransmitter systems. Schmitz et al. (1996) have demonstrated that secretion of melatonin, which is produced by the pineal gland and serves to initiate sleep, is inhibited by alcohol, and remains inhibited during abstinence in chronic drinkers. Recent research in humans and animals has demonstrated that the hippocampus is particularly sensitive to the deleterious effects of alcohol. In an extensive review of the literature, White, Matthews, and Best (2000) conclude that moderate alcohol intake disrupts hippocampal function in a number of ways, leading to impairment in learning and memory.

As a medicine for sleep, alcohol could not be more disastrous. Tolerance to it can develop so fast that even after a single large dose, there are withdrawal symptoms (the mirror image of tolerance) such as hypersensitivity to sound and light. But because of these acute tolerance and mini-withdrawal phenomena, someone taking "a couple stiff ones" just to get to sleep is likely to wake up "wired" after just a few hours, unable to get back to sleep. If this person then drinks more to get back to sleep, he or she is on the path to alcohol dependence and abuse.

Practical Considerations in the Use of Alcohol. Unfortunately, because alcohol is available, affordable, accepted by the military culture, and legal, many young men and women with stress injuries use it to alleviate stress and/or to try to fall asleep, not realizing that neurochemically, alcohol not only prevents them from healing, but exacerbates their symptoms. For many, it is far easier to pick up a bottle at the 7-day store or the PX than it is to walk into the medical treatment facility and ask for help. Frequently, these stress-injured service members are involved in alcohol related incidents, including fights, intoxicated driving charges, and episodes of domestic violence, with consequential loss of rank, income, family, or worse.

Preventing and recognizing deleterious self-medication with alcohol by stress-injured warfighters is the shared responsibility of many. First and foremost, commanding officers are responsible for making it safe for service members to seek help for their combat/operational stress injuries. Commanding officers must educate their NCOs about how to recognize when troops are suffering from combat/operational stress, and when they may be abusing alcohol to try to cope. But military service members are trained to "suck it up," and to their credit, are skilled at remaining professional, even when they may be enduring extraordinary personal hardship. Sometimes the first sign of trouble is involvement in an alcohol-related incident. In those cases, it is important that commanders consider whether or not the soldier or Marine has deployed, and to offer the opportunity for treatment along with nonjudicial punishment.

NCOs have their fingers on the pulses of their troops, though, far more than their commanding officers. NCOs must ensure that they do not promote and condone the machismo of heavy drinking, especially among those who have served in combat zones. NCOs must reinforce the importance of taking responsibility, not only for physical fitness and readiness, but of mental fitness and readiness as well.

Medical personnel, especially primary care providers in military treatment facilities, must screen for potential combat/operational stress injuries and concomitant alcohol abuse. Every clinical encounter is an opportunity to educate military members about the common symptoms of combat/operational stress and the deleterious effects of alcohol in the recovery process, and to offer treatment.

NOVEL CLASSES OF MEDICATION

Above we have reviewed the medications currently available in the pharmacologic armamentarium for prescription to the stress injured. In addition, entirely new classes of medication are under development, with novel mechanisms of action, which may prove effective in the treatment and/or prevention of stress injuries in the future. Malberg and Schechter (2005) describe how the strides made in our understanding of the neurobiology of stress and depression have opened up the potential of developing new drugs that increase the activity of trophic factors in the brain, such as BDNF, and inhibit the activity of damaging neurochemicals and neurohormones, such as CRF. Readers can go to the U.S. National Institutes of Health Website (http://www.clinicaltrials.gov) to learn about novel drugs that are currently in various phases of clinical trial, which hold the potential for safe, rapid, effective treatment, and/or prevention of stress injuries.

CONCLUSION

The remarkable advances in the fields of neurobiology and psychopharmacology have begun to make it possible for us to offer today's stress-injured service members something that has not been available for past generations of warriors: a rational approach to restoring some of the biological disruptions that can result from the stress of warfighting. Medication certainly cannot treat all of the wounds sustained from combat or operational stress, and not every person exhibiting stress symptoms needs to take a medication. But because medication can provide significant symptomatic relief from suffering, and can contribute to the restoration of normal neurobiological function, it should always be considered as part of a comprehensive treatment plan for stress-injured service members. Thanks to the painstaking work done by dedicated experts in the field, guidelines are available to assist clinicians in prescribing safe, effective medication regimens. These options will only continue to improve, as more data become available from current and future research. It is important that clinicians and commanders work together to help reduce the stigma long associated in the military with taking psychiatric medication, to ensure that warfighters have access to all of the best available treatment options.

REFERENCES

American Psychiatric Association. (1980). *Diagnostic and Statistical Manual of Mental Disorders* (3rd ed.). Washington, DC: Author.

American Psychiatric Association. (2000). *Diagnostic and Statistical Manual of Mental Disorders* (4th ed., text rev.). Washington DC: Author.

Brady, K., Pearlstein, T., Asnis, G. M., Baker, D., Rothbaum, B., Sikes, C. R., et al. (2000). Efficacy and safety of sertraline treatment of posttraumatic stress disorder: A randomized controlled trial. *Journal of the American Medical Association, 283,* 1837–1844.

Bremner, D. J., Krystal, J. H., Southwick, S. M., & Charney, D. S. (1996a). Noradrenergic mechanisms in stress and anxiety: I. Clinical studies. *Synapse 23*(1), 28–38.

Bremner, D. J., Krystal, J. H., Southwick, S. M., & Charney, D. S. (1996b). Noradrenergic mechanisms in stress and anxiety: II. Clinical studies. *Synapse 23*(1), 39–51.

Brenner, G. M., & Stevens, C. W. (2006). *Pharmacology* (2nd ed.). Philadelphia: Saunders Elsevier.

Cooper, J., Carty, J., & Creamer, M. (2005). Pharmacotherapy for posttraumatic stress disorder: Empirical review and clinical recommendations. *Australian and New Zealand Journal of Psychiatry, 39,* 674–682.

Davidson, J. R. T. (2004). Remission in post-traumatic stress disorder (PTSD): Effects of sertraline as assessed by the Davidson Trauma Scale, Clinical Global Impressions and the Clinician-Administered PTSD scale. *International Journal of Psychopharmacology, 19,* 85–87.

Department of Veterans Affairs & Department of Defense. (2004). *VA/DoD Clinical Practice Guideline for the Management of Post Traumatic Stress.* Washington, DC: Office of Quality and Performance. Available at http://www.oqp.med.va.gov/cpg/PTSD/PTSD_Base.htm

Devane, C. L. (1995). Comparative safety and tolerability of selective serotonin reuptake inhibitors. *Human Psychopharmacology: Clinical and Experimental, 10*(S3), S185–S193.

Domino, E. F. (1999). History of modern psychopharmacology: A personal view with an emphasis on antidepressants. *Psychosomatic Medicine, 61,* 591–598.

Durmer, J. S., & Dinges, D. F. (2005). Neurocognitive consequences of sleep deprivation. *Seminars in Neurology, 25,* 117–129.

Foa, E. B., Keane, T. M., & Freidman, M. J. (Eds.). (2000). *Effective Treatments for PTSD.* New York: Guilford Press.

Friedman, M. J. (2001). Allostatic versus empirical perspectives on pharmacotherapy for PTSD. In J. P. Wilson, M. J. Friedman, & J. D. Lindy (Eds.), *Treating Psychological Trauma & PTSD.* New York: Guilford Press.

Friedman, M. J., Davidson, J. R. T., Mellman, T. A., & Southwick, S. M. (2000). Pharmacotherapy. In E. B. Foa, T. M. Keane, & M. J. Friedman (Eds.), *Effective Treatments for PTSD.* New York: Guilford Press.

Friedman, M. J., & Schnurr, P. P. (1995). The relationship between trauma, posttraumatic stress disorder, and physical health. In M. J. Friedman, D. S. Charney, & A. Y. Deutch (Eds.), *Neurobiological and Clinical Consequences of Stress: from Normal Adaptation to Posttraumatic Stress Disorder.* Philadelphia: Lippincott-Raven.

Gangwisch, J. E., Malaspina, D., Boden-Albala, B., & Heymsfield, S. B. (2005). Inadequate sleep as a risk factor for obesity: Analyses of the NHANES I. *Sleep, 28,* 1289–1296.

Geracioti, Jr., T. D., Baker, D. G., Ekhator, N. N., West, S. A., Hill, K. K., Bruce, A. B., et al. (2001). CSF norepinephrine concentrations in posttraumatic stress disorder. *American Journal of Psychiatry, 158,* 1227–1230.

Hajak, G., Muller, W. E., Wittchen, H. U., Pittrow, D., & Kirch, W. (2003). Abuse and dependence potential for the non-benzodiazepine hypnotics zolpidem and zopiclone: A review of case reports and epidemiological data. *Addiction, 98,* 1371–1378.

Harvey, A. G., Jones, C., & Schmidt, D. A. (2003). Sleep and posttraumatic stress disorder: A review. *Clinical Psychology Review, 23,* 377–407.

Husum, H., Mikkelsen, J. D., Hogg, S., Mathe, A. A., & Mork, A. (2000). Involvement of hippocampal neuropeptide Y in mediating the chronic actions of lithium, electroconvulsive stimulation, and citalopram. *Neuropharmacology, 39,* 1463–1473.

Kaplan, H. I., & Sadock, B. J. (Eds.). (1985). *Comprehensive Textbook of Psychiatry/IV.* Baltimore: Williams & Wilkins.

Kelly, D. (1985). Pharmacology of stress: Beta-blockers in anxiety. *Stress Medicine, 1,* 143–152.

Kostis, J. B., & Rosen, R. C. (1987). Central nervous system effects of beta-adrenergic-blocking drugs: The role of ancillary properties. *Circulation, 75,* 204–212.

Landolt, H. P., & Gillin, J. C. (2001). Sleep abnormalities during abstinence in alcohol-dependent patients: Aetiology and management. *CNS Drugs, 15,* 413–425.

Lavie, P. (2001). Sleep disturbances in the wake of traumatic events. *New England Journal of Medicine, 345,* 1825–1832.

Loscher, W. (1998). Pharmacology of glutamate receptor antagonists in the kindling model of epilepsy. *Progress in Neurobiology, 54,* 721–741.

Malberg, J. E., & Schechter, L. E. (2005). Increasing hippocampal neurogenesis: A novel mechanism for antidepressant drugs. *Current Pharmaceutical Design, 11,* 145–155.

Marshall, R. D., Beebe, K. L., Oldham, M., & Zaninelli, R. (2001). Efficacy and safety of paroxetine for chronic PTSD: A fixed-dose, placebo-controlled study. *American Journal of Psychiatry, 158,* 1982–1988.

McDermott, C. M., LaHoste, G. J., Chen, C., Musto, A., Bazan, G. N., & Magee, J. C. (2003). Sleep deprivation causes behavioral, synaptic, and membrane excitability alterations in hippocampal neurons. *Journal of Neuroscience, 23,* 9687–9695.

Mellman, T. A., Kumar, A., Kulick-Bell, R., Kumar, M., & Nolan, B. (1995). Nocturnal/daytime urine noradrenergic measures and sleep in combat-related PTSD. *Biological Psychiatry. 38.* 174–179.

Neylan, T. C., Lenoci, M., Franklin, K. W., Metzler, T. J., Henn-Haase, C., Hierholzer, R. W., et al. (in press). Guanfacine does not improve symptoms of posttraumatic stress disorder. *American Journal of Psychiatry.*

Nicholi, Jr., A. M. (Ed.). (1988). *The New Harvard Guide to Psychiatry.* Cambridge, MA: Belknap Press of Harvard University Press.

Nikisch, G., Agren, H., Eap, C. B., Czernik A., Baumann, P., & Mathe, A. A. (2005). Neuropeptide Y and corticotrophin-releasing hormone in CSF mark response to antidepressive treatment with citalopram. *International Journal of Neuropsychopharmacology, 8,* 403–410.

Pitman, R. K., Sanders, K. M., Zusman, R. M., Healy, A. R., Cheema, F., Lasko, N. B., et al. (2002). Pilot study of secondary prevention of posttraumatic stress disorder with propranolol. *Biological Psychiatry, 51,* 189–192.

Raskind, M. A., Peskind, E. R., Hoff, D. J., Hart, K. L., Holmes, H. A., et al. (2006, in press) A parallel group placebo controlled study of prazosin for trauma nightmares and sleep disturbance in combat veterans with posttraumatic stress disorder. *Biological Psychiatry.*

Raskind, M. A., Peskind, E. R., Kanter, E. D., Petrie, E. C., Radant, A., Thompson, C. E., et al. (2003). Reduction of nightmares and other PTSD symptoms in combat veterans by prazosin: A placebo-controlled study. *American Journal of Psychiatry, 160,* 371–373.

Robert, S., Hamner, M. B., Kose, S., Ulmer, H. G., Deitsch, S. E., & Lorberbaum, J. P. (2005). Quetiapine improves sleep disturbances in combat veterans with PTSD: Sleep data from a prospective, open-label study. *Journal of Clinical Psychopharmacology, 25,* 387–388.

Sattler, R., & Tymianski, M. (2000). Molecular mechanisms of calcium-dependent excitotoxicity. *Journal of Molecular Medicine, 78,* 1432–1440.

Schmitz, M. M., Sepandj, A., Pichler, P. M., & Rudas, S. (1996). Disrupted melatonin-secretion during alcohol withdrawal. *Progress in Neuro-psychopharmacology and Biological Psychiatry, 20,* 983–995.

Schoenfeld, F. B., Marmar, C. R., & Neylan, T. C. (2004). Current concepts in pharmacotherapy for posttraumatic stress disorder. *Psychiatric Services, 55,* 519–531.

Southwick, S. M., Bremner, J. D., Rasmusson, A., Morgan, III, C. A., Arnsten, A., & Charney, D. S. (1999). Role of norepinephrine in the pathophysiology and treatment of posttraumatic stress disorder. *Biological Psychiatry, 46,* 1192–1204.

Stahl, S. M. (2000). *Essential Psychopharmacology of Depression and Bipolar Disorder.* New York: Cambridge University Press.

Stahl, S.M. (2002). *Essential Psychopharmacology of Antipsychotics and Mood Stabilizers.* New York: Cambridge University Press.

Stein, M. (2006). *Pharmacoprevention of adverse psychiatric sequelae of physical injury.* Presented at the 21st Annual Meeting of International Society for Traumatic Stress Studies, Toronto, November 2006.

Stein, M. B., Kline, N. A., & Matloff, J. L. (2002). Adjunctive oanzapine for SSRI-resistant combat-related PTSD: A double-blind, placebo-controlled study. *American Journal of Psychiatry, 159,* 1777–1779.

Stoschitzky, K., Sakotnik, A., Lercher, P., Zweiker, R., Maier, R., Liebmann, P., et al. (1999). Influence of beta-blockers on melatonin release. *European Journal of Clinical Pharmacology, 55,* 111–115.

Vaiva, G., Ducrocq, F., Jezequel, K., Averland, B., Lestavel, P., Brunet, A., et al. (2003). Immediate treatment with propranolol decreases posttraumatic stress disorder two months after trauma. *Biological Psychiatry, 54,* 947–949.

van Zwieten, P. A., van Meel, J. C., de Jonge, A., Wilffert, B., & Timmermans, P. B. (1982). Central and peripheral alpha-adrenoreceptors. *Journal of Cardiovascular Pharmacology,* 4(S1), 19–24.

Wadsworth, E. J. K., Moss, S. C., Simpson, S. A., & Smith, A. P. (2005). SSRIs and cognitive performance in a working sample. *Human Psychopharmacology: Clinical and Experimental, 20,* 561–572.

Warner-Schmidt, J. L. & Duman, R. S. (2006). Hippocampal neurogenesis: Opposing effects of stress and antidepressant treatment. *Hippocampus, 16,* 239–249.

White, A. M., Matthews, D. B., & Best, P. J. (2000). Ethanol, memory, and hippocampal function: A review of recent findings. *Hippocampus, 10,* 88–93.

12

The Royal Marines' Approach to Psychological Trauma

CAMERON MARCH AND NEIL GREENBERG

HISTORICAL DEVELOPMENT

The link between the effects of warfighting and psychological trauma has been known for centuries. The effects can trace their lineage from the campaigns of the ancient Roman and Greek armies to the punitive World Wars of the last century. However, from the 1914–1918 Great War, and latterly, since the Vietnam War, an ever-increasing level of attention has been paid to the corrosive effect of trauma on serving military personnel and veterans. Looked at in depth, the way that trauma has been perceived by the military has not followed a consistent or even path. The reality is that the perception of stress within the military has been a mixture of stops, starts, false turns, and disinterest. To quote Ben Shephard (2000) from his excellent book, *A War of Nerves*, the military treatment of stress is "at first denied, then exaggerated, then understood, then ignored" (p. 53). There is much truth in his statement; as we militarily campaigned into the latter part of the 20th century, there was little put in place to deal with the effects of traumatic stress. Indeed, the expectation was that you "drove or cracked on" and ensured that you completed the task. The essential ethos for the U.S. and British Marines was: "Mission, Men, Self"—always in that priority.

There is much to commend this thought process; after all, the role of the military is to fight, or, to quote British military historian Richard Holmes, "[t]o take the bayonet to Her Majesty's enemies." It is this requirement, which requires robust personnel to achieve it, that poses the inherent dilemma for senior ranking military officers. In years gone by, put simply, the chain of command was under the clear impression that introducing trauma support policies would in turn open the floodgates and provide the shirker, the timid, and the malingerer to exhibit stress reactions in order to avoid doing their duty.

In general terms, people who volunteer for service with at-risk professions (including the military, police, emergency service, and frontline journalists) are usually aware that they may potentially put themselves into harm's way. Thus, as a self-selecting group they are more likely to have effective coping skills and display above average levels of resilience and fortitude. Against this background, such organizations have a moral responsibility to look after their personnel and meet their duty of care. Closely linked to this requirement is the economic imperative to keep highly trained and valued people in service. The failure to address work-generated psychological distress can be economically devastating. Furthermore, in the absence of a credible protocol, personnel are likely to be less effective at work and less likely to remain within the parent organization.

Not addressing the issue of duty of care can have substantial and costly legal ramifications. All organizations, including the military, have a legal requirement to look after their employees both physically and psychologically as far as reasonably practicable. Under British law this interaction is known as the "master and servant" relationship. In the U.K., organizations are also bound by the overarching requirements of the Health & Safety at Work Acts. It is easy, therefore, to come to the conclusion that the "do nothing" option has ceased to become a viable option at all. For the U.K. Ministry of Defence (MoD), this was profoundly brought into focus by the advent of the "MoD PTSD Case" (Applegate, 2003). It may be useful to examine this case in more detail.

THE MOD PTSD CASE

The joint-action case was bought against the MoD by a number of ex-military personnel. The basis of their claim was three-fold: first, that the MoD had not prevented them from getting posttraumatic stress disorder (PTSD); second, that the MoD had not detected their problems at an early stage; and third, that the MoD had not treated them effectively after they had subsequently developed PTSD. Counsel for the claimants argued that by nature of their employment, the MoD knew that their personnel could be exposed to traumatic stressors. Their case was that in view of this the MoD should have been aware and have done more to assist personnel in the preparation for and the aftermath of challenging situations. The case was opened in February 2002 and the findings were presented some 15 months later.

It is always difficult to distill many pages of legal summing up into a simple paragraph, but in essence the judgment was broadly in favor of the

MoD. This was on the basis that the MoD systems of practice (late 1960s until 1996) were sufficient and in line with accepted practice. However, the judge in his judgment also made a number of other telling points. He stated that posttraumatic stress reactions were an organizational issue for the military to address, and that there should be sufficient training for military personnel managers to be able to exercise their duty of care. Although the judge ruled that the MoD had discharged its duty of care for the time period in question, it was clear that this was a wakeup call for the MoD and that they have in future to keep up with current evidence-based interventions. The issue of postincident management is a dynamic one, and trauma risk management (TRiM) was discussed (briefly) during the case as a model that was likely to be of use after traumatic events. One of the other findings from the case was that the judge made it clear that there was no absolute duty on individuals toward their comrades during war. Therefore, the traditional and pivotal buddy-buddy system that has underpinned the U.K. Armed Forces is based on a purely voluntary basis and the individual does not have a duty of care to his colleague. A MoD appraisal document suggested that as a result of the case, military units should ensure that NCOs and officers were skilled in the detection of adverse posttraumatic psychological issues (but did not indicate how those skills were to be acquired), and that military culture should become more accepting that posttraumatic stress reactions were a real issue.

TRAUMA RISK MANAGEMENT (TRIM)—THE BACKGROUND

At this juncture, and against the background detailed in earlier paragraphs, it may be useful to examine the approach made by the Royal Marines on this subject. The Royal Marines provide Britain's amphibious landing force and are a major component in the country's Special Forces. Since the conclusion of World War II to the present day there has only been 1 year (1968) that the Corps of Royal Marines have not been engaged in active service. The roles of the Corps range from direct warfighting operations to peacekeeping and humanitarian deployments. It was against this background that in 1996 the Royal Marines reviewed its stance on the introduction of a stress control policy. Despite the concerns relating to mental fortitude as discussed above, it was that felt that something positive needed to be done. The drive to "do something" and actively support our personnel came from then Colonel (now Major General) David Wilson. David Wilson was a highly experienced combat Royal Marines Command, who, at that time, was employed out of his more normal operational field, as the Head of Royal Marines Personnel Branch at Headquarters Royal Marines in Portsmouth. While not sure which stress protocol to adopt, he felt intuitively that action was needed to address this potential problem. As a result, orders were issued and budgets allocated to progress matters further.

It was from this decision that TRiM was born. The origins of the system presented in this chapter were drawn together over a period of years, the original idea being devised by Major Norman Jones and Captain (rtd) Peter Roberts, two community psychiatric nurses who had vast experience in the British Army Medical Services. As it developed over the subsequent years, the system was adapted and implemented within the Royal Marines under the guidance of Major Cameron March and Surgeon Commander Neil Greenberg.

Trauma risk management or TRiM is a maturation of the initial before-during-after (BDA) system that Jones and Roberts had pioneered. The strategy was wrought from experiences during humanitarian and peacekeeping operations, as well as warfighting. After all such operations, it has become accepted practice that some form of psychological support should be available for service personnel. In many ways this military culture is a reflection of the culture of the wider civilian community. As such, until the year 2000, the U.K. military operated under the policy that if a unit experienced a traumatic event, then single-session psychological debriefings would be undertaken. However, as scientific literature began to suggest that formal debriefings such as critical incident stress debriefing (CISD) were of questionable effectiveness (Rose, Bisson, & Wessely, 2003), and may have the potential to cause additional distress, it was clear that other ways of managing the aftermath of potentially traumatic events (PTEs) needed to be found. The evidence available at that time did, however, suggest that non-CISD interventions, based on both practical support and multiepisodic interviews, might have benefit after potentially traumatic events (Bisson, 2003; Van Emmerik et al., 2002).

As the Royal Marines used and refined their TRiM model, in 2000, things were also moving within the MoD. Based on emerging scientific literature, the Military Surgeon General (the Head of Medical Services for the British Armed Forces) banned the practice of routine debriefings for military personnel (Revision of Surgeon General's Policy Letter 7/95, 2000). As well as altering the previous policy by banning psychological debriefing, the British military had other practical factors to consider when selecting a replacement trauma intervention protocol. These factors derive from logistical and geographical constraints, the paucity of readily available psychological assets, and its credibility among its target audience. When viewed in the round, one is steered to the view that an effective traumatic stress management strategy should be initially delivered by known peer group personnel embedded with their military units, rather than parachuting in outsiders (Jones et al., 2003). Experience gleaned over the last 9 years within the Royal Marines indicate that the employment of peer group personnel or stress practitioners have great utility in overcoming the stigma commonly associated with psychological distress (Greenberg, Cawkill, & Sharpley, 2005).

The strategy presented in this chapter is based on the use of peer group practitioners who have been trained in how to assist the command to deal with the psychological aspects of traumatic events. The strategy does not focus on forced emotional ventilation as a therapeutic measure; rather, it builds on everyday supportive peer group interaction and the concomitant social support which results. It is known that military personnel who do want to speak about their operational experiences prefer to speak to a peer rather than to other forms of support such as medical staff or managers (Greenberg et al., 2003), and thus a peer-delivered strategy is more likely to be acceptable than other forms of possible intervention.

PSYCHOLOGICAL MANAGEMENT FOLLOWING A TRAUMATIC EVENT

TRiM is not based on the medical management of stress or CISD (Mitchell, 1983). It is an informed, common sense, postincident management protocol implemented by trained individuals working to a strict code of practice. The protocol is not intended to deal with occupational stress. However, the embedded TRiM practitioners are by virtue of their training empathic, and have skills which may be able to assist with certain occupational stress issues. Naturally, this is not their prime function. Primarily their role is to manage those personnel who have been exposed to traumatic episodes.

Traumatic Stressors

By their very nature traumatic events can be very variable, and sometimes it is misleading to think purely about the event without considering actual consequent impact the event has had on the individual. By virtue of our antecedence, one's degree of support, one's coping and psychological defensive mechanisms, and simply the nature of our employment are all likely to affect the way military personnel perceive traumatic incidents. TRiM practitioners receive training to aid them in distinguishing between normal reactions and more extreme reactions. In some cases, it is a matter of the intensity and duration of the reaction rather than the characteristics of the reaction that dictates whether the individual may encounter subsequent formal psychological difficulties.

TRiM—The Three Pillars

TRiM stands on three primary pillars:

- **Education**—Education can be divided into pre- and postincident strategies. Education is given to all recruits on entering the Royal Marines and is built on during subsequent professional and promotion courses. In the aftermath of warfighting and peace keeping activities, operational "decompression" forms an essential element of our returning preparations. The exact nature of the decompression process will be dependent on the nature of the operation and available resources.

- **Risk Assessment**—The risk assessment process using the before–during–after matrix provides a central pillar of the strategy. Its primary aim is to gauge the amount of stress a person has assimilated. By measuring at predetermined times—3 days, 1 month, and in some cases 3 months—it is possible to assess an individual's recovery rate. Risk assessment can be conducted as a singleton or group process. The group is never more than eight people and all groups are conducted by at least two stress practitioners.
- **Mentoring**—Following the risk assessment process, contact is maintained with the individual who have experienced the process. The stress practitioner is in a position to assist and mentor in the months following the traumatic event.

THE BEFORE–DURING–AFTER MATRIX

The BDA grid (Figure 12.1) was developed from a technique called functional analysis (part of behavioral psychotherapy) by Peter Roberts and Norman Jones some years ago. It is a simple interviewing tool that can be used for groups or individuals. It enables TRiM practitioners to discuss the incident with the participants in a narrative format. It allows for a proper assessment of risk in order that an effective management

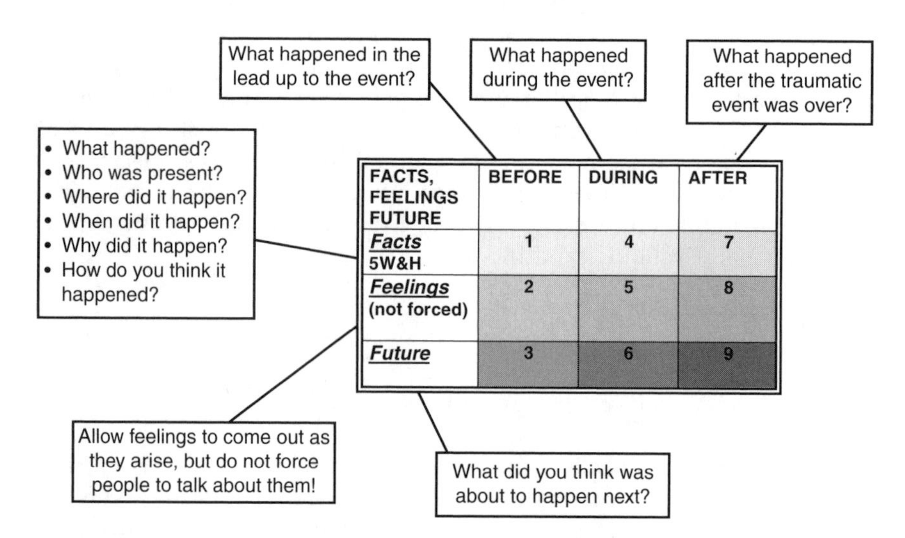

FIGURE 12.1 The Before-During-After (BDA) Matrix.

strategy can be initiated. Also, if required, early referral for treatment can be quickly instituted.

The Before–During–After Phases

By explaining clearly and adhering to the three distinct phases of BDA, the TRiM practitioner and the participants navigate themselves safely through the risk assessment process.

The Before Phase. Normally the conversation begins by discussing events that occurred perhaps 2 to 3 hours or the day before the event. This phase is important in the building of rapport between those assessing and those being assessed.

The During Phase. This is the heart of the assessment. Surprisingly the risk assessment is often the first time that those involved have met and discussed the event in a structured and semiformal manner! Normalization of reaction can also occur in this phase by recognizing that many other group participants are experiencing similar reactions.

The After Phase. This covers all the events relating to the conclusion of the incident to the present. It encompasses current coping patterns and/or substance misuse. At the conclusion of this phase, a number of simple coping strategies are discussed.

The Two TRiM Assessments

The TRiM module is based on two separate risk assessments. Each has a differing character and feel. As outlined, the initial risk assessment can be conducted as a singleton or a group. However, the 1-month risk assessment is always conducted on an individual basis. Experience has shown that this allows a more open and frank discussion.

The 3-Day TRiM Risk Assessment. The initial, 3-day TRiM risk assessment has two primary functions. First, it allows for a simple needs assessment for those who have been exposed to a traumatic event. Second, it sets a baseline by which the second TRiM risk assessment, implemented at 1 month, can be compared. The TRiM practitioner should pay attention to how the participant is coping. If the individual continues to be effectively engaged with work, colleagues, peers, and family, then it is likely that little, if any, further intervention is required at this stage.

The 1-Month TRiM Assessment. The 1-month TRiM assessment is vitally important. It is carried out in a different format from the earlier assessment. To avoid retraumatization the narrative tool is not employed. Stress practitioners, working collaboratively with the individual, compare and contrast initial scores and gauge the persons' coping mechanisms. If they are making progress their efforts are validated, but if they continue to experience psychological distress they may be considered to be at risk and access to further formal help may be appropriate.

The TRiM Scoring System

Each risk factor (Figure 12.2) is scored on a 0-1-2 scale, giving a maximum score of 20. A score of 0 indicates that nothing is present. Score 1 is appropriate when the risk factor is present to a moderate degree. A score of 2 suggests that the risk factor is strongly present. The recorded scores should not be considered in isolation at the 1-month assessment but the two scores are compared and contrasted.

TRiM Risk Factors

One of the central objectives of this protocol is to assess how much stress a person has assimilated in the aftermath of a traumatic incident. Currently, there is no definitive profile to identify who will develop psychological problems following a PTE. However, there is a growing body of research evidence that indicates that certain factors may reliably predict if a person may develop problems. In this section we attempt to define the risk factors employed in this protocol.

The 10 risk factors used in the risk assessment tool are not exhaustive (Figure 12.2). They do, however, represent a useful cross-section of the findings from previous research studies. There are also other risk factors which may predict subsequent development of posttraumatic syndromes. These include the level of education of the victim, ethnicity, IQ, and posttraumatic heart rate.

The following sections are abridged amplifying notes for the ten risk factors listed in Figure 12.2.

Risk factor 1—The person felt that they were out of control during the event.

Those who experience the stressor as uncontrollable or unpredictable are more likely to experience psychological problems (Foa, Zinbarg, & Ruthbaaum, 1992).

NB: The event is invariably out of control; this factor relates to a person's perception of his personal level of control during the event.

Risk factor 2—The person felt that their life was threatened during the event.

Individual reactions to trauma have also been shown to be important, and individuals who felt that their life was threatened have been shown

	Risk Factor			
		0	1	2
1	The person felt that they were out of control during the event			
2	The person felt that their life was threatened during the event			
3	The person blamed others for what happened			
4	The person feels ashamed about their behaviour during the event			
5	The person perceives the event as serious (likely to involve death, serious injury or near miss)			
6*	The person experienced acute stress following the event			
7	The person is having problems with day to day activities			
8	The person has been involved in previous traumatic events			
9	The person has poor social support, (Family, Friends, Unit Support)			
10	The person has been drinking alcohol excessively to cope with their distress			
	Total (max = 20)			

FIGURE 12.2. Initial and one-month risk assessment

to have an increased risk of developing future psychological problems (Brewin et al., 1996; Udwin et al., 2000).

NB: The main point relating to this factor is the significant threat to a person's personal safety.

Risk factor 3—The person blamed others for what happened.

Blame often directed at an organization or an employer. People can articulate high levels of anger. In many cases the anger is unrealistic and everyone, including the risk assessor, can be a target.

Risk factor 4—The person feels ashamed about their behavior during the event.

Personnel who feel a strong sense of shame after an event are also at increased risk of developing longer-term psychological problems (Brewin et al., 1998).

NB: Although shame is often unrealistic, it is often very powerfully felt.

Risk factor 5—The person perceives the event as being serious (likely to involve death, serious injury, or a near miss).

Studies of posttraumatic stress disorder suggest that the severity of the traumatic event may affect outcome, so that a high-intensity, long-duration event can be associated with the development of posttraumatic illnesses (Kessler et al., 1995).

Risk factor 6—The person experienced acute stress following the event.

People who have been exposed to trauma and move on to develop an acute stress reaction are more likely to have longer-term psychological difficulties (Brewin et al., 1998; Yehuda, 2004). Acute stress reactions are

TABLE 12.1
Acute Stress Matrix

Acute Stress Aid Memoir

1. Upsetting thoughts or memories about the event that have come into their mind against their will.
2. Upsetting dreams about the event.
3. Acting or feeling as though the event were happening again.
4. Feeling upset by reminders of the event.
5. Bodily reactions (such as fast heartbeat, stomach churning, sweatiness, dizziness) when reminded of the event.
6. Difficulty falling or staying asleep.
7. Irritability or outburst of anger.
8. Difficulty concentrating.
9. Heightened awareness of potential dangers to oneself and others.
10. Being jumpy or being startled at something unexpected.

discussed in detail below. The presence of this risk factor may be the most powerful individual predictive risk factor.

NB: Refer to the acute stress checklist above (Table 12.1). This ten-item checklist is taken from research undertaken by Brewin and colleagues (2002).

Risk factor 7—The person is having problems with day-to-day activities.

Evidence from numerous studies have shown that people who are not able to maintain functional effectiveness in their day-to-day life remain at increased risk of having the longer-term problems (Lecrubier, 2004).

NB: This concerns how people function in their lives, occupationally and socially. It includes significant deviations from normal adaptive routines.

Risk factor 8—The person has been involved in previous traumatic events.

Those who have experienced previous significant trauma are at increased risk of going on to develop posttraumatic illness when exposed to additional future traumatic events (McFarlane, 1988).

NB: The importance of this factor is strengthened when this and previous trauma events share similar and significant key points, and when it occurred in the recent past.

Risk factor 9—The person has poor social support (family, friends, unit support).

One pivotal finding derived from a large body of research is that accessible social support is only beneficial if it is perceived as being useful. This is particularly so when associated with lower levels of psychological difficulty following a traumatic event (e.g., Green, Grace, & Gleser, 1985; Norris et al., 2002). When the subject remains isolated from their social support, they remain at risk of developing psychological problems.

NB: The key point here is whether people are taking strength and support from whatever source is suitable to them.

Risk factor 10—The person has been drinking alcohol excessively or using nonprescribed medication to cope with their distress.

There is significant evidence that suggests that alcohol misuse is associated with poor psychological coping strategies. This misuse of alcohol may signify impending posttraumatic stress disorder illnesses (Lecrubier, 2004; Zweben, Clark, & Smith, 1994).

NB: The key point here is to check out how someone's use of alcohol has altered since the event, not just a check on unhealthy drinking habits and lifestyle issues.

TRiM Risk Assessment—The Structure

Confidentiality is emphasized (see below). Participants are told that there will be no extensive notetaking, tape recording, or other detailed recording of the proceedings. However, it should be emphasized that a trauma risk assessment form will be completed. It may be necessary to record *brief* points notes during the interview. In addition, the interviewer may wish to seek advice about subsequent action from managers and other key personnel. The outcome of the assessment should be discussed with the participant(s), positive coping strategies are emphasized, and agreement is gained should there be a need to take matters further. If during the course of the assessment the stress practitioner becomes substantially concerned that the individual has the potential to harm themselves or others, the stress practitioner is taught that he needs to discuss his concerns with someone else (such as a medic or his chain of command). However, he or she will always discuss any actions with the individual before embarking on this course of action and this is emphasized during the initial confidentiality discussion.

Following the initial risk assessment, appropriate management strategies should be considered depending on the initial risk assessment score. With the agreement of the individual and the chain of command, the practitioner ensures that potentially distressed individuals are able to access social support (peers, team, friends, and family) while protected from excessive stress as much as reasonably possible. However, all TRiM practitioners are told about the benefits of continued active employment and the potential risk of separating a possibly distressed individual from his peers. TRiM aims to conserve the fighting force wherever possible and not to create casualties by overmedicalization of normal human distress. TRiM practitioners may also be in a position to offer appropriate practical advice and mentor distressed individuals by virtue of their training and organizational experience.

SUMMARY

Within the military forces of the Western nations, the nature of warfighting has changed over recent years. The isolation caused by dispersment policies, the changing nature of the battle space, and the enhanced lethality of weapon systems have all potentially ratcheted up the stress levels felt by military personnel. The "for Queen and country whatever the costs" mentality has gone; soldiers are more questioning about the types of wars they are required to fight. Throughout history we have been faced with the choice of "psychologically good wars" and "psychologically bad wars." Psychologically good wars have clearly defined aims and timelines, they enjoy broad public support, and troops are engaged in combat with a clearly defined enemy. An added bonus in this definition is that your side is victorious. Conversely, a psychologically bad war has ill-defined aims, no end state, a faceless enemy, and poor public support. In the armed forces of the Western world, the laws and ethics of armed conflict are scrutinized in the closest detail. Through a combination of litigation and fear, Western culture is now extremely risk-aversive. Correspondingly, there is an apprehension that realistic training, when combined with an awareness of traumatic stress, will weaken the moral and fighting fabric of the military.

We believe that there is merit in allowing the chain of command at the battalion, ship, or squadron level to be involved with management of traumatic stress and that stress should not be made a primarily medical concern. Psychology and medical personnel are there to support command structures and not to replace them where stress is concerned. Empowerment of the command may be accomplished by the deployment of an embedded peer group–delivered system such as the TRiM model. We believe that active consideration of the potential for longer-term psychological disability by the command using a nonmedical approach is likely to effectively manage a problem that otherwise has the potential to be denied and avoided by both the distressed individual and the organization. We further believe that for all parties it is important to keep the corrosive effect of traumatic stress in perspective. For instance, after the first Gulf War, only 3% of British troops deployed in theater went on to develop PTSD (Ismail et al., 2002).

To requote Ben Shephard's words, "at first denied, then exaggerated, then understood, then ignored." We hold the conviction that exaggerating the scale of the problem is as deleterious as attempts to ignore it. The chain of command should actively encourage fortitude and resilience among its service men and women, and senior management should strive to promote the ethos of being "strong in stress." However the flipside to the promotion of resilience is to ensure that effective and credible trauma control systems are put in place which assist individuals to readily access services when their "bank account of courage" becomes expended. We contend that the TRiM concept supports this aspiration, and is perhaps best illustrated by the words of a highly experienced Royal Marine Warrant Officer who commented on TRiM in the aftermath of the warfighting operation in Iraq during Operation Telic (the 2003 Iraq War): "The guys sometimes see it as a bit fluffy, but actually, they really like to know it's there." For too long now the stigma preventing distressed service personnel from asking for help and the barriers which prevent care from being rendered have had deleterious effects for both individuals and the wider military. We believe that an effective peer delivered support program such as TRiM can be an important tool to prevent those who are "down" from also being "out."

REFERENCES

Applegate, D. (2003). *Lessons learnt from the PTSD group action, service personnel board.* Paper by the Stress Project Leader. London: Ministry of Defence.

Bisson, J. (2003). Early intervention following traumatic events. *Psychiatric Annals, 33,* 37–44.

Breslau, N., Chilcoat, H. D., Kessler, R. C., Davis, G. C. (1999). Previous exposure to trauma and PTSD effects of subsequent trauma: results form the detroit area survey of trauma. *Am J Psychiatry 156,* 902–907.

Brewin, C., Andrews, B., & Rose, S. (1998). *A Preventative Programme for Victims of Violent Crime.* NHSE Research and Development Programme Final Report.

Brewin, C. R., Rose, S., Andrews, B., Green, J., Tata, P., McEvedy, C., et al. (2002). A brief screening instrument for post-traumatic stress disorder. *British Journal of Psychiatry, 181,* 158–162.

Brewin, C. R., Dalgleish, T. and Joseph, S. (1996). A Dual Representation Theory of Posttraumatic Stress Disorder. *Psychological Review, 103,* 670–686.

Foa, E. B., Zinbarg, R., & Ruthbaaum, B. O. (1992). Uncontrollability and unpredictability in posttraumatic stress disorder: an animal model. *Psychological Bulletin, 112,* 218–238.

Greenberg, N., Maingay, S., Iversen, A., Palmer, I., Hull, L., Unwin, C., et al. (2003). Perceived psychological support of UK military Peacekeepers on return from deployment. *Journal of Mental Health, 12,* 565–573.

Greenberg, N., Cawkill, P., Sharpley, J. (2005). How to TRiM away at post traumatic stress reactions: Traumatic risk management—now and the future. *Journal of the Royal Naval Medical Service, 91,* 26–31.

Green, B. L., Grace, M. C., & Gleser, G. C. (1985). Identifying survivors at risk: Long-term impairment following the Beverly Hills Supper Club fire. *Journal of Consulting and Clinical Psychology, 53,* 672–678.

Jones, N., Roberts, P., & Greenberg, N. (2003). Peer-group risk assessment: A post-traumatic management strategy for hierarchical organizations. *Occupational Medicine, 53,* 469–475.

Kessler, R. C., Sonnega, A., Bromet, E., Hughes, M., & Nelson, C. B. (1995). Posttraumatic stress disorder in the National Comorbidity survey. *Archives of General Psychiatry, 52,* 1048–1060.

Ismail, K., Kent, K., Brugha, T., Hotopf, M., Hull, L., Seed, P., et al. (2002). The mental health of UK Gulf war veterans: Phase 2 of a two phase cohort study. *British Medical Journal, 325,* 576.

Lecrubier, Y. (2004). Posttraumatic stress disorder in primary care: A hidden diagnosis. *Journal of Clinical Psychiatry, 65*(Suppl. 1), 49–54.

Macklin, M. L., Metzger, L. J., Litz, B. T., et al. (1998). Lower pre-combat intelligence is a risk factor for posttraumatic stress disorder. *J Consult Clin Pyshcol, 66,* 323–326.

McFarlane, A. C. (1988). Posttraumatic morbidity of a disaster. A study of cases presenting for psychiatric treatment. *Journal of Nervous and Mental Disease, 174,* 4–14.

Mitchell, J. (1983). When disaster strikes.... the critical incident stress debriefing procedure. *Journal of Emergency Medical Services, 8,* 36–39.

Norris, F. H., Friedman, M. J., Watson, P. J., Byrne, C. M., Diaz, E., & Kaniasty, K. (2002). 60,000 disaster victims speak: Part 1. An empirical review of the empirical literature, 1981–2001. *Psychiatry, 65,* 207–239.

Revision of Surgeon General's Policy Letter 7/95—Stress-Related Disorders. Ministry of Defence. London.. October 2000.

Rose, S., Bisson, J., & Wessely, S. (2003). Psychological debriefing for preventing post traumatic stress disorder (PTSD) (Cochrane Review). Cochrane Library, 1. Oxford: Update Software.

Shalev, A. Y., Pitman, R. K., Orr, S. P., Peri, T., & Brandes, D. (2000). Auditory startle in trauma survivors with PTSD: A prospective study. *American Journal of Psychiatry, 157,* 255–261.

Shephard, B. (2000). *War of Nerves, Soldiers and Psychiatrists 1914–1994.* London: Jonathan Cape.

Udwin, O., Boyle, S., Yule, W., Bolton, D., & O'Ryan, D. (2000). Risk factors for long-term psychological effects of a disaster experienced in adolescence: Predictors of post traumatic stress disorder. *Journal of Child Psychology and Psychiatry, 41,* 969–979.

van Emmerik, A. A., Kamphuis, J. H., Hulsbosch, A. M., &Emmelkamp, P. M. (2002). Single session debriefing after psychological trauma: a meta-analysis. *Lancet* Sep 7; 360(9335):766–71.

Yehuda, R. (2004). Risk and resilience in posttraumatic stress disorder. *Journal of Clinical Psychiatry, 65*(Suppl. 1), 29–36.

Zweben, J. E., Clark, H. W., & Smith, D. E. (1994). Traumatic experiences and substances abuse: Mapping the territory. *Journal of Psychoactive Drugs, 26,* 327–344.

13

The Operational Stress Injury Social Support Program: A Peer Support Program in Collaboration Between the Canadian Forces and Veterans Affairs Canada[1]

STEPHANE GRENIER, KATHY DARTE, ALEXANDRA HEBER, AND DON RICHARDSON

The views expressed in this chapter are solely those of the authors. They do not necessarily reflect the policy or the opinion of any agency, including the Government of Canada and the Canadian Department of National Defence and Veterans Affairs Canada.

INTRODUCTION[2]

The Creation of Operational Stress Injury Social Support

Canada has a proud tradition of participating in United Nations (UN) peacekeeping missions. Since 1947, 125,000 Canadian military peacekeepers have participated in UN and North Atlantic Treaty Organization (NATO) peace enforcement and humanitarian activities (http://www.forces.gc.ca/site/operations/past_ops_e.asp). Like many other NATO countries, Canada experienced a significant increase in operational tempo following the end of the Cold War. Canadians have served in the first Gulf War, Somalia, the former Yugoslavia, Rwanda, East Timor, Haiti, and Cambodia, and are currently deployed to Afghanistan. The last 15 years have, therefore, produced seasoned military veterans who have experienced wars and conflicts around the globe.

UN and humanitarian missions to places such as Rwanda, Somalia, and the former Yugoslavia closely resemble traditional warfare (Litz, 1996). Posttraumatic stress disorder (PTSD) is one of the significant operational stress injuries (OSI) resulting from exposure to combat (Hoge et al., 2004; Kulka, 1990; Lipton, 1986). During these operations, military members may be exposed to profoundly disturbing events that include witnessing extreme atrocities, death of children, comrades being killed and wounded, as well as feelings of responsibility for the death of a friend from the local population (American Psychiatric Association, 2000). Studies have found that the rate of PTSD among peacekeepers varies from 2.5% to 20% (Birenbaum, 1994; Passey and Crockett, 1999; Litz et al., 1997; MacDonald et al., 1998; Passey and Crocket, 1995; Richardson, 2002; Statistics Canada, 2002; Weisaeth et al., 1996). Other studies of peacekeeping have demonstrated that the rate of PTSD among peacekeepers varies from 8% to 20% (Birenbaum, 1994; Litz et al., 1997; Passey, 1995).

Rwanda was a particularly difficult deployment for Canadian soldiers. In 1994, Lieutenant Colonel Stephane Grenier was part of a small group of Canadian military members deployed as part of the United Nations military force in Rwanda. There he witnessed one of that century's most horrific genocides, where close to 1 million people were slaughtered during a civil war that lasted less than 5 months. The experience of this deployment was instrumental in the development of the Operational Stress Injury Social Support (OSISS) program.

LIEUTENANT COLONEL GRENIER'S EXPERIENCE

After almost 10 months of service in Rwanda, I returned home to a country that for the most part had forgotten what had happened in Rwanda. Having experienced firsthand the atrocities during that country's civil war, and the aftermath of a society of displaced people and refugees, my return to Canada to my previous life was difficult, and I was invaded with a strong feeling that I did not belong. Nevertheless, as with most soldiers returning from difficult deployments, the disturbances in sleep, the recurring nightmares and flashbacks, were considered normal. Although my spouse regularly had to wake me from nightmares during the night, and she noticed significant changes in my behavior, we did not realize that these experiences were likely the first signs of what would become a debilitating mental health condition.

Six months after my return from Rwanda, I had trouble being around my coworkers and family. Somehow nothing I was experiencing back home fit with who I had become. As time went on, I started to isolate myself. I thought that I was the only one feeling this way, and I was terribly ashamed that I was not coping. I sought medical attention. However, I did not find it helpful. After showing up at the National Defence Medical Centre following a night where I had come close to committing suicide, I was eventually referred to a psychiatrist for an assessment. Having never been in therapy before, I had little idea how talking could help with the many feelings and experiences I was having. Of course, I had no difficulty in articulating clearly the difficulties I was experiencing at work and the many conflicts I was experiencing with everyone around me. Being a military officer, I felt compelled to minimize the impact of my tour in Rwanda. Even in the relative safety of the doctor's office, I did not feel I could admit that I felt as if all of this was caused by what I had experienced in Rwanda. I focused on conflicts I was having in the workplace. Probably partly due to this, I was told I had a personality disorder and a problem with anger. Feeling frustrated and misunderstood after this first encounter with the mental health community, I went back home and disposed of all my newly prescribed medications by flushing them down the toilet. I also made the decision to get over my problems on my own.

For 2 years, I managed to mask the condition that was slowly worsening by becoming a workaholic. But although I learned to mask my symptoms at work, I was unable to hide my condition from my family. As time went on, the undiagnosed and untreated injury I had sustained in Rwanda was only getting worse. Over time it became impossible for me to mask the problems and behaviors at work. By then I was employed as a media liaison officer for the Canadian Forces, responding to media queries of national scope on behalf of the Canadian military. My behavior was becoming increasingly aggressive with my superiors and antisocial with my colleagues.

As a result of this out-of-character behavior, I was forced by my superiors to seek medical attention. Because I feared losing my career, when I saw my doctor I minimized my symptoms to receive a "clean bill of health." To cope, I once again immersed myself in work. To avoid working in Canada, and having to face my chronic sense of alienation, I volunteered to travel to remote locations such as Cambodia, Lebanon, the Persian Gulf, and Haiti. Like many other veterans affected by what they had to do and witness overseas, anywhere in the world was now more comfortable than home. As I often heard other veterans say, "I was addicted to adrenaline."

This obsession with work eventually led to a promotion and a posting. It was only after being posted to Toronto in an Army Headquarters, where my level of activity came to a virtual standstill, that I was forced to confront the demons I had been avoiding for years. Having sought medical treatment several times in the past, my negative experiences, combined with the fear that I would be branded as weak or as a malingerer, motivated me once again to try to deal with this on my own. Then, one of my superiors, Colonel Chris Corrigan, the Chief of Staff of the Headquarters and my boss, took me aside and acknowledged that I had endured a significant amount of hardship in the past years and that my behavior seemed to be inconsistent with the reputation I had as an armored corps officer. The understanding and empathy that Colonel Corrigan showed that day was probably the most significant event that had occurred since my return from Rwanda. He provided me with the confidence I needed to seek help and tend to my injuries. As I slowly started to recover from posttraumatic stress disorder, I began thinking of my experience with Colonel Corrigan and wondered how many other veterans were not as fortunate as me and how many of them were still trying to deal with these demons on their own.

On the family front, my tolerance level as a father was not what it should have been. I slowly retreated, and thereby influenced my family to also become reclusive. With the clarity of hindsight I now know why. I did not feel comfortable around others; I felt we had nothing in common. The only people I enjoyed being with were veterans of the mission I had been on. I felt no one was interested in what had happened overseas. People would rarely ask about my tour. I remember clearly listening to endless conversations around the dinner table with my brother-in-law and sister-in-law after they returned from a week of vacation in the Caribbean. As I listened, I wondered why no one had ever asked me about my 10 months in Rwanda. It felt like I had so much to say but no one to say it to. As for my spouse, I later realized how difficult it had been for her to stay behind and watch the news every night, wondering if I was still alive. This had affected her profoundly, and no doubt explained why she was not eager to discuss Rwanda with me. She had her own demons to face. For myself, I felt that I simply did not fit into society anymore and somehow everyday that went by provided yet another example that I did not belong. For a few years, I had resisted attending the traditional family summer gathering at my aunt's cottage. But after some convincing I had given in and decided to see relatives I had not seen in years. Close to 40 people were enjoying a beautiful summer day, swimming, waterskiing, and walking about as I was sitting on the end of the dock. One of my young cousins approached me. She was about 7 years old at the time, and she tapped on my shoulder and asked me gently if I was still "crazy from the war in Rwanda." I remember going through a range of emotion. I realized that she had not come up with this thought on her own. She had likely heard the adults talking about me. Although no one ever mentioned the word Rwanda to me, somehow it had come to this 7-year-old's attention that I was not well because of Rwanda. I found it very odd, as I had never outwardly exhibited signs of my suffering to these people. Yet they knew. But no one approached me to ask me how I was doing. Worse yet, no one asked my wife how she was doing. Everyone knew that something was wrong, but the stigma was such that it paralyzed everyone in the family. It kept them from providing that natural support to a loved one.

As the years went by I was driven to search for a solution to the stigma, shame, and isolation victims of posttraumatic stress disorder endure. I wanted to remove the barriers that often prevent soldiers from seeking help. I reviewed the literature and it became clear to me that social support was key to the recovery process. However, most of the research in this field examined the effects of a lack of social support, rather than on the positive effects social support could have on military members recovering from war-related mental health conditions. The literature was clear on the association between the absence of social support and the subsequent development of PTSD (Brewin, Andrews, & Valentine, 2000). However, there was limited research on the effects of enhanced social support on treatment outcomes. Through reviewing the available research, and talking to many veterans, it became clear to me that if the military was to provide the best care for serving and retired members, more was needed than the traditional medical model of care.

In March 2001, a Canadian infantryman who had served in the Balkans and in Africa earlier in the 1990s drove his sport utility vehicle through the headquarters of the Edmonton, Alberta army base in the late hours of the night. It was this desperate act that provided me with the impetus to create a peer support program. The media quickly concluded that the soldier was suffering from posttraumatic stress disorder. I became interested in how the Canadian military had treated this individual. My Commander at the time, knowing of my interest in this subject, allowed me to visit the individual. On arriving in Edmonton, I signed the individual out of the psychiatric ward of the local hospital and spent the day with him, discussing his ordeal. We talked about how the Canadian military could have better assisted him. I discussed my vision for a program to provide peer support to those suffering from PTSD, and sought his opinion on whether or not such a program would have helped him. Without a doubt this soldier's positive response to my query motivated me to put forward a recommendation to launch what has now become a nationwide peer support program for the Canadian military.

Three weeks later, I was tasked by Lieutenant General Christian Couture, who was responsible for Canadian Military Personnel at the time, to develop the concept of a peer support program for those suffering from deployment-related stress disorders. In May 2001, I began to work officially as the program manager and made the decision to focus on the importance of terminology and its impact. From the outset, I recognized that soldiers like me do not want to be labeled as "mentally ill." I believed that diagnostic terms such as posttraumatic stress disorder led to further stigmatization within the soldiering community.

During combat, soldiers can be physically injured. This kind of injury is considered honorable. However, others, like me, can be injured psychologically. The difference is, the type of injury we sustained is not visible. Because of this, these psychological injuries are viewed as less honorable. When I consulted my colleagues and a number of veterans on the value of referring to these problems as injuries of combat situations, much like the physical wounds that others sustain, their reaction was unanimous. They supported the creation of a term that would help classify psychological injuries the same way as physical ones. Inspired by a paper authored by Dr. Allan English, reviewing the historical and contemporary interpretations of combat stress reaction (English, 1999), the term operational stress injury was therefore created and became part of our new program's name. Since then, the term has been officially defined by the Department of National

Defence & Veterans Affairs Canada (2002) and endorsed by the Canadian Forces Surgeon General.

Definition

An operational stress injury is any persistent psychological difficulty resulting from operational duties performed by a Canadian Forces member. The term OSI is used to describe a broad range of problems which usually result in impairment in functioning. OSIs include diagnosed medical conditions such as anxiety, depression, and PTSD, as well as a range of less severe conditions, but the term OSI is not intended to be used in a medical or legal context. From the onset of the program, some clinicians warned me of the dangers of involving people suffering from operational stress injuries in helping others suffering from the same conditions. Warnings of negative health outcomes for those employed to support others, combined with the fear that peers would attempt to "become clinicians or rescuers," motivated me to add the words "social support" to the name of this new program. The program became known as the "Operational Stress Injury Social Support," or OSISS. To illustrate some of the resistance experienced in the early stages, I particularly remember an animated discussion during which I had presented the notion that my employees would "listen, assess, and refer" as part of their functions. Some clinicians took exception to my suggestion that nonclinicians would be permitted to use the term "assess and refer." I remember putting an end to this debate by quoting the Oxford Dictionary and pointing out the fact that this word was not for the exclusive use of clinicians and that we would ensure through proper training that OSISS Peer Support Coordinators understood clearly the limitations of their particular assessment.

Six months after being assigned the task to develop the concept of peer support into a full-fledged program, the vision had been translated into a business case and was presented for endorsement to the senior decision-making body in the Canadian Forces, Armed Force Council. In October 2001, Armed Force Council approved the OSISS program and mandated the program to not only provide peer support across the country, but to look at shifting the attitudes toward operational stress injuries in the Canadian Forces by raising awareness and understanding and creating acceptance that these nonphysical injuries are honorable wounds.

Over the years since the OSISS program was launched, there are countless success stories telling of the need to formalize peer support in large organizations such as the Canadian military and veteran populations to better serve Canadian Forces members, veterans, and their families affected by operational stress. There has been countless research reporting on the negative effects of isolation and loss of peer support from clients and consumers of mental health services. However, there is very little research on the phenomenon that occurs when peer support is organized, formalized, and proactively injected into an injured population such as the military and veterans populations affected by OSIs. I have theorized that although peer support can be very positive, in many cases, especially within a military population suffering the effects of OSIs, left to its own, peer support could be, or become, very destructive and even an obstacle to recovery.

In the late 1990s and early 2000, as I was researching the question of peer support, I discovered how many veterans were supporting themselves socially; and while some entertained positive and supportive relationships, many on the other hand did not. It was not uncommon to hear of small groups of veterans getting together to consume excessive amounts of alcohol or illicit drugs. One group I came in contact with in 2000 told me that one night, a group of veterans decided to play poker using their medication as poker chips, for a few good laughs. The evening ended on a relatively positive note; however, some veterans left with the wrong medication. One veteran in particular had a very negative reaction and was admitted to hospital after he decided to consume the medication he had "won" rather than his own. After more than 5 years of successful operations, it is our firm belief that OSISS is tangible evidence that through a formal government-driven peer support program, it is possible to remove the potentially negative elements found within the fabric of natural support and ensure that all support received and given by peers remains positive and conducive to recovery.

PEER SUPPORT

Over the past 20 years, community mental health has expanded the traditional medical model to mobilize community resources for the benefit of clients. This has meant that the highly specialized skills of a few professional groups treating those with mental health problems, while still critical in the care of the mentally ill, are no longer the pivotal activity. What has emerged is a client or client-centered approach, where the needs identified by the client become a central focus of care.

A vital part of this expansion of mental health philosophy has been the inclusion of peers or consumers, who have themselves suffered a mental illness, as part of the frontline care team, and as active participants in overall mental health program planning and development (Davidson et al., 1999; Dixon, Krauss, & Lehman, 1994). Their inclusion has not come without some struggle. There has been skepticism and distrust on both sides. Peers, often angry and disillusioned with the gaps in the mental healthcare system, and resentful of the role they were often expected to play as the passive recipients of the doctor's or therapist's treatment, wanted a voice in the care they received, and in how the priorities for that care were decided. Clinicians, on the other hand, were concerned about boundaries between themselves and the peers they might work with. They were unsure what they could expect from the self-help model, and how resilient a peer who had formerly been a client could be in working with other clients. They had trouble with the relationship shift from being the caregiver to becoming a colleague. It was likely boundaries that could become blurred, and it was not always clear what to do when this happened.

The authors presented a workshop at the International Society for Traumatic Stress Studies (ISTSS), 21st Annual Meeting, November 2005 entitled, *Overcoming Stigma and Delivering Care: Combining Peer Support and Treatment in the Canadian Forces*. The audience was asked to describe what they imagined would be barriers or challenges to integrating peer support workers into their clinical practice. One honest and brave clinician stood up and said, when describing the traditional model of care, "It's my turf, and the power relationship is very much shifted to me. In [a peer support model] we try to intentionally cultivate cohesion and bonding among the veterans, so that for one thing, the power relationship can be shifted back to them, where, in my view, it belongs. But I would not pretend that I don't get scared, sometimes" (Heber et al., 2005).

The concerns and reluctance of professionals was sometimes seen as resistance, and as an expression of negative attitudes toward the self-help movement (Hodges, 2003; Labonte, 1989). Professionals were criticized for not being interested or motivated to build a cohesive service system that included self-help or peer support services (Emerick, 1990; Hodges et al., 2003).

At the same time, clinicians who were open to trying this model began to write about the unique benefits of peer support to their clients and to treatment outcomes (Carling, 1995; Dixon et al., 1994; Felton et al., 1995; Solomon, 2004). Benefits of peer support which they described include:

1. First-hand knowledge of systems in which professionals often have had no experience. A good example of this in the military mental health clinics is experience with military life, the chain of command, and the process of release. Most providers of mental health services in the Canadian military are now civilians, who can find themselves at a disadvantage when trying to understand and empathize with the particular work-related situations facing their clients. One telephone call to a peer support colleague often quickly clears up this confusion.
2. Another benefit of peer support workers is engagement of clients into care. As illustrated by the following anecdote, soldiers suffering from OSIs are often suspicious of the system and reluctant to seek help. However, many clients who do not trust the military system or the mental health professional will trust a fellow injured soldier.

   In the early stages of implementation of the OSISS program, Michael Spellen, one of the newly hired peer support coordinators (PSCs), was called by a psychologist in Western Canada who was struggling with a veteran who was severely isolated and refused to come to town for treatment. The veteran had attempted suicide several times, on one occasion attempting to shoot himself with a rifle at a hunting camp. The psychologist thought that perhaps Mike might be able to work with the veteran and motivate him to get into care. She told Mike that the veteran had been decorated with a medal for bravery after a peacekeeping tour of Bosnia. He had developed an operational stress injury, then took his voluntary release from the military, changed his name, and moved out into the bush of Western Canada, far away from all friends and family. For the first few months the veteran would not take Mike's calls or visits. But Mike persisted. The veteran eventually began to see Mike, and after many encounters with individual peer support, he started to attend regular treatment and group therapy. Once Mike got to know him, he asked the veteran why he had changed his name and moved so far away. The veteran replied that after he developed an operational stress injury, he felt ashamed and wanted to disappear. He moved away and changed his name so that he would not dishonor his regiment.

3. The personal experience which the peer support worker shares with the clients (such as suffering the debilitating symptoms, experiencing the stigma of an OSI, the ambivalence about accepting therapy, and experiencing both the benefits and side effects of medication) makes the peer support worker an invaluable teacher for the mental health team. He or she can sensitize the clinicians to these experiences and thereby influence how they provide care.
4. Peer support workers can serve as role models to their peers and therefore provide a powerful sense of hope. As well, they can be role models to the professional team, providing clinicians with a potent example of how clients can succeed, despite having suffered from severe and debilitating symptoms.

The mental health self-help movement is built on the assumption that one does not need to rely solely on professionals to overcome one's afflictions or injuries. Self-help groups "emphasize self-definition of needs, voluntary participation, and autonomy," and tend to be nonhierarchical in structure (Dixon et al., 1994). The importance of those with mental health problems being active participants in both their individual recovery and in organizing the services delivered to them is integral to this model. The shared decision making between the consumers and providers of mental health services parallels the trends occurring in other areas of modern medical care (Dixon et al., 1994). "Peer support," "consumer-driven services," "social support," and "recovery model" are all terms used to describe this growing self-help movement.

There are generally four models of formalized peer support described in the literature.

1. **Mutual Support Groups**. These are voluntary, informal groups, often conducted on a drop-in basis, which are led by peers and focus on a particular condition or life stage. Bereavement groups and groups for cancer survivors are two examples. These groups usually aim to help people overcome the stigma attached to their condition, to feel supported and accepted by others sharing similar challenges in life, and thereby to gain more self-acceptance. Another common goal of mutual support groups is to engage in advocacy. Through these mutual endeavors, rather than through professional interventions, the group strives to help its members. Structure and governance vary widely, and often, as these groups mature, their focus evolves from one of individual and family support to action-oriented advocacy (De Sousa & Leung, 2002; Schubert & Borkman, 1991; Wintersteen & Young, 1988).
2. **Consumer or Peer-Run Services**. Peers are paid employees of the program and do not expect and are not allowed to receive support or other assistance from those served by the program. They usually aim to provide a supportive setting, and to engage in more formal interactions with the peers they serve. The fiduciary responsibility of the peer support workers toward the peers they serve becomes an increasing factor in this model, compared with mutual support (Davidson, 1999; De Sousa & Leung, 2002).
3. **Peers as Part of the Mental Health Team**. Peers are paid employees, and work as providers of support within the professional setting. The peer support workers (PSWs) are usually hired and trained by mental health professionals and work alongside them as colleagues. The advantages of this model are that it can provide more opportunity for the PSW to be a visible role model for the peers being served, and the PSW can reach a larger number of clients. As well, it offers the PSW greater potential to influence reforms in the way clinical and support services are designed and delivered. Therapeutic boundaries are less ambiguous as the PSWs behavior is more likely to be guided by conventional professional practices (De Sousa & Leung, 2002). Disadvantages include the potential value of peer support being diminished by the prevailing medical or clinical paradigms of the setting (Davidson, 1999). Peers can be "absorbed" into the clinical culture, thereby neutralizing their greatest effect.
4. **Workplace Embedded Peers**. This model is often used in workplaces where the normal demands of the job can put an employee at increased risk of developing a physical or psychological injury. Careers in firefighting, policing, and the military are examples of such professions. Certain employees receive special training to provide support and counseling to their colleagues when a job-related traumatic incident occurs. Often these peer support workers provide this service in addition to carrying out their regular (nonsupport) duties. This model is distinct from the previous three. Here, one

is a "peer," not because of a shared disability or type of suffering, but because one shares an occupation or workplace with those one supports or counsels. Critical incident stress debriefing and critical incident stress management programs use this model, with the peer counsellor sometimes pairing up with a mental health professional, to provide debriefings or defusing and follow-up support to his colleagues (Mitchell, 1988). Advantages of this model include the proximity (both physically and psychosocially) of the PSC to those he or she supports. Challenges include the add-on nature of the role, where the PSC usually carries this job in addition to his or her regular workload. As well, since this person may not have experienced the injury or condition from which his or her peer suffers, the ability to be an effective role model, "engager" of reluctant clients, and educator of professional staff may be diminished. In some ways, when this peer says to his or her suffering colleague, "I know you can get better, and though you don't want to take medications I really believe they'll help you," his or her credibility may be diluted if they have no firsthand experience on which to draw.

In the end, OSISS was designed using principles from all four of the models described above. However, its structure and functioning most closely resembles the model of consumer or peer-run services.

THE OPERATIONAL STRESS INJURY SOCIAL SUPPORT PROGRAM—A PEER SUPPORT MODEL

Dealing with the issue of operational stress injuries within the military clearly extends beyond the realm of medical treatment alone. Although healthcare professionals play a critical role in the delivery of healthcare to military members and veterans, there was a need to address the social and educational aspects associated with operational stress injuries.

To address this issue, and recognizing that a strong social support network is a key determining factor in returning Canadian Forces members and retirees suffering from operational stress injury to good health and helping them stay healthy, the OSISS program was initiated in the spring of 2001 at the Department of National Defence (DND). Soon thereafter, Veterans Affairs Canada (VAC) became a partner in this program. Under this program, military veterans who have suffered operational stress injuries themselves will be available to offer assistance to serving and former members who are currently affected by operational stress injuries.

The accepted World Health Organization's definition of health includes physical, mental, and social well-being. Social support is defined as "the resources provided by other persons" (Cohen, & Syme, 1985, p. 4), and is often used in a broad sense as "any process through which social relationships might promote health and well-being" (Cohen, Underwood, &

Gottlieb, 2000, p. 4). Kaniasty (2005) states that social support is referred to as social interactions that provide individuals with actual assistance and embed them into a web of social relationships that are caring and available when needed. This definition points to three major facets of social support: perceived support (social resources that people perceive to be available when needed); received support (social support that is actually received); and social embeddedness (the quality and type of relationships with others) (Kaniasty, 2005).

Social support is important for people coping with trauma. Kaniasty (2005) reports that reviews of the literature frequently report the limitations of supportive relationships during stressful times but have generally concluded that social support is beneficial to psychological well-being and physical health. What happens after a trauma has been shown to have an impact on whether a person develops PTSD. Brewin, Andrews, and Valentine (2000) conducted a meta-analysis of studies of PTSD risk and protective factors and found life stress subsequent to trauma and social support from others at the top of the list of predictive factors.

There are two main theories of how social support works: the stress buffer model and the main effect model. The stress buffer model has a direct effect on people's health, through a buffering effect, moderating the impact of stress on health by preventing responses to stressful events that are adverse to good health. Believing that others will provide resources when needed may redefine the potential for harm by a situation, and prevent the individual from appraising the situation as highly stressful (Cohen, Underwood, & Gottlieb, 2000). Having someone to talk to about problems has been found to help in preventing maladaptative responses to stressful events (Lepore et al., 1996). Social relationships can also have a main effect on health. Cohen et al. (2000) reports that "[t]hose who participate in a social network are subject to social controls and peer pressures that influence normative health behaviors" (p. 11). Integration in a social network is also seen as a source of positive affect, providing a sense of stability, purpose, belonging, and security, as well as recognition of self-worth. These positive psychological states can result in reducing psychological despair (Thoits, 1985, as cited in Cohen et al., 2000) as well as result in a greater motivation to care for oneself (Cohen & Syme, 1985, as cited in Cohen et al., 2000).

When people undergo stressful life events, they typically orient themselves to and seek out similar peers who are able to help them integrate their new identities; for example, first-time parenthood, recent retirees who are first in their social network to leave the workforce, or adults returning to school in midlife. De Rosenroll (1994) defines peer support as "an umbrella term used to describe a sanctioned program where individuals receive appropriate training and supervision so that, formally and informally, they can directly and indirectly offer assistance in a variety of ways to individuals who, based on their situational defined similarities, would refer to themselves as peers" (p. 32). This definition clearly describes the

OSISS program whereby it is a sanctioned program where individuals receive appropriate training so that they can offer support and assistance to their peers.

The main intervention in the OSISS program is the peer support coordinator and the family peer support coordinator (Family PSC). These support coordinators (SC) are provided with knowledge, information, and practice tools needed to establish functional helping relationships. Helping is about learning and relearning. The helping relationship with the PSC is meant to guide the peer and engage him or her in a process of exploring, understanding, and activity in accordance to his or her own free will, and to help the peer make changes in his or her situation. All the PSCs employed in this program have an operational stress injury such as PTSD, and the Family PSCs have lived with or supported a military member or veteran suffering from an OSI.

OSISS PSCs are paid federal government employees hired after a careful selection process which includes a medical screening. All potential candidates for the position of PSC are initially referred to the DND Medical Advisor to OSISS for first level of screening. The Medical Advisor then communicates with the health professional who made the referral to discuss the potential candidate, review the requirements of the job, and determines if the candidate has sufficiently recovered from his or her injury to be employed within OSISS. A decision is then made to proceed to the next step whereby the DND Medical Advisor meets with the individual to further discuss the interest of the candidate, as well as discuss the work of OSISS. If the individual is still interested, a formal medical screening form is completed and signed by the candidate's treating clinician. In most cases, these individuals are referred to the OSISS management team, a multidisciplinary team which includes psychiatry, nursing, and social work for review and discussion. Without the first level of screening carried out by the DND Medical Advisor, candidates are not referred to the OSISS management team for further consideration and eventual selection. Names of individuals who successfully completed this first step are then scheduled for a formal interview process in accordance with the Public Service Employment Act. Family peer support coordinators do not have a diagnosed OSI and are not required to have a medical screening.

Prior to beginning their job, the PSC and Family PSC must attend mandatory Peer Support Training. This course has been specifically designed for the OSISS Program and the critical skills development portion of the 2-week initial course is delivered by mental health staff including psychiatrists, psychologists, clinical nurse specialists, and social workers from the only Veterans Affairs Canada hospital remaining in Canada, Ste. Anne's Hospital, Sainte-Anne-de-Bellevue, Quebec. The peer support training course material includes knowledge and skills development in peer support, helping relationships, conflict resolution, understanding and respecting boundaries, active listening, problem solving, crisis management, suicide intervention, volunteer management, group work, as well as emphasizing the importance of respecting program boundaries and self-care. Employees receive information on programs, benefits, and services, as well as policies. They are given information concerning the many agencies and services with which they will need to work. They are briefed on the resources within the federal government and in their communities to which they will refer their peers. The program also provides ongoing professional development to the PSCs and Family PSCs through quarterly workshops. The curriculum for this training has been developed over a period of 5 years by a dedicated multidisciplinary team who balanced the needs of the program with the skills brought forward by seasoned veterans and their families to offer relevant and meaningful information and develop tangible skills rather than overwhelming employees with academic theories (Veterans Affairs Canada, 2006).

At the heart of any helping relationship are communication skills. The PSC learns how to conduct a supportive or therapeutic interview. Information about the peer is gathered, selected, and analyzed to make an assessment. The assessment drives all intervention strategies and is based on the information obtained. Information for the assessment comes from direct conversation (the interview process) about concrete issues or situations ("the story") and the process of the interview; how the conversations flows, what is said or not said, the tone of voice, or silence. It is a complex "dance" and must be handled with care if it is to be an effective helping process.

Through OSISS training, SCs are taught that helping networks have four major functions: to act as buffers between the individual and the source of stress; to provide practical and emotional support; to provide information on resources; and to act as referral agents. As referral agents, helping networks will address needs defined by the network members, not by others. They will help their peers to increase self-esteem ("helper therapy"), and learn new skills and knowledge about accessing resources. They will empower them to think differently about themselves, and to act on those differences. They are also sensitized that as SCs they are in the "zone of helpfulness" when they are attentive to the peer's needs, and guides the peer toward the appropriate resources. A support coordinator is outside the "zone of helpfulness" when he or she denies, minimizes, or ignores the peer's needs or suffering. The SC also falls outside the "zone of helpfulness" when he or she begins to treat the peer (conducting therapy sessions, for example), rather than referring the peer to the appropriate resources.

Empathy is a necessary quality in every helping relationship. The helper uses knowledge and understanding to help the other person. Empathy requires the capacity to feel an emotion deeply and yet maintain "separateness" from it so that knowledge and skill can be put to good use. The SC must suspend all preconceived expectations of resolving the peer's problems. The best they can do is to be empathic and apply their knowledge and skill. Good working relationships are built on trust, caring, sincerity, acceptance, respect, and absence of judgment. Uncontrollable events call for increase in emotional support. Controllable events call for the need for instrumental support. When the SC has not experienced the same life event as the intended support recipient, it is much harder for the recipient to gain the needed emotional support or information about ways of coping. Patterns of preference are empathy and understanding communicated by those who have "walked in the same shoes."

On completion of their training, the SCs focus their energy on getting started a community-based network of both resources and "clients" referred to as "peers." This can take up to 9 months but as the network is developed, the SC will provide initial support to the extent possible to all individuals who seek out the services of OSISS. This support will be offered until such time as the SC has recruited enough peers into the network to begin developing a formal program with trained peer support volunteers. Once at this stage of development, normally after 1 year of operations, volunteers undergo a screening process. The selection process varies for OSISS Public Service employees and the OSISS volunteers. The volunteers are not subject to a formal interview process. They are recruited and selected by the PSC or Family PSC as being a suitable and appropriate volunteer following an in depth interview with the PSC or Family PSC. Subsequent to this interview, additional OSISS volunteer policy criteria must be met, such as a mandatory police check and an official letter of endorsement from the treating psychiatrist and/or psychologist for all those who have an OSI. Once all the criteria has been met in accordance with OSISS policy, all OSISS volunteers are required to attend a mandatory 3-day Peer Support Training program provided by mental health professional staff at VAC's Ste. Anne's Hospital in Sainte-Anne-de-Bellevue, Quebec. This Volunteer Peer Support Training is exclusively dedicated to skills development in providing peer support and less focused on programs, policies, and administrative functions. The ultimate goal of each SC is to recruit volunteers, who are empowered to support others in a safe way, and delegate and coordinate this support within their community.

The PSC, responsible for the military member and veteran peer support, and the Family PSC, responsible for the families of military members and veterans with an OSI, have independent and separate roles and networks within the same program; however, the same theory and goal are applied (Table 13.1). The prime role of an OSISS SC is to increase the level of social support to serving military members and veterans who have experienced symptoms consistent with operational stress injuries, as well as their families. The goal of the SC is not to replace anyone in the helping community, but rather to complement existing services available from the Department of National Defence, Veterans Affairs Canada, and the local community, and to empower Canadian Forces members and veterans to assist in reducing OSI symptoms, improve functional status and occupational performance, and improve quality of life. The main role of the OSISS PSC and Family PSC as outlined in Table 13.1 is to provide one-on-one assistance; organize and conduct group work; recruit and manage volunteers; and provide outreach briefings. Activities involved in this role

TABLE 13.1
The Roles of an OSISS Peer Support and Family Peer Support Coordinator

Peer Support Coordinator	Family Peer Support Coordinator
Focus on military members and veterans	Focus on families of military members and veterans with an OSI
Provide one-on-one assistance: listen, assess and refer	Provide one-on-one assistance: reach out, inform, and connect
Organize and conduct peer support groups	Organize and conduct psycho-education groups
Select, train, and manage volunteers	Select, train, and manage volunteers
Provide program outreach briefings	Provide program outreach briefings

include, but are not limited to, such things as creating a peer support network; creating an environment for individuals who contact the PSC and Family PSC; assisting individuals to make contact with appropriate DND and VAC offices, as well as the community; providing general information on VAC, DND, and community resources, and assisting with finding the resources; and conducting follow-up with their peers.

The relationship between the SC and the peer is fundamentally unequal, as it is designed to address the needs of the client. This inequality, however, is more theoretical, given that the ones offering this support are nonclinicians and are individuals who have "walked in the shoes" of their peers. A lot of emphasis is placed on the term "peer" to foster a helping relationship where there is no power differential. The helping relationship seeks to develop and maintain autonomy and competence in the peer client. It provides emotional, informational, and esteem support, as well as tangible aid and social integration. Essentials of a professional peer support relationship include trust, clear expectations, responsibilities and limits, and a caring attitude, while maintaining autonomy of the peer being served. Table 13.2 provides some general tips for maintaining a professional relationship with a peer client.

The OSISS program was created and is led by a military member suffering from an OSI. However, the program has a strong multidisciplinary

TABLE 13.2
General Tips for a Professional Relationship

If you feel you are getting too close, i.e., "caught up in the situation," discuss it with your manager or management.
If you are unclear or uncomfortable, speak to your manager or management.
It is human nature to feel closer to some than others. However, make sure the time, care, and concern shown is equal among all peer-client cases.
You are responsible for creating the professional relationship with each peer-client.

management team where clinical considerations are always considered prior to making any and all decisions. The management structure includes a co-manager provided by Veterans Affairs Canada and the management team includes mental health advisors and specialists, including a psychiatrist, nurses, social workers, and former military service members.

The success of OSISS rests on several important pillars, but one of its strongest remains the volunteer component of the program. They are mainly veterans suffering from an OSI and family members who have experienced firsthand what it is like to live with and support a military member or veteran suffering from an OSI. Respecting strict guidelines and principles of volunteerism, including the Canadian Code for Volunteer Involvement (Volunteer Canada, 2000), OSISS provides a unique and safe opportunity to help others with an OSI. This not only allows OSISS to reach a greater number of injured personnel, but it plays a role in helping the volunteer to reestablish self-confidence and a sense of self-worth.

It is recognized that OSISS peer support workers and peer volunteers may be at risk of relapse by being exposed to the kind of work they do. Therefore, during the official peer support training a great deal of emphasis is placed on boundaries and the limits of involvement in peer client cases, as well as the importance of a strict regimen of self-care that is monitored biweekly by a clinical psychologist.

The SCs need to be clear on their roles and expectations, as well as boundaries in their working relationship with peers, in order to protect and preserve the integrity of the helping relationship. When one person entrusts his or her welfare to another, there is always an inherent power differential. The peer–SC relationship is never a relationship of equality. The SC has the power of knowledge, training, experience, and authority in relation to the peer he or she serves. Peers should reasonably expect that those who help them would do so in a safe way.

The OSISS Peer Support Training includes training in maintaining appropriate boundaries to help manage the power differential and allow for a safe connection between the SC and the peer receiving help, based on the peer's needs. Boundaries help to differentiate oneself from others and maintain a sense of self. They help to establish the appropriate limits of the interactions. Boundaries help to protect the peer and to prevent the "offered help" from becoming compromised. Boundaries create clarity, safety, and predictability for both the peer being served and the SC. They help to define the relationship with the peer as professional, not personal.

Support coordinators must understand the difference between personal and professional relationships. Peers seek out someone from within their own community, believing the SC will better understand them. Despite the fact that dual and overlapping relationships are unavoidable in the military community, they usually complicate the helping relationship. It is always the SCs' responsibility to monitor the relationship and to ensure that the best interest of the peer is the priority. The program must not become primarily a way for SCs to gain personal recognition and fulfillment. The SCs must avoid becoming too directive in their interventions and must always remember that what worked for them is not necessarily the solution for the peer they are supporting. Finally, SCs must be willing to refer a peer to a health professional.

A major concern of all involved in the OSISS program has been the issue of boundary violations. This occurs when standard practices are not followed, resulting in the confusion of roles and expectations. The SC's own personal interests must never compete with those of the peer being helped. There may be no clear or obvious answer to questions, such as: Is this in my peer-client's best interest? Whose needs are being served? Will this have an impact on the service I am delivering? And does this contravene my code of ethics? Within the OSISS program, a healthy culture of respect of these boundaries has been established. However, in the early stages of program implementation, OSISS management was required to take corrective actions. As time goes by, SCs police themselves and remind each other as colleagues of the importance of boundary respect. This phenomenon was fostered and supported by management and took approximately 3 years to occur, but once established, it became a very effective way to solidifying service delivery and retain credibility within the helping community. Support coordinators who have difficulty respecting boundaries do not remain within the OSISS program. In most cases, they realize for themselves that perhaps this work is not for them and choose a different career path even before management has to take formal corrective action.

It is paramount that the SCs clearly understand their role and staying within the confines of that role. They must be aware of their own "triggers" and remember that they are there for the peer and not for themselves. Sometimes a peer's behavior evokes in the SC feelings relating to unresolved situations in their own life, causing the helper to respond to the peer client in a nonobjective way. When this occurs, SCs are trained to understand that they must discuss and review the situation with their health professional resources.

It is also recognized that individuals assisting those who have been traumatized may be at risk of becoming negatively affected by their work. This phenomenon has been called "compassion fatigue" (Figley, 1995), "secondary victimization" (Figley, 1982), "secondary traumatic stress" (Figley, 1985; Stamm, 1995, 1997), "vicarious traumatization" (McCann & Pearlman, 1990; Pearlman & Saakvitne, 1995), or simply "burnout" (Miller, Stiff, & Ellis, 1988). Burnout is a combination of emotional, mental, and physical exhaustion caused by the cumulative strain of working with emotionally demanding situations where one gradually wears down over time. Vicarious trauma, compassion fatigue, or secondary traumatization occurs when the SC unwittingly takes on the reactions of the peer he or she is supporting and revisits his or her own traumatic life events. It is a process in which the SC's inner experiences are negatively transformed through empathic engagement with the peer's trauma. Contributing factors to the occurrence of vicarious traumatization include sharing many similarities (such as age, gender, and profession) with the peer-client; physical and emotional fatigue; over involvement in the case; unresolved personal issues; and lack of appropriate support and supervision for the SC.

In the OSISS program, self-care is what you do for yourself. It is recognizing your own limits and being kind to yourself. It is understanding what you need and making sure your needs are met at work and at home. Self-care is utilizing your team of colleagues and consultants. It is staying involved in your personal relationships, and it is respecting the choices of others.

Self-care is of utmost importance for the well-being of the peer and family support coordinator. The Peer Support Training includes a module on self-care and has built in a number of self-care mechanisms. These include the requirement for a prehiring medical screening by the treating clinician for those who have an OSI, and for an annual follow-up assessment. As well, SCs are expected to maintain a regular health regimen. They receive training in self-care. They are given linkages in their local community to mental health professionals for advice and guidance, and they participate in a biweekly self-care teleconference with a clinical psychologist.

One of the requirements for individuals who have an OSI to work for this program is to remain in a therapeutic relationship with a mental health clinician. They must also be linked to a mental health professional in their own community so that they have daily contact if they need support, guidance, advice, or direction concerning their work, as well as their own health and well-being. They are also required to connect with a clinical psychologist at Ste. Anne's Hospital who discusses self-care with the SC by teleconference biweekly. This teleconference focuses on the well-being of the SCs and not on their case work.

The OSISS program was subjected to a formal evaluation. The Department of National Defence and Veterans Affairs Canada's (2005) *Interdepartmental Evaluation of the OSISS Peer Support Network* found that "a strong social support network is a key determining factor in returning Canadian Forces members and retirees suffering from Operational Stress Injury to good health and helping them stay healthy" (pp. i–ii). The Peer Support Network was found to be the only common and continuous formal social support capability that a CF member and/or retiree suffering from OSI experiences in his or her recovery and/or transition from regular military service to retirement. Overall, the evaluation found that the Peer Support Network had been successfully implemented and is contributing to effectively meeting the social support needs of CF members and veterans with operational stress injuries. It acknowledged the challenges with implementation, and report that early program issues and risks have been and are being addressed, including an improved PSC selection process and enhanced safeguards to maintain the health of each PSC. Recommendations were made to advance some key areas leading to enhanced program effectiveness.

RESEARCH COMPONENT

It was recognized early in the development of the OSISS program of the potential negative health impact on the peer support coordinator in working with peers who also suffer from a similar operational stress injury. It is also recognized that individuals assisting those who have been traumatized may be at risk of becoming negatively affected by their work. Therefore, management instituted a mechanism to research the impact on those who provide the social support on an ongoing basis. Subsequently, the research component of the OSISS program investigates the health and social outcomes of those Canadian Forces veterans who are providing social support to their military peers. The OSISS PSCs (paid employees) are asked to voluntarily complete a series of four standardized questionnaires prior to starting their work, 6 months later, and then repeating the scales at 1-year intervals.

To measure symptoms of PTSD, the PTSD Checklist–Military Version (Weathers, Litz et al., 1993) is utilized. The PCL-M is a 17-item, DSM-IV–based PTSD symptom measurement tool assessing the extent to which symptoms of PTSD have been experienced over the previous month. Although the main reason for using this measure was to assess changes over time, we also used the cut-off score of 50 to establish the presence of probable PTSD. This was consistent with the authors of the PCL-M Checklist (Weathers, Litz et al., 1993), who found that a cut-off score of 50 yielded a sensitivity of 0.82 and specificity of 0.83 in a combat veteran sample. Using Weathers et al.'s (1993) cut-off score of 50 with civilian motor vehicle accident victims and sexual assault victims, Blanchard, Jones-Alexander et al. (1996) demonstrated good sensitivity (0.78), specificity (0.86), and overall diagnostic efficiency (0.83). A PCL-M cut-off score of 50 was also used in a study by Forbes et al. (2003) in cases of combat-related PTSD.

To assess for symptoms of depression, the Center for Epidemiological Studies-Depression Scale (CES-D) is used. The CES-D (Radloff, 1977) is a 20-item self-report instrument for depression. It is a Likert scale using four points. A cut-off score of 16 indicates high-end depressive symptoms. Excellent reliability (internal consistency was 0.84–0.90) and good test-retest reliability (0.51 at 2-weeks and 0.67 at 4-weeks) has been shown (Radloff, 1977). A 0.88 internal consistency was found by Knight, Williams et al. (1997). Good construct validity is reported, with moderate correlations with the Hamilton Clinician's Rating Scale and the Raskin Rating scale (0.44–0.54) on initial administration, and higher after 4 weeks of treatment (0.69–0.75).

To measure alcohol abuse the Alcohol Use Disorders Identification Test (AUDIT) (Allen et al., 1997; Babor et al., 1992, 1995; Claussen and Aasland, 1993; Saunders et al., 1993;) was administered. The AUDIT consists of 10 questions that measure the quantity and frequency of alcohol consumption while simultaneously assessing harmful and hazardous drinking. As these three domains are included in the DSM-IV diagnosis for alcohol use disorders and dependence, this instrument is useful for detecting clinically significant alcohol misuse. To define cases of problematic alcohol use, the recommended cut-off score of 8 was utilized (American Psychiatric Association, 2000).

There has been a growing body of research examining the impact of psychiatric illnesses on health-related quality of life (HRQoL). A number of studies have demonstrated a decrease in HRQol in individuals suffering with PTSD (Cordova et al., 1995; Schnurr, Friedman, & Green, 1996; Schonfeld et al., 1997). To assess the HRQoL in individuals suffering with an operational stress injury such as PTSD, working with a similar population, the Medical Outcomes Study (MOS) 36-item Short-Form Health Survey (Ware et al., 1993) was also utilized. The SF-36 Health Survey measures impairments in eight domains, or subscales: four of which relate to mental health, vitality (VT; a measure of energy level and fatigue), social functioning (SF), role emotional (RE; difficulties in work or daily activities due to emotional problems) and mental health (MH; anxiety and depression); four of which relate to physical health, physical functioning (PF), role physical (RP; difficulties in work or daily activities due to physical problems), bodily pain (BP), and general health (GH). For each of these subscales, scores range from 0 to 100, with higher scores indicating better functioning.

It is clear that this research would not be able to determine the positive benefits on individuals of working in a peer support program; however, we will remain hopeful that this research may contribute to the body of knowledge advocating that with appropriate training, medical screening, medical monitoring and follow-up, and self-care, the potential negative impacts of working in peer support can be mitigated. Preliminary results have demonstrated that the health and well-being of the PSCs have not been negatively affected by working in the peer support program.

STIGMA AND OPERATIONAL STRESS INJURIES

In ancient Greece, people would prick the skin of their slaves with a sharp instrument to demonstrate ownership and to signify that those so marked were unfit for full citizenship. The Greek word for prick is *stig*, and the resulting mark is a *stigma* (Stuart, 2004). Today, a stigma is a symbolic mark of social disapproval or disgrace (Dovidio, Major, & Crocker, 2000; Goffman, 1963). Those who are stigmatized are viewed as inferior or weak or damaged. Mental illness is highly stigmatized in our society (Sartorius, 2004). Some authors have pointed out that stigma alone "adds a dimension of suffering to the primary illness—a second condition that may be more devastating, life-limiting and long-lasting than the first" (Schulze & Angermeyer, 2003). According to the Canadian Mental Health Association, our society "feels uncomfortable about mental illness. It is not seen like other illnesses such as heart disease or cancer.... People [may]... believe that an individual with a mental illness has a weak character or is inevitably dangerous. Mental illness can be called the invisible illness. Often, the only way to know whether someone has...a mental illness is if they tell you. The majority of the public is unaware of how many mentally ill people they know and encounter every day" (Canadian Mental Health Association, 2005).

The invisibility of mental health problems is a double-edged sword. It allows those suffering to do so in private and to maintain their confidentiality. Often early on in treatment, this invisibility is called on as a positive coping strategy, to help decrease the client's sense of being exposed and vulnerable. However, the atmosphere of disapproval, and often disgust, that surrounds mental illness serves to encourage the person to keep his suffering a secret. By doing so, the person's sense of shame is reinforced. And the stigmatized and erroneous attitudes of those around him go unchallenged, and therefore do not change. Stigmatized views of mental illness are shared by everyone in our society. This includes the client suffering from such a disorder, as well as those who are treating him.

The Canadian Mental Health Association and Health Canada report that while 1 in 5 Canadians will suffer from a mental illness at some point in their life, almost 50% will never receive treatment, because of the stigma attached to their condition (Underwood, 2005).

Within certain social groups or workplaces, such as the military, stigmatization of mental illness can be profound (Hoge et al., 2004). This has been well documented in the Croatia Board of Inquiry (Croatia BOI, 2002), the Canadian Forces Ombudsman's report on the systemic treatment of CF members with PTSD (Marin, 2001) and the Ombudsman's report entitled, *Off the Rails: Crazy Train Mocks Operational Stress Injury Sufferers* (Marin, 2003). It would be hard to overestimate the effects of stigma on a military member. There are few activities in a person's life that convey a greater sense of self or self-worth than one's career. "Work influences how and where one lives, it promotes social contact and social support, and it confers title and social identity" (Stuart, 2004, p. 102). The military member's identity is closely tied to his work. For him, living in a closed and controlled environment, where one's neighbors, friends, and work colleagues are often the same people, the effects of stigma in the workplace can rapidly become universal.

Because of the importance of factors such as team cohesion, a willingness to follow orders, and a preparedness to put oneself in harm's way, military training tends to discourage attitudes and behaviors seen as rebellious, unstable, or morally feeble. As long as mental health problems are considered signs of rebellion, instability, and moral weakness, the stigma attached to these disorders will continue to permeate the views held by the troops and the chain of command. Clinical experience shows that some of the most difficult and demoralizing issues for members with a mental illness or injury are: feelings of failure and shame; a sense of being betrayed and abandoned by the military; guilt over one's own previous stigmatized views of mental illness; fear of harming one's career and anger over the effects on one's career; lack of faith that the mental health system can help; and embarrassment and a desire to hide from one's colleagues, one's boss, or from all things military. These issues are all rooted in stigma.

CHANGING ATTITUDES—THE OSISS SPEAKERS BUREAU

Another component of the OSISS mandate is to develop the methodology to effect an institutional cultural change regarding the realities of operational stress injuries. The speakers bureau was created to provide professional development briefings that can assist Canadian Forces personnel to better understand operational stress injuries and help address the stigma of mental illness.

Although it is clear that many of the theaters of operations Canadian Forces personnel serve in expose our men and women to traumatic events and wide-scale tragedy, the trauma they endure may not be the only cause of their OSIs. It may be a major contributing factor in a cocktail of elements that together cause some Canadian Forces members to develop full-blown OSIs. Research has shown that social support is one of the most effective prevention strategies, which can moderate the effects of trauma and have a positive impact on mental health outcomes (Brewin, Andrews, & Valentine, 2000; Stephens, Long, & Miller, 1997). But what happens when this support is not positive? We have come to realize that much of what constitutes the very fabric of social support extends well beyond what we, as a military institution, can control. Outside of the workplace, we have very little control over how our members draw support from friends, neighbors, family, extended family, etc. However, within the Canadian Forces, our members draw this support from each other and it is an institutional obligation for any large organization to ensure that the support they draw and provide to each other within the context of OSIs is positive rather than negative.

In many military organizations, the chains of command often associate good morale and esprit de corps with behaviors marked by excessive bravado, machismo, and rowdy camaraderie. These types of behaviors often include cruel and insensitive comments that further stigmatize the soldier with the invisible wounds of an operational stress injury. Although it is fully understood and accepted that the behaviors outlined above are certainly part and parcel of a healthy military unit, members within each unit need to understand the nature of OSIs through formal education and professional development, in order to properly support their colleagues who have difficulties. As reported in the Croatia Board of Inquiry (BOI) and the Ombudsman's Report (Croatia BOI, 2002; Marin, 2001), military members have a tendency to demonstrate behaviors and attitudes which are counter productive and detrimental to those who suffer from OSIs. An attitude can be defined as a learned, global evaluation of a person, place, or issue that influences thought and action (Perloff, 2003), and as such it can be altered or unlearned. Attitudes serve many functions within the context of military culture; the current general attitudes toward OSIs serve a social acceptance function where members generally feel that they will be more accepted by their peer group if they demonstrate bravado and harbor negative attitudes toward OSIs and those who suffer from them. Unfortunately, this situation causes many of the injured members to feel ostracized and marginalized. For others, witnessing this phenomenon becomes a disincentive to seeking treatment. Their condition often worsens as a result.

OSIs such as PTSD often present with comorbidities of depression and addictions (Forbes et al., 2003; Kessler et al., 1995), which may significantly affect treatment response. More than 50% of PTSD clients have symptoms of a major depressive disorder (Kessler, 1995), but in the veteran population, possibly due to delayed treatment, the percentage may be much higher (Keane & Wolfe, 1990; Richardson, 2002). When depression and anxiety present simultaneously, there is an increased likelihood of treatment resistance and greater functional and psychosocial impairment (Barlow et al., 1986; Enns et al., 2001; Regier et al., 1998; Radloff 1977; Weathers, Litz et al., 1993; Passey and Crocket, 1995; Blanchard, Jones-Alexander et al., 1996; Weisaeth, Mehlum et al., 1996; Knight, Williams et al., 1997; Ravindran, Judge et al., 1997; Rudolph, Entsuah et al., 1998). Also, comorbid depression significantly increases suicide risk (Kaufman & Charney, 2000). PTSD is commonly associated with other conditions, such as panic disorder, social phobia, and obsessive-compulsive disorder (Kessler et al., 1995; Statistics Canada, 2002). Alcohol abuse or other substance abuse affects over 50% of those with PTSD (Kessler et al., 1995). In the uninitiated military chain of command these comorbidities are often viewed as behavioral and disciplinary matters that are punishable under military justice, which only serves to further injure the affected member. The speakers bureau serves as one tool to educate military members at all levels of the chain of command about OSIs and how in the military culture at present as behavioral or functional changes in their work environment.

It was recognized that the lack of understanding about mental health issues is not unique to the military, but is present in all of Canadian society. In 2003, the Canadian Psychiatric Research Foundation launched a shocking multimedia advertising campaign aimed at destigmatizing mental health in Canada (http://www.cprf.ca). The CPRF's campaign won a United Nations Department of Public Information award in May 2004. The OSISS Speakers Bureau, inspired in part by this campaign, developed didactic material in the same manner used in advertising. The goal was to "sell" a new idea.

What is remarkable is that as an organization whose function it is to ultimately send people into harm's way, where, as Breslau et al. (1991) explain, the members are far more likely to be exposed to significant traumas than the general public, one would expect the military to be well acquainted with psychological difficulties members can develop when faced with events that are considered beyond the realm of normal experience. To develop an appropriate intervention within the military institution the OSISS program first approached Defence Research and Development Canada to conduct a literature review (Thompson & McCreary, 2003) and analyze how effective our strategy of involving veterans to educate military members would be. It was our belief that in order to reach and convince the military chain of command as well as the troops that operational stress injuries were not imaginary and that most who suffer from them do not fake their condition, we needed to use seasoned veterans to communicate this message.

Although still in its infancy, the speakers bureau now delivers half-day awareness training using predesigned educational modules to all new recruits in the Canadian Forces, and 90-minute professional development periods to formed military units and personnel around the country. The modules include an historical overview of OSIs, a clinical module, two moderated group discussion workshops where scenarios are presented, and, at the end, practical tips and lessons learned are shared. The participative approach and the careful selection of credible veterans as speakers are to this date the cornerstone of the program. Although the speakers bureau was designed to change attitudes within the military community toward those suffering from operational stress injuries, a byproduct of openly addressing the issue of OSIs has been a positive impact on treatment seeking among the injured population. In the words of a military member who attended such a presentation when the presenter "stood up and talked about how he had PTSD, and he was so willing to be open about it, and when he described his problems and how he had been thinking and behaving, I could really identify with him, and this completely broke down my denial."

OSISS is currently developing more advanced training which will be delivered in increments as military members gain experience and seniority within the military institution. It will include detailed discussions on leadership responsibilities and awareness of departmental policies pertaining to OSIs, to ensure the members' actions are consistent with the departmental ethos, objectives, and philosophy as they move up in rank.

COMBINING PEER SUPPORT AND CLINICAL TREATMENT

Despite the creation of special clinical programs in Canada such as the Operational Trauma and Stress Support Centres (OTSSC) by the Department of National Defence, followed a few years later by the creation of Operational Stress Injury Clinics by Veterans Affairs Canada, it is still common for CF members to first present for treatment several years after their traumatic deployment. Although already secretly suffering from an

OSI, many of these soldiers have gone on to complete several more tours to very active and war-torn parts of the world.

This raises the question of why many military members wait so long to get help for these very debilitating and distressing symptoms. As was previously reviewed in this chapter, common barriers to seeking treatment include fear of loss of military career, the stigma attached to having psychological problems, feelings of shame, fears of being considered weak or a burden by the chain of command, and fears of being considered a faker or freeloader by the institution or by one's friends. Hoge et al. (2004) found that American soldiers returning from active duty in Afghanistan and Iraq cited similar barriers to seeking care. Soldiers' symptoms, such as avoidance and isolation, lack of energy and motivation, decreased self-esteem (feelings that one is not worthy of help), and little hope for the future can also prevent them from coming forward.

As well, there are structural factors in the military organization that present barriers to care. It is now generally accepted that for members with PTSD to attempt a meaningful recovery, their family must be included in, and provided with, necessary treatment as well (Galovski & Lyons, 2004; Shehan, 1987). However, although the Canadian military member is provided with complete healthcare services within the military, his or her spouse and children must access most of their medical and mental health care through the civilian healthcare system. This approach fragments the care these families receive. Although PTSD and other OSIs are conditions that affect all members of a family, the OTSSCs can provide family or couple assessment and therapy to a limited degree when it is deemed to be in the best interests of the military member (RX, 2000). Spouses are offered emotional support and education about operational stress injuries under this provision. However, many of these families have much greater need for services than the OTSSC or the local base mental health centers can provide.

The mental health and medical services offered to serving members are discontinued once the member releases from the military. He or she then becomes a client of Veterans Affairs Canada for any ongoing treatment benefits. This state of affairs further bifurcates and fragments the care of members suffering from OSIs. Having recognized this as a problem, the Canadian Forces and Veterans Affairs Canada have struck joint committees and organized projects to explore avenues for providing seamless care to members before and after release from the military (Lt. Colonel H. Matheson, personal communication, January 2006).

One of the benefits of the support provided by OSISS is that it crosses the above-mentioned divides. OSISS is mandated to support and care for both serving military members and veterans, as well as the families of those who suffer from OSIs. OSISS does not discriminate, and peer support coordinators can provide a constant link in the continuum of care for the CF member after release from the Forces.

As the OSISS program was created and the new peer support coordinators were trained and started working with military members and veterans across Canada, the clinicians assessing and treating operational stress injuries started having contact with them, often through their clients. It became increasingly common for clients to mention, during their appointment with their doctor or social worker, that they had started attending the peer support group organized in their local community by the OSISS peer support coordinator, or that they'd had coffee with the PSC last week and found it helpful to talk to someone who had been through similar experiences. Or perhaps the client had attended a briefing on PTSD and a PSC had been one of the speakers and had talked about his or her personal experiences struggling to put his or her life back together. In other, more serious cases, an OSISS worker would be directly involved in providing critical crisis support to a suffering member.

Sergeant Jones's case is an example of this. This 33-year-old member, married father of three children, had been deployed overseas to war zones four times in the past 8 years. He was first seen in the OTSSC 4 months after his last tour of duty. He presented severely depressed and met criteria for PTSD. He had been having suicidal thoughts on and off for several months. Because he was posted to a base that was a 3-hour drive from the OTSSC, the client was only seen once every month by his psychiatrist, and he was followed by a community psychologist and the general practitioner at his home base. The client told his doctor that he had met with the OSISS worker on his base, who had invited him to a group meeting, but the client found the group triggered too many memories and created a great deal of anxiety for him. The OSISS worker had therefore suggested they continue to meet one-on-one as often as the client wanted, just to chat. Sergeant Jones found this helped, even though he didn't talk much, and because of his poor concentration, couldn't often remember what he and the OSISS worker had spoken about. With the client's permission, the psychiatrist contacted the OSISS worker, who was quite enthusiastic about collaborating with the treatment team in the care of this man.

At one point early in his treatment, Sergeant Jones's symptoms increased and he began ruminating about how useless he felt, thinking he could no longer be a good soldier, husband, or father. He started having increasing thoughts of killing himself. His psychiatrist decided to admit him to hospital. The OSISS worker accompanied Sergeant Jones and his wife to the emergency department. During his hospital stay, the worker visited him daily and took him off the ward for brief outings as soon as the client was no longer a risk to himself. After discharge, when he returned to see his psychiatrist in the OTSSC, Sergeant Jones reported that the time the OSISS worker had spent with him had had a profound effect on him. He said that he had been surprised and grateful that the OSISS worker had not abandoned him when he became very ill. The nonjudgmental support of this worker had remained an important constant in his life as he struggled with his symptoms and impulses.

The collaboration between OSISS and clinical staff around the country has continued to expand and become more formalized. In some clinics, the OSISS worker calls the OTSSC clinicians with questions or concerns about accessing care for a peer, and they will collaborate around care for the client/peer. In others, the OSISS worker has been integrated into the group treatment programs. In some OTSSCs, providing information about OSISS and the peer support coordinators' contact numbers to each new client has become standard procedure.

CONCLUSIONS

With careful program planning and implementation of program policies and guidelines, combined with strict recruiting practices and procedures, training, medical screening, medical monitoring and follow-up, as well as a strict self-care regime, the OSISS program has shown that that it is possible to employ individuals who suffer from operational stress injuries such as PTSD to assist those who suffer from similar conditions. For organizations such as the military, where employees are at significant risk of being exposed to traumatic events, it is important that the support members provide to and draw from each other is guided and structured. OSISS has demonstrated that it is possible to exercise leadership in this field, to ensure that the social support given to serving military members and veterans is positive and beneficial.

Research has explored how the erosion of social support can contribute to the development of mental illness for those who have been traumatized. The OSISS program is a template for restoring this vital ingredient into the lives of our wounded soldiers.

ACKNOWLEDGMENT

Many people have been instrumental in the creation and success of the OSISS program and we wish to acknowledge the extensive contribution and dedication of Jim Jamieson, Canadian Forces Medical Advisor to the OSISS program, and the staff of the Veterans Affairs Canada, Ste. Anne's Hospital, Sainte-Anne-de-Bellevue, Quebec, and specifically Johanne Isabel, Francine Gagnon, and Juan Cargnello.

Most important, we want to acknowledge that this chapter would not be possible without the courage of those who have suffered the stresses and traumatic experiences of military operations. These many individuals speak out about their own personal experiences and struggles through personal testimonies, treatment, peer support groups, or through their own individual efforts to help others on their road to recovery and rehabilitation. Their courage and generosity to their fellow peers are admirable. For this, we express our gratitude.

IN MEMORY

This chapter is dedicated to the memory of the late Lt. General (ret) Christian Couture, who died January 28, 2006. His vision and unfailing support and commitment to the establishment of the OSISS program is the inheritance of all Canadian Forces members, veterans, and their families affected by operational stress injuries.

REFERENCES

Allen, J. P., Litten, R. Z., Fertig, J. B., & Babor, T. (1997). A review of research on the Alcohol Use Disorders Identification Test (AUDIT). *Alcoholism, Clinical and Experimental Research, 21,* 613–619.

American Psychiatric Association. (2000). *Handbook of Psychiatric Measures.* Washington, DC: Author.

Babor, T. F., Bohn, M. J., & Kranzler, H. R. (1995). The Alcohol Use Disorders Identification Test (AUDIT): Validation of a screening instrument for use in medical settings. *Journal of Studies on Alcohol, 56,* 423–432.

Babor, T. F., de la Fuente, J. R., Saunders, J., et al. (1992). *AUDIT, the Alcohol Use Disorders Identification Test: Guidelines for Use in Primary Health Care (WHO Publication No. PSA/92.4).* Geneva: World Health Organization.

Barlow, D. H., DiNardo, P. A., Vermilyea, B. B., Vermilyea, J., & Blanchard, E. B. (1986). Comorbidity and depression among the anxiety disorders. Issues in diagnosis and classification. *Journal of Nervous and Mental Disease, 174,* 63–72.

Birenbaum, R. (1994). Peacekeeping stress prompts new approaches to mental-health issues in Canadian military. *Canadian Medical Association Journal, 151,* 1484–1489.

Blanchard, E. B., Jones-Alexander, J., et al. (1996). Psychometric properties of the PTSD Checklist (PCL). *Behaviour Research and Therapy, 34,* 669–673.

Breslau, N., Davis, G. C., Andreski, P., & Peterson, E. (1991) Traumatic events and posttraumatic stress disorder in an urban population of young adults. *Archives of General Psychiatry, 48,* 216–222.

Brewin, C. R., Andrews, B., & Valentine, J. D. (2000). Meta-analysis of risk factors for posttraumatic stress disorder in trauma-exposed adults. *Journal of Consulting and Clinical Psychology, 68,* 748–766.

Canadian Mental Health Association. (2005). *Stigma and mental illness.* Retrieved March 2006 from http://www.mentalhealthworks.ca

Carling, P. J. (1995). *Return to Community: Building Support Systems for People with Psychiatric Disabilities.* New York: Guilford Press.

Claussen, B., & Aasland, O. G. (1993). The Alcohol Use Disorders Identification Test (AUDIT) in a routine health examination of long-term unemployed. *Addiction, 88,* 363–368.

Cohen, S., & Syme, S. L. (Eds). (1985). *Social Support and Health.* New York: Academic Press.

Cohen, S., Underwood, L. G., & Gottlieb, B. H. (2000). *Social Support Measurement and Intervention.* London: Oxford University Press.

Cordova, M. J., Andrykowski, M. A., Kenady, D. E., McGrath, P. C., Sloan, D. A., & Redd, W. H. (1995). Frequency and correlates of posttraumatic-disorder-like symptoms after treatment for breast cancer. *Journal of Consulting and Clinical Psychology, 63,* 981–986.

Davidson, L., Chinman, M., Kloos, B., Weingarten, R., Stayner, D., & Tebes, J. K. (1999). Peer support among individuals with severe mental illness: a review of the evidence. *Clinical Psychology: Science and Practice, 6,* 166–187.

Department of National Defence (2000). *Final Report: Board of Inquiry Croatia*. Ottawa: Department of National Defence. Retrieved April 2006 from http://www.dnd.ca/boi/engraph/home_e.asp

Department of National Defence & Veterans Affairs Canada. (2002). *Operational Stress Injury Social Support Policy*. Retrieved March 2006 from http://www.osiss.ca

Department of National Defence & Veterans Affairs Canada. (2005). *Interdepartmental Evaluation of the OSISS Peer Support Network* (CRS No. 1258-138). Ottawa, Ontario: DND Chief Review Services.

De Rosenroll, D. A. (1994). Toward an Operational Definition of Peer Helping. The *Peer Facilitator Quarterly*, 12(1), 30–32. De Sousa, L., & Leung, D. (2002). A vision and mission for peer support: Stakeholder perspectives. *International Journal of Psychosocial Rehabilitation, 7*, 5–14.

Dixon, L., Krauss, N., & Lehman, A. (1994). Consumers as service providers: The promise and the challenge. *Community Mental Health Journal, 30*, 615–625.

Dovidio, J. F., Major, B., & Crocker, J. (2000). Stigma: Introduction and overview. (pp. 1–28) In T. F. Heatherton, R. E. Kleck, M. R. Hebl, & J. G. Hull (Eds.), New York: Guilford Press. *The Social Psychology of Stigma*.

Emerick, R. E. (1990). Self-help groups for former patients: Relations with mental health professionals. *Hospital and Community Psychiatry, 41*, 401–407.

English, A. D. (1999). Historical and contemporary interpretations of combat stress reaction. *Canadian Forces Journal, 1*, 37.

Enns, M. W., Swenson, J. R., McIntyre, R. S., Swinson, R. P., & Kennedy, S. H.; CANMAT Depression Work Group. (2001). l. Clinical guidelines for the treatment of depressive disorders. VII. Comorbidity. *Canadian Journal of Psychiatry, 46*(Suppl. 1), 77S–90S.

Felton, C. J., Stastny, P., Stern, D. L., Blanch, A., Donahue, S. A., Knight, E., et al. (1995). Consumers as peer specialists on intensive case management teams: Impact on client outcomes. *Psychiatric Services, 46*, 1037–1044.

Figley, C. R. (1982). *Traumatization and comfort: Close relationships may be hazardous to your health*. Keynote presentation at the Conference for Families and Close Relationships: Individuals in Social Interaction, Texas Tech University, Lubbock, Texas, February.

Figley, C. R. (1985). The family as victim: Mental health implications. *Psychiatry, 6*, 283–291.

Figley, C. R. (Ed.). (1995). *Compassion Fatigue: Coping with Secondary Traumatic Stress Disorder in Those Who Treat the Traumatized*. New York: Brunner/Mazel.

Forbes, D., Creamer, M., Hawthorne, G., Allen, N., & McHugh, A. F. (2003). Comorbidity as a predictor of symptom change after treatment in combat-related posttraumatic stress disorder. *Journal of Nervous and Mental Disease, 191*, 93–99.

Galovski, T., & Lyons, J. A. (2004). Psychological sequelae of combat violence: A review of the impact of PTSD on the veteran's family and possible interventions. *Aggression and Violent Behavior, 9*, 477–501.

Goffman, E. (1963). *Stigma: Notes on the Management of Spoiled Identity*. Englewood Cliffs, NJ: Prentice Hall.

Heber, A., Grenier, S., Richardson, D., Darte, K., Burr, M., & Dallaire, R. (2005). *Overcoming stigma and delivering care: Combining peer support and treatment in the Canadian Forces*. Presented at the International Society for Traumatic Stress Studies 21st Annual Meeting, Toronto, Ontario.

Hodges, J. Q., Markward, M., Keele, C., & Evans, C. J. (2003). Use of self-help services and consumer satisfaction with professional mental health services. *Psychiatric Services, 54*, 1161–1163.

Hoge, C. W., Castro, C. A., Messer, S. C., McGurk, D., Cotting, D. I., & Koffman, R. L. (2004). Combat duty in Iraq and Afghanistan: Mental health problems, and barriers to care. *New England Journal of Medicine, 351*(1), 13–22.

Kaniasty, K. (2005). Social support and traumatic stress. *National Center for PTSD Research Quarterly, 16*(2), 128.

Kaufman J., & Charney D. (2000). Comorbidity of mood and anxiety disorders. *Depression and Anxiety, 12*(Suppl. 1), 69–76.

Keane, T. M., & Wolfe, J. (1990). Comorbidity in post-traumatic stress disorder: An analysis of community and clinical studies. *Journal of Applied Social Psychology, 20,* 1776–1788.

Kessler, R. C., Sonnega, A., Bromet, E., & Nelson, C. B. (1995). Posttraumatic stress disorder in the National Comorbidity Study. *Archives in General Psychiatry, 52,* 1048–1060.

Knight, R. G., S. Williams, et al. (1997). Psychometric properties of the Centre for Epidemiologic Studies Depression Scale (CES-D) in a sample of women in middle life." *Behaviour Research and Therapy, 35,* 373–380.

Kulka, R. A., Schlenger, W. E., Fairbank, J. A., Hough, R. L., Jordan, B. K., Marmar, C. R., et al. (1990). *Trauma and the Vietnam War Generation: Report of Findings from the National Vietnam Veterans Readjustment Study.* New York: Brunner/Mazel.

Labonte, R. (1989). Community and professional empowerment. *Canadian Nurse, 85,* 22–28.

Lepore, S. J., Cohen Silver, R., Wortman, C. B., & Wayment, H. A. (1996). Social constraints, intrusive thoughts, and depressive symptoms among bereaved mothers. *Journal of Personality and Social Psychology, 70*(2), 271–282.

Lipton, M. I., & Schaffer, W. R. (1986). Post traumatic stress disorder in the older veterans. *Military Medicine, 151,* 522–524.

Litz, B. (1996). The psychological demands of peacekeeping. *National Center for PTSD Clinical Quarterly, 6*(1).

Litz, B. T., Orsillo, S. M., Friedman, M., Erhlich, P., & Batres, A. (1997). Post-traumatic stress disorder associated with peacekeeping duty in Somalia for U.S. military personnel. *American Journal of Psychiatry, 154,* 178–184.

MacDonald, C., Chamberlain, K., Long, N., Pereira-Laird, J., & Mirfin, K. (1998). Mental health, physical health and stressors reported by New Zealand Defence Force peacekeepers: A longitudinal study. *Military Medicine, 163,* 477–481.

Marin, A. (2003). *Off the Rails: Crazy Train Float Mocks Operational Stress Injury Sufferers,* March 2003. http://www.ombudsman.forces.gc.ca/reports/annual/2002-2003_e.asp#crazy

Marin, A. (2001*). Special Report: Systemic Treatment of CF Members with PTSD. Report to the Minister of National Defence,* September 2001. http://www.ombudsman.forces.gc.ca/reports/special/PTSD-toc_e.asp

McCann, I. L., & Pearlman, L. A. (1990). Vicarious traumatization: A framework for understanding the psychological effects of working with victims. *Journal of Traumatic Stress, 3,* 131–149.

Miller, K. I., Stiff, J. B., & Ellis, B. H. (1988). Communication and empathy as precursors to burnout among human service workers. *Communication Monographs, 55,* 336–341.

Mitchell, J. T. (1988). Development and functions of a critical incident stress debriefing team. *Journal of Emergency Medical Services, 12,* 43–46.

Passey G., Crocket D. (1995). *Psychological consequences of Canadian UN peacekeeping in Croatia and Bosnia.* International Society for Traumatic Stress Studies. Boston.

Pearlman, L. A., & Saakvitne, K. W. (1995). Treating therapists with vicarious traumatization and secondary traumatic stress disorders. In C. R. Figley (Ed.), *Compassion Fatigue: Coping with Secondary Traumatic Stress Disorders in Those Who Treat the Traumatized* (pp. 150–177). New York: Brunner/Mazel.

Perloff, R. (2003). *The dynamics of persuasion: Communication and attitudes in the 21st Century.* Mahwah, NJ. Publisher.

Radloff, L. S. (1977). The CES-D Scale: A self-report depression scale for research in the general population. *Applied Psychological Measurement,* 1, 385–401.

Ravindran A. V., Judge R., et al. (1997). A double-blind, multicenter study in primary care comparing paroxetine and clomipramine in patients with depression and associated anxiety. Paroxetine Study Group. *Journal of Clinical Psychiatry, 58,* 112–118.

RX.(2000). *RX 2000: A Prescription for Health Care Reform in the Canadian Forces.* Canadian Forces Health Services, Department of National Defence, Ottawa, Ontario. Retrieved from http:/www.forces.ga.ca/health/news_pubs/engraph/HCReform_home_e.asp? Lev1=4&Lev2=6&Lev3=/

Regier, D. A., Rae, D. S., Narrow, W. E., Kaelber, C. T., & Schatzberg, A. F. (1998). Prevalence of anxiety disorders and their comorbidity with mood and addictive disorders. *British Journal of Psychiatry. Supplement, 34,* 24–28.

Richardson, D. (2002). *Rates of PTSD and co-morbidity in Canadian peacekeepers.* Paper presented at the Annual Meeting of the International Society for Traumatic Stress Studies, Miami, Florida.

Rudolph, R. L., Entsuah, R., et al. (1998). A meta-analysis of the effects of venlafaxine on anxiety associated with depression. *Journal of Clinical Psychopharmacology, 18,* 146–44.

Sartorius, N. (2004). The World Psychiatric Association global programme against stigma and discrimination because of stigm. In A. H. Crisp (Ed.), *Every Family in the Land.* (pp. 373–375) London: Royal Society of Medicine Press.

Saunders, J. B., Aasland, O. G., Babor, T. F., et al. (1993). Development of the Alcohol Use Disorders Identification Test (AUDIT): WHO Collaborative Project on Early Detection of Persons with Harmful Alcohol Consumption–II. *Addiction, 88,* 791–804.

Schnurr, P. P., Friedman, M. J., & Green, B. L. (1996). Posttraumatic stress disorder among World War II mustard gas test participants. *Military Medicine, 161,* 131–136.

Schonfeld, W. H., Verboncoeur, C. J., Fifer, S. K., Lipschutz, R. C., Lubeck, D. P., & Buesching, D. P. (1997). The functioning and well-being of patients with unrecognized anxiety disorders and major depressive disorder. *Journal of Affective Disorders, 43,* 105–119.

Schulze, B., & Angermeyer, M. C. (2003). Subjective experiences of stigma: A focus group study of schizophrenic patients, their relatives and mental health professionals. *Social Science and Medicine, 56,* 299–312.

Shehan, C. L. (1987). Spouse support and Vietnam veterans' adjustment to post-traumatic stress disorder. *Family Relations, 36,* 55–60.

Schubert, M. A., & Borkman, T. J. (1991). An organizational typology for self-help groups. *American Journal of Community Psychology, 19,* 769–787.

Solomon, P. (2004). Peer support/peer provided services underlying processes, benefits, and critical ingredients. *Psychiatric Rehabilitation Journal, 27,* 392–401.

Stamm, B. H. (Ed.). (1995). *Secondary Traumatic Stress: Self-care Issues for Clinicians, Researchers and Educators.* Lutherville, MD: Sidran Press.

Stamm, B. H. (1997). Work-related secondary traumatic stress. *National Center for PTSD Research Quarterly, 8*(2).

Statistics Canada. (2002). Canadian community health survey: Mental health and well-being (No. 82-617-XIE). Retrieved April 2006 from http://www.statcan.ca

Stephens, C., Long, N., & Miller, I. (1997). The impact of trauma and social support of post-traumatic stress disorder: A study of New Zealand police officers. *Journal of Criminal Justice, 25*(4), 303–314.

Stuart, H. (2004). Stigma and work. *HealthcarePapers, 5,* 100–111.

Thompson, M. M., & McCreary, D. R. (2003). *Attitudes and Attitude Change: Implications for the OSISS Speakers Bureau Programme.* Toronto: Stress & Coping Group Defence R&D Canada.

Underwood, N. (2005). *Suffering in silence: Stigma, mental illness and the workplace.* Canadian Health Network, October 15, 1–3.

Veterans Affairs Canada. (2006). *Peer Helpers Training Manual.* Sainte-Anne-de-Bellevue. Quebec: Ste. Anne's Hospital.

Volunteer Canada. (2000). *Canadian Code for Volunteer Involvement.* Ottawa: Volunteer Canada (ISBN 0-9680701-2-4). www.volunteer.ca.

Ware, J. E., Snow, K. K., Kosinsk, M., & Gandek, B. (1993). *SF-36 Health Survey: Manual and Interpretation Guide.* Boston: New England Medical Center.

Weathers, F. W., Litz, B. T. , et al. (1993). *The PTSD checklist: Reliability, validity, & diagnostic utility*. Annual Meeting of the International Society for Traumatic Stress Studies, San Antonio, Texas.

Weisaeth, L., Mehlum, L., et al. (1996). Peacekeeper Stress: New And Different? *NCP Clinical Quarterly, 6*(1).

Wintersteen, R. T., & Young, L. (1988). Effective professional collaboration with family support groups. *Psychosocial Rehabilitation Journal, 12,* 19–31.

ENDNOTES

1. A portion of this chapter was presented at the NATO Research and Technology Organization, Human Factors and Medicine Panel Symposium in Brussels, Belgium, April 24–26, 2006.
2. This section on the background on the creation of OSISS is written by Lt. Colonel Stephane Grenier.

14

Spirituality and Readjustment Following War-Zone Experiences

KENT D. DRESCHER, MARK W. SMITH, AND DAVID W. FOY

Religious beliefs and practices (spirituality) aid many people in developing personal values and beliefs about meaning and purpose in life. They can provide an avenue for coping with difficult life events including trauma. Mental health professionals increasingly recognize spirituality as a primary human dimension, and a potentially robust area of research. The military has a long tradition of providing for the spiritual needs of its troops through chaplains representing many faith traditions. However, the direct spiritual consequence of participation in war has only recently begun to be studied, as has the potential role spirituality may play as a healing resource for those recovering from war-zone trauma.

Researchers and theorists about the effects of trauma have suggested that traumatic events frequently call into question existential and spiritual issues related to the meaning of life, self-worth, and the safety of life (Janoff-Bulman, 1992). For those whose core values are theologically grounded, traumatic events often give rise to questions about the fundamental nature of the relationship between the creator and humankind. The question of how belief in a loving, all-powerful God can be sustained when the innocent are subjected to traumatic victimization has been labeled "theodicy" by philosophers. Frequently called "the problem of evil," theodicy poses the question: If God is all-powerful, and God is all-good, how does God allow evil to exist in the world? Historically, varied solutions have been proposed to the theodicy question, including solutions that diminish God (i.e., God is not all-powerful, God is not all-good, God does not exist), or that diminish evil (i.e., it is a punishment for sin, it may bring about some greater good), and perhaps individual solutions that diminish the self (e.g., self-blame, rage, loss of meaning, purpose, or hope). However, theodicy is not a philosophical question to trauma survivors—it is real, tangible, and can be an obstacle to full recovery.

The purpose of this chapter is to provide an overview of the ways that war generally, and the current wars in Afghanistan and Iraq specifically, may affect the spirituality of returning troops. We hope to review the empirical literature related to trauma and spirituality among veterans of previous wars, share the anecdotal experiences of a chaplain who has directly debriefed many returning troops, and offer suggestions gleaned from clinical work among veterans with posttraumatic stress disorder (PTSD), of ways to help veterans utilize healthy spirituality in their trauma recovery.

Reflections on the Spiritual Effects of the War on Returning Troops: Commander Mark W. Smith (Navy Chaplain)

Many Marines are reluctant to go to the chaplain after a dramatic or traumatic event such as war. They are afraid it might look as if they couldn't handle the pressure and needed the chaplain. That is the advantage of requiring every returning Marine—gunnery sergeant through general in a particular force—to schedule a personal debrief with the chaplain. They can come into the outer office complaining that they don't need this, then tell their whole story to the chaplain once the counseling office door is closed.

I was that chaplain for the First Marine Expeditionary Force Command Element when they returned from Iraq in the spring of 2005. As part of the "Warrior Transition" program, I individually debriefed some 200 officers and senior enlisted personnel in a 45-day period. Their appointments ranged from 10 minutes to 1 hour and 15 minutes. Most were in the 30- to 45-minute range. The longer appointments were usually the Marines who most boldly declared their lack of need of debriefing by the chaplain. These appointments were long not because I had to work so hard to get the Marines to admit they had emotions—the length was usually because I asked them how they were doing, and they had a lot more to say than they thought they did.

Did some of what they talked about include spiritual issues? I haven't yet decided if it was surprising that so many gruff, tough Marines had sincere spiritual issues. You might think they would feel obligated to talk spirituality since they were in the chaplain's office. But very few of these discussions had the sound of someone trying to make the chaplain feel needed. The Marines appeared to be grappling with the real meaning of life, and their own place in this world. Can't get much more spiritual than that.

It seems to me that these Marines fell into two categories in relation to how they perceived they were responding to their experiences of trauma in war: the "Never-recovers" and the "Nothing-wrongs."

The Never-recovers: When the counseling office door closed, the fears and concerns began to pour out. Often Marines would say, "I'm not really having any problems, but I did see…." And the story would take them deep into the basic questions of life, and their own painful thoughts. They actually often seemed relieved to talk with someone other than a mental health clinician or an official member of the chain of command. They were glad to be able to talk honestly with no fear of their words ever getting out to the rest of the command. When these Marines talked about their fears, one of the chief concerns was that they would never be the same. Probably true. But they also feared that meant they would never get better. They would never recover from this.

Very few Marines actually found themselves angry with God. Very few blamed God for what they had seen or for what was going on around them. In fact, many actually had some newfound appreciation for the faith of others, but still retained the right to hate those who sought to destroy under misguided interpretations of valid religious expressions. And this appreciation of other religions caused a number of Marines to say that they planned to get more involved in their own faith group's practices. They might never be the same, but maybe they could live a better life anyway, despite the fact they feared the hate would never go away, or the compassion would never return.

Another characteristic of some of the Never-recovers: they were tired of talking about their war experiences. They had done enough of that, they had processed it, and wanted to be done with it. What they really wanted to talk about was the pain they were having right now. They didn't seem to want to consider that the two were still related.

The Nothing-wrongs seemed to come in two variations: those in denial, and those who actually grew through their experiences.

The Marines in denial continued to maintain nothing was wrong with them despite the painful stories and torturous symptoms they said they were living with. A number of them did say they appreciated that the military took their personhood seriously enough that they wanted everyone to see the chaplain. They appreciated that the focus was not just medical or cover-your-tail-in-case-something-bad-happens when a Marine returns. They were hinting they wanted to go deeper. We usually did. Sometimes they left still maintaining they didn't need any help. But that was pretty rare.

Many other Marines talked about not being traumatized because their faith sustained them. They had strong beliefs about their place in this world and their hope in an afterlife. They frequently believed that God would take care of them, but surprisingly that was not a naive belief that they would be safe, but that they would either be protected or God would comfort them even as they were called home (to heavenly places) or allowed to suffer. Their theology after war was better than the average Marine often manages. They were spiritually stronger than they had been before.

There were also a number who decided it was time to grow up and get more serious about their faith. They needed to decide what they believed and then get more involved in the practice of that faith. Nearly all were in agreement that their spiritual selves were important and had been affected by this deployment.

I also debriefed several chaplains and a number of medical personnel. These debriefs were both one-on-one, and in two retreats aimed at helping caregivers deal with the extra burden of trauma they encountered while trying to take care of others. Participants in these retreats completed an anonymous survey on spirituality. Here are some of my observations from that experience.

The retreats included 31 recent returnees from Iraq, split between chaplain staff and medical staff. They were surveyed on some of their reactions since returning. Some of the questions were specific toward ministry, others aimed at spirituality issues in general, and others directly targeted some of the human spirit connection points for spirituality. Whether we are looking at cognitive, behavioral, or relational definitions of spirituality, spiritual issues all seem to involve these human connection points.

Nearly all participants, including the medical personnel, strongly agreed that spirituality is important. Additionally, a large majority of the respondents showed a high level of agreement that several spiritual aspects of their lives had been affected. Three were strongest: (1) their faith had been challenged, (2) they had found new purpose, and (3) their spiritual religious practices had changed. None of these changes are necessarily negative; in fact, they are probably another example of adversity providing the opportunity to grow in positive ways. Other responses may indicate when the changes were negative. There was nearly universal agreement that they were suffering spiritual burnout and were emotionally drained. There were also a significant number who felt they had lost some of their "sense of call"—usually a term used by ministers to describe why they became ministers, but here used by medical personnel as well.

A majority of these caregivers also reported damage to parts of their humanity that could arguably be the connection points for spirituality in the human makeup: (1) loss of their creativity, (2) having greater difficulty expressing and receiving love, and (3) having greater difficulty expressing themselves or understanding others. Creativity, ability to love and be loved, and language are some of the quintessential aspects of being human—in other words, part of the spirit of being human.

Other spiritual connection points were seen as damaged by a significant minority of respondents: (1) having trouble seeing themselves as others see them, (2) loss of their sense of place in this world, (3) loss of the ability to make choices for themselves, and (4) loss of their appreciation of beauty. These could be identified as the human distinctives of self-transcendence, autonomy, and aesthetics.

These returning combatants and caregivers showed a high awareness of spiritual needs, newfound theological understandings, and clear damage to elements of spirituality within themselves. Areas such as creativity, the ability to love and be loved, advanced language, self-transcendence, autonomy, and aesthetics were clearly affected. My impression of our times together lead me to think that a significant number of these Marines were coming to see great potential for growth.

Admittedly, these interviews and retreats were in close proximity to the participants' return from war. True damage caused by exposure to the trauma of war will undoubtedly not be fully realized or diagnosed until much later, but it does add credence to the common wisdom that there are no atheists in a foxhole.

HOW SPIRITUALITY MAY PROMOTE TRAUMA RECOVERY

Though not all these areas have been researched, there are several important clinical themes among trauma survivors that potentially involve religion and spirituality. For example, anger, rage and a desire for revenge may be tempered by forgiveness, spiritual beliefs, or spiritual practices. Feelings of isolation, loneliness, and depression related to grief and loss may be lessened by the social support of religious participation (McIntosh, Silver, & Wortman, 1993). Spirituality, as it is frequently experienced in community settings, places survivors among caring individuals who may provide encouragement, emotional support, as well as possible instrumental support in the form of physical or even financial assistance in times of trouble. Recovery of meaning in life may be achieved through changed ways of thinking and involvement in meaningful caring activities or through religious rituals experienced as part of religious/spiritual involvement. In addition, traumatic experiences may become a starting point for discussion of the many ways in which survivors define what it is to have "faith." Finally, religion and spirituality may be associated with beliefs about healthy lifestyles and may keep people from engaging in unhealthy coping behaviors. This may, for instance, decrease survivors' risk for substance abuse and social isolation in the aftermath of trauma. It also may provide stress reduction through practices such as prayer and meditation.

REVIEW OF RECENT RESEARCH ON THE SPIRITUALITY-TRAUMA LINK AMONG COMBAT VETERANS

On the positive side, spirituality may help combat veterans achieve posttraumatic growth (Linley & Joseph, 2004) that could lead to benefits, such as increased resilience in the face of future life challenges, increased meaning or purpose, and strengthened capacity to utilize positive coping resources amid crises. However, surviving trauma may also be associated with a shift to more negative beliefs about the safety, goodness, and meaningfulness of the world (Janoff-Bulman, 1992), negative views of one's relationship with God/deity (i.e., beliefs that God is punishing me, or has abandoned me) (Pargament, Koenig, & Perez, 2000), loss of core spiritual values, and estrangement from or questioning of one's spiritual identity (Decker, 1993; Drescher & Foy, 1995; Falsetti, Resick, & Davis, 2003; Wilson & Moran, 1998). Additionally, several authors (Gorsuch, 1995; Pargament et al., 2003) have suggested that unhealthy aspects of spirituality might actually lead to worse clinical outcomes.

An early study (Green, Lindy, & Grace, 1988) found increased religious coping and attempts to assign meaning to war-zone events in military combat veterans. Additionally, a study from a residential PTSD treatment program found strong religious/spiritual distress (i.e., abandoning faith in the war zone, difficulty reconciling war-zone events with faith) in a high percentage of military veterans (Drescher & Foy, 1995). To date, dimensions of spirituality and their relationships to clinical outcomes among veterans treated for PTSD have not been examined.

Several more recent studies have identified both positive and negative associations between spirituality and war-zone trauma or related PTSD. Witvliet and colleagues (2004) identified two dimensions of spirituality, i.e., lack of forgiveness and religious coping (both positive and negative) that were associated with PTSD and depression severity in an outpatient sample of veterans treated for PTSD. Further, another recent study (Fontana & Rosenheck, 2004) found a significant structural equation model pathway between war-zone trauma, change in religious faith, and increased utilization of VA mental health services for veterans being treated for war-zone–related PTSD. Specific types of war-zone experiences (killing others, failure to save the wounded, etc.) were directly and indirectly (mediated by guilt) associated with reduction in comfort derived from religious faith. Both guilt and reduced comfort from religious faith were shown to be associated with increased use of VA services (Fontana & Rosenheck, 2004).

In a recent study of women veterans, those who reported being sexually assaulted (23% of the sample) while in the military were found to have poorer overall mental health and higher levels of depression than veterans who did not report being assaulted (Chang, Skinner, & Boehmer, 2001). The study also found that more frequent religious participation among the sexually assaulted women was associated with lower depression, higher overall mental health scores, consistent with a buffering effect for religious participation on mental health.

Taken together, these studies raise several key considerations for professionals interacting with military service personnel returning from combat deployment. First is the potential that trauma exposure may lead to a loss of faith. Spiritual tensions that arise for many combat veterans attempting to come to terms with their war-zone experiences may reduce their use of spiritual resources as part of reentry, and may in turn lead to worse psychiatric symptoms and higher medical service utilization. Additionally, it is important to stay alert for signs of "negative religious coping" (e.g., God has abandoned me, God is persecuting me) or negative attributions about God, as these can be associated with more severe PTSD and depression in some veterans. Finally, difficulties with forgiveness and higher levels of hostility or guilt may be associated with more severe problems later on. It is notable that much of our current knowledge about relationships between trauma and spirituality comes from studies conducted years after those traumatic experiences occurred. It will be important to continue this line of research with individuals returning from the present conflict, soon after their actual combat experiences.

REASONS WHY SOLDIER REACTIONS TO THIS WAR MAY BE MORE VARIED

There are a number of reasons why the current war might provoke more varied spiritual reactions than previous wars. First, the personal characteristics of the soldiers, and the context in which they serve, are different. This war is being conducted by an all-volunteer military with extensive use of National Guard and military reserve troops. Among personnel in the present war, greater variability in age, gender, and avenue of deployment (reserves or National Guard) exists than in previous wars. Because there is no military draft, experience levels of troops in the war zone may be somewhat higher as well. As has been true in previous wars, perception and impact of war-zone experiences among officers and senior enlisted soldiers may be different from those of more junior personnel. In the current conflict, repeated deployments of uncertain duration have created significant stress. This is particularly true for reserves and National Guard troops, who left careers and businesses behind, and for whom supportive resources may be lacking upon return home. Homecoming experiences may be another source of differential spiritual impact for returnees from the current war. Though there is active and vocal opposition to the current war in some segments of the United States and even more largely abroad, there seems to be an awareness, even among those opposing the war, of mistakes made in previous wars. Even those in opposition seem to be making active attempts to express support and concern for returning personnel—something that was not always true for returning Vietnam veterans.

Trauma exposure is another area of difference from experiences in previous wars, which may contribute to differential spiritual impact. Aside from initial battles in the first weeks of the war and sporadic intensive battles within constrained geographic areas (e.g., Al Fallujah), a great number of the life-threatening experiences individuals are exposed to appear somewhat random. Many deaths have occurred from improvised explosive devices, rocket propelled grenades, and suicide bombers. As a result, those who, in previous wars, might have been considered noncombatants (e.g., truck drivers) are now subject to high risk of traumatic exposure and injury.

GUIDELINES FOR INCORPORATING SPIRITUALITY INTO TRAUMA RECOVERY

Much of the clinical work that led to the development of the suggestions that follow has been done within a PTSD treatment program that utilizes a group therapy format. It is important to note that the principles discussed here are also appropriate for use with individuals. However, there are a number of potential advantages to using a group format. Groups provide the opportunity to learn from the experiences and thoughts of others. Discussing issues with other veterans who have similar experiences helps veterans counter the idea that "I am the only one with problems like this." Groups also provide a wider range of feedback, which is often better received, because the feedback comes from peers rather than from staff. Finally, group interaction builds actual connections and friendships among group members, which carries with it both immediate and potential long-term benefit. The authors recognize that many of the clinical opportunities helpers will have with returnees from the war will occur one on one, and have tried wherever possible to tailor suggestions to be useful in both individual and group contexts.

Another important question is who most appropriately should provide the services to address the interaction between spirituality and trauma. Traditionally, chaplaincy and mental health have operated somewhat independently, and not always collaboratively. Both disciplines have unique strengths and potentially serious limitations. Chaplains frequently receive little more than basic training in clinical skills and lack specialized knowledge of interventions for specific mental health disorders. Equally true is that mental health providers usually receive little or no training in how to address spiritual/religious issues. The clinical suggestions in this chapter were developed and originally implemented within a mental health context. They, however, explicitly do not attempt to answer or resolve theological or religious questions which arise for veterans; rather, these interventions are conducted from a motivation enhancement perspective. We hope to increase veterans' openness to spiritual exploration and to remove barriers that have prevented them from seeking to utilize spirituality in their recovery from trauma. We encourage veterans to seek out additional support from chaplains to talk through specific issues or questions in greater individual detail. The ideal intervention might be collaborative in nature where both disciplines make contributions to the intervention process.

To institute a spiritual component into efforts supporting recovery from trauma, one critical ground rule must be maintained: the experience must be experienced as safe. Safety in a clinical experience involving spirituality has two distinct components: (1) intentional awareness and acceptance of diverse spiritual experiences, and (2) mutual respect for the views of others and openness to new learning. It is very important for clinical staff to model these characteristics and to verbalize these ideals repeatedly. Safety is necessary for helpful discussion of sensitive and delicate issues and creates an environment that allows for honest, vulnerable self-disclosure on the part of participants.

The tone and content of conversation needs to be fully inclusive so that participants can feel comfortable with, and benefit from, the experience.

To participate fully, participants need to experience the conversation as a place where feelings related to the existential impact of trauma can be expressed regardless of one's individual beliefs about religion or God. Intentional awareness of diversity means there can be no assumptions that individuals share common beliefs or religious traditions. Helpers need to be careful in their use of language to ensure that the way they speak about spiritual issues is not heard by clients as biased or advocating a particular spiritual perspective. If helpers choose to self-disclose information about their own spiritual history, that choice should be both intentional and directly based on the clinical needs of the client.

Arising out of these core values flow "group rules" about using "I" statements when speaking about one's personal views and beliefs, and a proscription against "proselytizing" or speaking to persuade others that one's beliefs are correct or "true." Ultimately, this need for inclusiveness extends even to definition of terms. For example, we have selected a definition of spirituality that does not require belief in God or a "higher power" and can even accommodate active hostility toward religion in all its forms. Spirituality is defined for clinical purposes as "an individual's understanding of, experience with, and connection to that which transcends the self" (Drescher, 2006, p. 337). The idea of spirituality as connection with something beyond self allows a given individual to define their personal spirituality as relationships with friends or family, or connection with nature, if connection with God or a higher power is not a personal option.

When religion is discussed, emphasis should be placed on aspects that varied spiritual traditions share, rather than on those which separate. Acknowledgment of the varied contributions of each religious/spiritual tradition and culture represented should be made whenever possible and a tone of acceptance set by facilitators. As with all clinical activity, an environment should be established that allows for appropriate emotional expression and self-disclosure.

It should be explicitly stated that helping conversations of the sort we are describing are not designed to teach spirituality; rather, they should be seen as providing a safe space in which to discuss the possibility that spirituality might be a recovery resource. One might view this as motivation enhancement toward reconsidering spirituality as a potential healing resource for veterans following war-zone experiences. The overarching goal of such conversations would be to allow individuals to reconsider the role that spirituality might play following trauma.

Specific Suggestions for Topics/Activities That Address Spiritual Needs

Redefine Spirituality. Many people do not think very often about how they define spirituality. Engaging in a discussion of what an individual sees as core elements to a definition can be useful in helping that person realize that he or she can actually reconsider views which may have been learned in childhood. Defining spirituality as "connecting to something outside the self" frees each individual to define that connection for him/herself. We encourage individuals to engage in a journey of a new discovery of what spirituality might now mean for their lives.

Group Exercise to Encourage Self-Disclosure and Relationship Building. In order to feel comfortable speaking about issues that can be very personal, it is very helpful for participants to get better acquainted. One exercise we have found useful, in helping facilitate both relationship and the realization of how life experiences and spirituality have been related over time, is called a spiritual autobiography. Individuals are asked to describe their spiritual journey from childhood to the present using a timeline chart. This highlights key experiences and decisions which were made regarding their religious faith and illustrates the context in which they occurred. This exercise allows clients to clarify and see more objectively their current religious beliefs and practices and reflect on directions they would like to pursue. Autobiographies are presented in turn by group members during sessions and help to identify and begin discussion of relevant themes and issues.

Encourage Involvement in Community. As humans, we are primarily social creatures. However, trauma and PTSD frequently impair relationships and distance survivors from potential support systems. Defining spirituality as "connection" fosters a reconsideration of that distancing process. Our culture is highly individualistic, and one problem as society has begun to see spirituality as an individual endeavor is that the community aspect of spirituality, which is inherent in most religious traditions, sometimes gets lost. We encourage veterans to seek out healthy, supportive communities, whether religious or not. Examples of these communities obviously include churches or Alcoholics Anonymous meetings, but could also include nonprofit helping organizations, service clubs, meditation groups, and even sports teams. We also gently confront trauma victims who seem intent on pursuing spirituality in total isolation and solitude, and encourage them to consider the possible benefits of incorporating community and relationship as aspects of their spiritual experience.

Incorporate Spiritual Practices. Spiritual activities should be described as being both inward and outward focused. A variety of inward experiential exercises involving meditation, breathing, guided imagery, and silent prayer are appropriate. Exercises should include a relaxation component which will build on existing stress management skills and which will contribute in a positive way to coping with war-zone–related stress. Activities should be drawn from a variety of religious traditions. In addition, outside "practice" of prayer and meditation exercises experienced during group sessions is encouraged.

From an outward perspective spiritual practice should include service and work on behalf of others. Nearly all religious traditions encourage service as a form of spiritual practice. One of the benefits of volunteering is engagement in the lives of others, which for a person suffering from PTSD addresses the tendency toward withdrawal and social isolation. Service for others also is a way of creating personal meaning and living a life that matters to others. Trauma victims are sometimes quite self-focused because of the damage they perceive has been done to them. Engaging with others who also have significant needs helps to broaden a victim's focus of attention, and helps them recognize they are not alone and that they can actually provide benefit to others. This can have very positive effects on the self-esteem of both the helper and of those who are served.

Examine Potentially Harmful Spiritual Attributions. Several studies (Gorsuch, 1995; Pargament & Brandt, 1998; Witvliet et al., 2004) have indicated that negative religious coping (i.e., negative attributions about God), such as "God has abandoned me," "God is punishing me," or anger at God, is associated with a number of poor clinical outcomes. We find it useful to talk about these data with veterans and to share ways to alter these viewpoints. Group interaction around these issues can be particularly helpful, as simply discussing the issue and hearing differing viewpoints voiced by other veterans can be helpful for those who are seemingly "stuck" in these negative ways of viewing their situation.

Address Important Existential Topics. Each session can include discussion of important existential issues frequently neglected in day-to-day life. Though helpers need to be informed about how various faith traditions have wrestled with these issues, it should be left to individual clients to struggle with each topic, developing their own individual solutions.

Theodicy—the Problem of Evil. This term comes from the Latin *théos díe*, meaning justification of God. The term was coined by the philosopher Leibniz, who in 1710 wrote an essay attempting to show that the existence of evil in the world does not conflict with belief in the goodness of God (Leibniz, 1890). Simply stated, theodicy poses the question: If God is all-powerful, and God is all-good, how does God allow evil to exist in the world? Historically, varied solutions have been proposed to the theodical problem, including philosophical solutions that diminish God (i.e., God is not all-powerful, God is not all-good, God does not exist), or that diminish evil (i.e., it is a punishment for sin, it may bring about some greater good), and perhaps personal nonphilosophical solutions that diminish the self (e.g., self-blame, rage, loss of meaning, purpose, or hope).

From a psychological perspective, Festinger's (1957) cognitive dissonance theory posits that individuals tend to seek consistency among their cognitions and experiences. When inconsistency exists between cognitions and experience, there is strong motivation for change, to eliminate the dissonance. In the case of a traumatic experience, the event itself cannot be changed, hence survivors must struggle to adapt their beliefs and attitudes to accommodate their experience in order to resolve the dissonance. Many trauma survivors, along with their families and friends, thus begin a lifelong journey toward making sense of their experiences.

Forgiveness. We have chosen to address the topic of forgiveness in two somewhat different ways. The first approach is to see forgiveness as something done in relation to a specific event. Thoresen, Harris, and Luskin (2000) define forgiveness as "the decision to reduce negative thoughts, affect, and behavior, such as blame and anger, toward an offender or hurtful situation, and to begin to gain better understanding of the offense and the offender" (p. 255). It has also been important to acknowledge that forgiveness does not include pardoning an offender, condoning or excusing an offense, forgetting an offense, or denying that an offense occurred. Rather, forgiveness involves choosing to abandon one's right to resentment and negative judgment, while nurturing undeserved qualities of compassion, generosity, and even love toward the offender (Enright & Coyle, 1998).

Within a military war-zone context, forgiveness sometimes becomes an issue of tension, in that it may suggest to veterans a pressure toward forgiving an enemy that killed your friends; forgiving the government that sent you into harm's way; forgiving people who perhaps didn't do their jobs effectively or who made mistakes; forgiving God who allowed all this to happen; and forgiving the self for perceived errors, mistakes, or lack of action. Though not all these issues are relevant for any given veteran, they are frequent areas of concern. Additionally, as veterans attempt to cope with the aftermath of war-zone experiences after returning home to families and friends, they frequently find the need for forgiveness or self-forgiveness in their relationships.

One issue that arises with veterans with war-zone trauma experiences is that perceptions and memories of these experiences, which are colored by strong emotions such as fear, rage, grief, guilt, and shame, seem to be particularly subject to cognitive distortion. The forgiveness process should not begin around distorted thinking. Rather, memories of things that happened around traumatic experiences should be examined carefully to look for distortions of belief, inappropriate assumptions or expectations, and illogical attributions about these traumatic events. After more reasoned, rational thoughts and attributions about the events have been attained, whatever remaining real blame or culpability that exists directed toward self or others can be addressed from the perspective of forgiveness.

PTSD symptoms themselves can sometimes become a barrier to forgiveness among veterans. In speaking with numerous veterans about difficulties they were having with forgiveness, reexperiencing symptoms are frequently cited as proof that forgiveness does not work. The authors have had to learn how to help disentangle the recognition that trauma experiences produce lasting intrusive memories from the actual forgiveness process.

The second way that forgiveness is addressed is by seeing it as a lifestyle. We talk about forgiveness as an attitude which exists at the opposite end of a continuum from the attitude of hostility. Hostility is described as an attitude that is closed to new experiences, pushes people away, expects the worst, and is harshly critical and judgmental of people and experiences. Forgiveness, on the other hand, hopes for the best, welcomes others, is open to new possibilities and new experiences, and is gently accepting and tolerant. Viktor Frankl, himself a trauma survivor, once said, "everything can be taken from a man but one thing: the last of the human freedoms—to choose one's attitude in any given set of circumstances, to choose one's own way" (Frankl, 1984/1946). In examining how one might move toward forgiveness, discussion centers are beginning to move away from an attitude of hostility and trying to take on some characteristics of an attitude of forgiveness. Ultimately, forgiveness is a choice. Forgiveness is presented as a positive choice, a choice for oneself, a choice that seeks health and wholeness, a choice that enhances supportive relationship with self and others.

Values for Living. Values are the ideas and beliefs that we hold as good, as important, as worthy of our time and energy. When speaking with veterans, the things they frequently mention as valuing the most include a sense of belonging; self-respect; inner harmony; freedom; family security; health; and enjoying life. A crucial question for all of us is to what degree our values are reflected in our day-to-day behavior. In other words, do we walk our talk? It is important for each of us to think about the degree to which our lives are authentic, such that how we spend our time accurately portrays what we hold to be important.

Finding Meaning and Purpose. We have found two separate means of addressing meaning in the context of spirituality and trauma. The first is the sense of personal meaning that one derives from one's own internal view of self. First we talk about meaning as "the story we tell ourselves about our life experiences." In this way, meaning can be construed as being the sum of our perceptions (i.e., sensory experiences, sight, sounds, touch), and our interpretation of those events. We point out that the problem with this equation is that both perception and interpretation are subject to error and cognitive distortion during experiences of stress or trauma. Frequently, the clients we treat carry with them extremely negative and distorted beliefs such as guilt, shame, or self-blame related to the trauma, which affects their perceptions of personal worth and value and of their efficacy in successfully recovering from their experiences. Finding ways to view their life experiences more accurately can be extremely important for those recovering from PTSD. In this context, recovery of meaning is a part of the cognitive restructuring process that trauma victims frequently do.

The second way that we look at meaning has to do with the sense of meaning that one derives from outside the self (i.e., from one's personal support system). In this context, we talk about finding meaning by "being meaningful," or creating a life where one "matters" to other people. Relevant to this discussion are the ways in which loss of a job or retirement, avoidance of social gatherings, and PTSD-related withdrawal and social isolation all serve to prevent one from making a significant positive impact in the lives of other people, and subsequently from receiving the positive regard and feedback that can allow a person to begin to feel better about themselves, and to see their lives as more meaningful and purpose filled. We encourage veterans to actively seek opportunities for service. Nonprofit service agencies as well as religious/spiritual communities are frequently looking for people with time on their hands, who can serve the community in significant ways. We point out that many spiritual traditions view service of others as a spiritual activity, where both the giver and receiver benefit greatly.

Utilize Simple Rituals. We have found it useful to incorporate simple rituals as a part of our group process, something which could potentially be part of individual-helping conversations as well. These can include readings of inclusive prayers or litanies from diverse spiritual traditions, or even poetry written by survivors. One challenge is that some religious perspectives can be inherently mutually exclusive. Selecting rituals based in specific religious traditions may be offensive for members of other traditions. To prevent this from being a problem, we carefully select resources with this in mind and often have crafted ritualized activities with no inherent religious connection. An example might be a closure exercise, such as holding hands and going around the circling with each member saying one positive word reflecting his or her best hope for that day. Rituals can be developed which reflect important clinical recovery themes such as forgiveness or grief/loss. Something as simple as holding open an empty chair in recognition of someone who did not return home can be extremely powerful. Traditional therapy utilizes mostly verbal processing. We have found that simple ritual seems to tap into a different type of processing, which is experiential in nature. These experiential exercises, though simple, can be experienced by members as both profound and emotional and seem to foster the experience of shared community.

We said from the outset that we approach spirituality as a resource in the recovery from trauma. We believe that the guidelines and suggestions addressed above specifically target a number of important PTSD symptoms. For example, the definition of spirituality as connection and the encouragement of group members to seek supportive healthy communities directly address the PTSD symptoms of isolation and social withdrawal. Self-forgiveness and an emphasis on compassion toward self address both guilt and shame, which though not formally a part of the diagnostic criteria of PTSD are certainly recognized as important clinical issues within certain PTSD populations. Forgiveness turned outward directly addresses anger and irritability which are PTSD symptoms, as well as chronic hostile attitudes that worsen social isolation and inhibit relationships with others. Inwardly directed spiritual practices such as mindfulness meditation can potentially have an effect on reducing hypervigilance and overall high levels of physiological arousal. Finally, rediscovery of meaning and purpose potentially enormously impacts at least two PTSD symptoms (i.e., foreshortened future and loss of interest in important activities). Taken together, we believe that finding ways to address spirituality in the context of trauma recovery can provide added benefit over treatment as usual.

In a number of ways, addressing spirituality supports healthy readjustment from war-zone trauma and recovery from PTSD. Defining spirituality as "connecting" confronts the tendency in PTSD toward isolation and withdrawal. Approaches addressing spirituality should emphasize the importance of building connections in creating for oneself a supportive healthy community.

REFERENCES

Chang, B.-H., Skinner, K. M., & Boehmer, U. (2001). Religion and mental health among women veterans with sexual assault experience. *International Journal of Psychiatry in Medicine, 31,* 77–95.

Decker, L. R. (1993). The role of trauma in spiritual development. *Journal of Humanistic Psychology, 33,* 33–46.

Drescher, K. D. (2006). Spirituality in the face of terrorist disasters. In L. A. Schein, H. I. Spitz, G. M. Burlingame, & P. R. Muskin, (Eds.). *Psychological Effects of Catastrophic Disasters: Group Approaches to Treatment.* (p. 940). New York: The Haworth Press.

Drescher, K. D., & Foy, D. W. (1995). Spirituality and trauma treatment: Suggestions for including spirituality as a coping resource. *National Center for PTSD Clinical Quarterly, 5*(1), 4–5.

Enright, R. D., & Coyle, C. T. (1998). Researching the process model of forgiveness within psychological interventions. In E. L. Worthington (Ed.), *Dimensions of Forgiveness.* Radnor, PA: Templeton Foundation Press.

Falsetti, S. A., Resick, P. A., & Davis, J. L. (2003). Changes in religious beliefs following trauma. *Journal of Traumatic Stress, 16,* 391–398.

Festinger, L. (1957). *A Theory of Cognitive Dissonance.* Stanford, CA: Stanford University Press.

Fontana, A., & Rosenheck, R. (2004). Trauma, change in strength of religious faith, and mental health service use among veterans treated for PTSD. *Journal of Nervous and Mental Disease, 192,* 579–584.

Frankl, V. E. (1984). *Man's Search for Meaning.* New York: Simon & Schuster. (Original work published 1946)

Gorsuch, R. L. (1995). Religious aspects of substance abuse and recovery. *Journal of Social Issues, 51,* 65–83.

Green, B. L., Lindy, J. D., & Grace, M. C. (1988). Long-term coping with combat stress. *Journal of Traumatic Stress, 1,* 399–412.

Janoff-Bulman, R. (1992). *Shattered Assumptions: Towards a New Psychology of Trauma.* New York: The Free Press.

Leibniz, G. W. (1890). *Philosophical Works* (G. M. Duncan, Trans.). New Haven, CT: Tuttle, Morehouse & Taylor.

Linley, P. A., & Joseph, S. A. (2004). Positive change following trauma and adversity: A review. *Journal of Traumatic Stress, 17,* 11–21.

McIntosh, D. N., Silver, R. C., & Wortman, C. B. (1993). Religion's role in adjustment to a negative life event: Coping with the loss of a child. *Journal of Personality and Social Psychology, 65,* 812–821.

Pargament, K., & Brandt, C. (1998). Religion and coping. In H. G. Koenig (Ed.), *Handbook of Religion and Mental Health* (pp. 112–128). San Diego, CA: Academic Press.

Pargament, K. I., Koenig, H. G., & Perez, L. M. (2000). The many methods of religious coping: Development and initial validation of the rcope. *Journal of Clinical Psychology, 56,* 519–543.

Pargament, K. I., Zinnbauer, B. J., Scott, A. B., Butter, E. M., Zerowin, J., & Stanik, P. (2003). Red flags and religious coping: Identifying some religious warning signs among people in crisis. *Journal of Clinical Psychology, 59,* 1335–1348.

Thoresen, C., Harris, A., & Luskin, F. (2000). Forgiveness and health: An unanswered question. In M. McCullough, K. Pargament, & C. Thoresen (Eds.), *Forgiveness: Theory, Research, and Practice* (p. 334). New York: Guilford Press.

Wilson, J. P., & Moran, T. A. (1998). Psychological trauma: Posttraumatic stress disorder and spirituality. *Journal of Psychology and Theology, 26,* 168–178.

Witvliet, C. V. O., Phillips, K. A., Feldman, M. E., & Beckham, J. C. (2004). Posttraumatic mental and physical health correlates of forgiveness and religious coping in military veterans. *Journal of Traumatic Stress, 17,* 269–273.

15

The Returning Warrior: Advice for Families and Friends

JUDITH A. LYONS

Your loved one has survived the war. You rejoice that they are coming home. You know they have experienced the horrors of war, but you are confident that the warmth of your love will sustain them and will sustain your relationship. You eagerly anticipate life and your relationship getting back to "normal" after their return.

Conceivably, it could work like that, with all the pieces falling effortlessly into place. In many—perhaps most—cases, readjustment requires more deliberate effort, flexibility, and sensitivity to the impacts of combat service.

This chapter is designed to provide guidance to families and friends of those returning from war, and to clinicians who are supporting them through the readjustment process. (Clinicians may also want to refer to Galovski & Lyons, 2004, a paper that addresses the same issues with a focus on the pathological end of the spectrum.) The chapter will address some of the challenges that emerge as veterans, friends, and families try to pick up where they left off. Some problems tend to be common across individuals so can be discussed as specific examples. Other difficulties are more idiosyncratic, but still tend to share some of the same underlying principles. By examining those principles, this chapter aims to take the reader beyond merely acknowledging that posthomecoming frictions are normal and predictable. The goal is to help identify specific situations that are likely to increase such frictions so that you can choose to avoid entering into such circumstances or can proceed with informed preparedness.

Armed with such insight, it is easier to avoid misinterpreting tensions as a sign of deteriorating commitment and caring, and recognize them instead as predictable responses to certain sets of cues that may have nothing to do with the quality of the relationship. The first half of this chapter presents tips derived from the author's two decades of work with veterans and their families in clinical and nonclinical settings. The second half of the chapter presents information on more formal clinical resources that may be needed if problems are severe or persistent.

THE WARRIOR

The term "warrior" is controversial, often deemed politically incorrect. However, it is deliberately used in this chapter to highlight that the experience of war does change a person. A persona and reaction style become engrained and remain forever a part of that individual—sometimes a minor component of the individual's character, deeply buried and rarely glimpsed; sometimes on the surface and "in your face," the defining essence of that individual for years to come.

To survive in a combat zone, a level of hypervigilance and suspicion is mandatory. The most innocuous-looking individual or item can prove to be the most deadly. Tendencies for sympathy and compassion are often used as lures to entrap the unwary—pick up the injured child and the booby trap goes off, try to get to the wounded comrade and find oneself in the crosshairs of the sniper. Losses are often inevitable. Choices may be limited to several unacceptable options, forcing the warrior into actions that run directly contrary to prior values and beliefs. The warrior must build a wall around tender emotions to be able to function in a calculated, all-about-business manner to stay alive and not jeopardize other comrades. A quiet moment to fully mourn a lost friend or the opportunity to stop to aid a wounded civilian are luxuries that are often not available when there are so many other demands at the same time. Working through exhaustion, filth, hunger, and thirst can become routine. To keep alive and perform combat duties successfully, the individual may have to remain in this combat mode 24/7, dozing only lightly, ready to pounce into attack mode at the slightest signal. After enough time and practice, this combat mode becomes second nature. The pattern does not fully shut off even after the person is home and safe.

RECOGNIZING AND ANTICIPATING COMMON PROBLEMS

The intensity of this combat-ready stance does diminish with time. However, certain circumstances can reactivate it to full intensity with little warning. Learning to anticipate such circumstances in advance and quickly recognize them when they occur without prior prediction can be a powerful asset for loved ones. The next three sections will present a variety of common situations in which the combat mode is reevoked and aspects of the warrior persona collide with the role of lover, family member, and friend.

Individual Relationships and Private Settings

Family and friends eagerly await the return of your loved one. You are excited to show off accomplishments, although perhaps with some uncertainty of how they will like certain changes or how the relationship will readjust. However, when your loved one arrives, his or her emotional focus seems scattered or erratic.

The returning warrior may feel relief and happiness to be home and joy at reuniting with loved ones. This may be mixed with insecurities about readjustment, guilt, or shame for combat decisions and actions, survivor guilt and concern for those still in the battle zone, and/or confusion regarding their own values and character. Things previously valued may now seem trivial, even irritating. Other things may take on increased significance, to a point that may seem irrational (never letting the gas gauge fall below half a tank, veering widely around seemingly harmless debris in the road, always having an escape route planned, insisting on always knowing the whereabouts of loved ones). Roles that shifted during separation may need to be renegotiated. Relationships that became idealized in memory during months of separation may seem tarnished in the light of reality. Under such pressures and with high expectations on both sides, the potential for unintended slights and open clashes is high (Knox & Price, 1995; Peebles-Kleiger & Kleiger, 1994).

One of the most significant areas of conflict pertains to talk about the war. There is almost invariably desynchrony in the warrior's readiness to discuss combat experiences and others' eagerness to inquire about such experiences. Particularly outside of military environments, well-meaning friends and family may jovially ask about kills and conquests. Those who are more accustomed to military culture or simply more sensitive in their approach may find that even modulated inquiries of "What was it like?" can spark angry replies. You soon learn not to ask, or to even avoid the topic at all costs. Weeks, months, or even years later when the warrior feels psychologically ready to share some of those war memories, the sad fact is that most people no longer want to listen. At that point, attempts the warrior makes to bring up the topic may be rebuffed with well-intended advice to "put that behind you" or less sympathetic commands to "get over it."

In addition to the issue of timing, there is another aspect of discussing war memories that damages many relationships. Do you harbor hurt or resentment that the warrior seems unwilling to share these stories with you? Many family members do not realize that the level of emotional confusion about events can take time to sort out before the stories are even in any organized fashion to tell. Warriors are likely to have some insecurity about whether they will still be viewed the same after they reveal what they have seen and done. (For an indepth discussion of the complex array of issues involved when a warrior confides atrocities, readers are referred to a classic 1974 paper by Haley.) Many warriors remain reluctant to share their stories with loved ones, not for lack of trust or intimacy, but because they do not want to infect loved ones with the same nightmarish images that haunt them. Such concerns reflect a real risk, as secondary traumatization is a phenomenon seen among therapists (Figley, 1995), partners (Dekel et al., 2005; Maloney, 1988), and children (Ancharoff, Munroe, & Fisher, 1998). Partners sometimes contribute to the tension by becoming jealous or resentful that the warrior shares these stories with a counselor instead, not realizing that it is precisely the *lack* of caring as much about the counselor or about the counselor's reaction that makes this possible.

The warrior may find it easier to share experiences with others who served in combat, even if they are strangers. The stories can be told in military shorthand without the necessity of explaining and elaborating. There is less fear of being misunderstood if the others also engaged in comparable experiences. This can facilitate conversations with family and friends who are also veterans. This would seem to be an advantage, but it does not always outweigh the challenges inherent in the warrior persona. Studies of father-son pairs who each experienced combat trauma have reported that sons of combat veterans adjusted less well to their own combat trauma than the sons whose fathers had not experienced war trauma (Rosenheck, 1985; Rosenheck & Fontana, 1998).

If you are not a veteran, one way you can prepare yourself for similar discussions—or, alternatively, reduce your own need to push for such a discussion—is to familiarize yourself with the combat events the warrior's unit might have experienced. Asking others who served in that unit would likely risk resentments and is generally not recommended. However, written information is available. After the Vietnam War, a book authored by the wife of a Vietnam veteran (Mason, 1990) served as a central reference for such information. Now, the Internet provides a wealth of information about contemporary war experiences. *Time* magazine (Bennett, 2005) reviewed various blogs by troops in Iraq and recommended several as offering a range of firsthand accounts of life in the war zone. A Web search for "Iraq blog" (or "Afghanistan blog," etc.) will pull additional/updated sites to peruse. Be prepared for crude language and explicit descriptions or pictures.

If you feel uncertain of how a discussion of war events might unfold, Johnson, Feldman, and Lubin (1995) offer very helpful case examples of ways to facilitate positive communication once the warrior is ready to disclose personal trauma experiences. Resist the desire to jump in and "fix" the hurt—focus on listening. Premature reassurances may be meant to comfort but can leave the warrior feeling that the degree of emotion or the complexity of the moral dilemma was dismissed.

Emotions about things unrelated to war have no immunity from the changes war imparts. After learning to shut off emotions to be able to carry out the cold priorities of combat, relearning to risk the emotional vulnerability that intimacy requires does not come easily. This may come across as uncaring and uninterested. Restless sleep may lead to separate bedrooms and further reduce intimacy. Try to be patient and not over-interpret such changes or take them too personally. Be forewarned, however, that such emotional numbing/withdrawal presents one of the biggest challenges to personal relationships (Frederikson, Chamberlain, & Long, 1996; Riggs et al., 1998; Wilson & Kurtz, 1997), including those with children (Ruscio et al., 2002).

It may come as a surprise to loved ones that the warrior's political, social, and religious views may have changed and may now contrast with those held by loved ones who remained stateside. Folks at home have not had access to the views and experiences of the war zone, whereas those overseas have been cut off from local developments and local opinions while away. Views on the war itself may be a particularly volatile subject. Loved ones may be baffled and irritated by the warrior's strident yet sometimes internally inconsistent views. Be aware that conflicted thoughts, like conflicted emotions, can take time to sort out. Some of the views the returning warrior espouses may be concepts that are being "tried on for size"—not necessarily the viewpoint the warrior will settle on after the sifting process is complete.

Problems in Public Settings

A huge welcome party is planned but the warrior acts reluctant to attend. Ditto the big weekend of the warrior's favorite sporting event. You ask what is wrong but get a vague reply. The warrior agrees to go, but then is irritable and disappears partway through the event. You ask what is happening.

Many group functions, such as parties, clubs, and most sports events, involve crowded venues with heightened levels of activity and noise. For someone still in combat mode, it can be exhausting to try to monitor everything going on in such a busy environment. Fireworks present a particular challenge, especially if the warrior is not in a position to observe them being lit (observation allows some ability to predict the flashes and booms). Quiet, stationary events (such as a classical concert, movie, or worship service) can also be problematic. A restless warrior's own fidgeting becomes conspicuous and a premature departure is disruptive to others. Thus, cut off from a timely escape route, such a setting can feel oppressive. In either case, the warrior can become frustrated by their own inability to relax and enjoy the activity, feeding the likelihood of irritability.

Restaurants may be somewhat easier to readjust to, but any public gathering spot still is recognized as a potential target. Thus, the warrior may be insistent on positioning him/herself to see everything that is happening. The warrior may not recognize that this seating arrangement is apt to leave a dining partner facing the wall on every outing.

Even when things are going smoothly, there can be potential for sudden volatility. A heightened emotional reaction to discussions of politics and questions or comments from others is common during the early days at home. With images of the combat zone still fresh, the newly returned warrior may have difficulty inhibiting a response when hearing others express "uninformed" views.

Fluctuations in Problems

It is normal for reminders of combat to stir a temporary increase in anxiety or moodiness. The anniversary of a battle and weather that is similar to the combat zone are common reminders. Current news events such as the rubble and evacuees from hurricanes and tsunamis may remind the warrior of the devastation of war. Personal life events can also serve as cues. A promotion can be disquieting for someone who felt responsible for an operation that failed under their command. Births and other children's milestones may also elicit ambivalence—pride and joy mixed with feeling undeserving of parenthood after being involved in deaths of other children or their parents.

One of the coping strategies many warriors adopt is to keep occupied so that their mind has no time to wander back to combat scenes and so that they are tired enough to sleep through the night without combat dreams. In many cases, this strategy serves very well until either a chronic illness or retirement leads to a more sedentary pace. In such cases, developing alternative ways to occupy time and thoughts is often all that is needed.

RESOURCES AND SERVICES

Counseling and mental health services are available to help shore up relationships that have been strained by deployment and war experiences. If the warrior remains actively involved in the military, a variety of family support services are usually available through their unit. The Department of Veterans Affairs also offers services at medical centers, community clinics, and Vet Centers. The following Web site provides a good starting point to learn about services in your area: http://www.va.gov/rcs/VetCenterDirectory.htm.

The extent of family services is expanding as the need becomes increasingly apparent. During the Gulf War in 1991, Reserve and Guard units offered deployment support groups on a scale not previously seen. Since that time, homecoming preparation and post-homecoming services have been expanded across branches of the military. Engaging families in the veteran's treatment was identified as one of seven Department of Veterans Affairs "priorities of quality" for clinical services (VHA Directive, 2001). The VA subsequently launched major initiatives to expand services for families.

Particular emphasis has been given to veterans who remain especially mired in the warrior role, i.e., those diagnosed with posttraumatic stress disorder (PTSD). PTSD is a psychiatric disorder in which a person frequently relives a horrific, life-threatening event via nightmares, repetitive thoughts and images, or even acting as if the event was happening all over again. One study (Riggs et al., 1998) found that only 30% of couples report relationship distress if the veteran does not have PTSD. However, if the veteran has PTSD then the odds reverse and only 30% of couples do *not* report distress. The severity of the PTSD symptoms is correlated with the severity of relationship distress. A similar pattern is reported by Dirkzwager et al. (2005), although the percentage of relationships classified as problematic is lower (16% in couples without PTSD, 26%–39% in couples with varying degrees of partial or full PTSD). The presence of PTSD is also associated with lower happiness and less life satisfaction (Jordan et al., 1992) and somatic and sleep-related complaints (Dirkzwager et al., 2005) in the veteran's spouse. Studies have also found children to be negatively impacted by a father's PTSD (Caselli & Motta, 1995; Davidson, Smith, & Kudler, 1989; Rosenheck, 1986; Rosenheck & Nathan, 1985), although other variables such as family violence (Harkness, 1991) or the father's participation in atrocities (Rosenheck & Fontana, 1998) are sometimes better predictors of problems. Some studies show that parents and/or siblings are less affected than partners by the warrior's PTSD (Dirkzwager et al., 2005; Lyons & Root, 2001). However, data from Dirkzwager and colleagues indicate a greater effect on mothers than fathers. Given that studies to date have focused almost solely on male veterans and female partners but have generally combined across genders when looking at parents and siblings, it is possible that the effects on other female family members have been underestimated. As more women are serving in combat zones, future studies can be expected to help elucidate gender-specific effects PTSD may have on various family relationships.

Effective therapies are available to treat PTSD. The *Journal of Clinical Psychiatry* published treatment recommendations compiled by a panel of experts (Foa & Davidson, 1999). These guidelines include a synopsis written specifically for clients and families. More technical information for clinicians is available in a textbook by Foa, Keane, and Friedman (2000). Many of the therapies used to treat PTSD are very effective, earning an "A" rating on the A–F scale employed in the text by Foa, Keane, and Friedman, indicating that the evidence supporting their use is from well-controlled, randomized clinical trials.

For more information about trauma, PTSD, and their effects, there are several good resources easily available. A 29-minute videotape, titled

PTSD: Families Matter (Abrams & Freeman, n.d.), depicts issues families encounter when a veteran has PTSD. It was developed by the VA's South Central Mental Illness Research, Education and Clinical Center and is available to licensed clinicians by contacting Michael.Kauth@med.va.gov. A video entitled *Living with PTSD: Lessons for Partners, Friends and Supporters* is accessible at the Website http://www.giftfromwithin.org. The National Center for PTSD (Carlson & Ruzek, 2005) offers an online fact sheet for families at http://www.ncptsd.va.gov/facts/specific/fs_family.html. A free 102-page booklet, titled *Veterans and Families' Guide to Recovering from PTSD* (Lanham, 2005), provides helpful information, including essays by veterans and family members and a resource directory.

There are often hurdles to surmount in accessing care (Glynn et al., 1999; Lyons, 2003). Specialized care may be a distance away. Scheduling and transportation can be problematic. There may be bureaucratic red tape to establish eligibility. There may be out-of-pocket costs. For the warrior, the mere idea of talking about the traumas can present the largest barrier. If PTSD symptoms are present, it is important to do whatever you can to help overcome these barriers. PTSD-related symptoms tend to expand if left untreated, as the traumatized person will go to greater and greater lengths to avoid reexperiencing the memories. This can lead to sleep avoidance, substance use, anger, and violence (to deflect deeper feelings of hurt and loss). Left untreated, the traumatized person can skip from job to job and relationship to relationship, trying to outrun negative thoughts and feelings.

Family and friends play a major role in recovery from PTSD. Negative family relationships account for nearly 20% of the variance in PTSD treatment outcome (Tarrier, Sommerfield, & Pilgrim, 1999). However, two particular aspects of PTSD are very easy for loved ones to misinterpret, and thus risk undermining treatment. First, it is common sense to think treatment is supposed to make someone better and that if symptoms get worse after beginning treatment, treatment should be discontinued. However, the treatment for PTSD often involves focus on the traumatic event so that the conflicted thoughts and feelings associated with it can be resolved. Thus, during the initial phases of treatment, increased combat-related thoughts, dreams, and anxiety are expected and can actually signify progress. The second common misinterpretation was mentioned previously but bears repeating in this context. When you are already frustrated that your loved one is more emotionally withdrawn and will not talk with you about what is bothering them, it can be upsetting to learn (or suspect) that they are baring their soul to a perfect stranger (the therapist). Keep in mind that reluctance to divulge horrible details may be due to the importance of your relationship—not an indication to the contrary. Being aware of these counterintuitive aspects of PTSD therapy may help you resolve any ambivalence you may have about the warrior seeking or continuing treatment.

Successful treatment is in the best interest of all concerned, as untreated or incompletely treated PTSD can remain chronic over many years. In such cases, the emotional drain on families can be extreme (Beckham, Lytle, & Feldman, 1996; Sautter et al., 2006). Recognizing the impact on loved ones, there are treatments available that include them as well. In a review of marital and family therapies being offered in the wake of trauma, Riggs (2000) identified two major approaches to PTSD-related family/marital treatment: systemic and support. Systemic approaches treat the relationship to reduce friction and/or strengthen bonds. Traditional marital and family therapies would be in this category. Support treatments are those that have the goal of increasing social support for the warrior and the warrior's treatment. Support treatments often include teaching about PTSD symptoms and helping family members develop ways to cope with the warrior's PTSD symptoms. In the A–F rating system used to evaluate the strength of the evidence in support of various therapies, Riggs rated marital and family therapy for PTSD an "E." This rating reflects Riggs' judgment that the interventions are derived from long-standing clinical practice by a certain groups of clinicians but are not in widespread use and have not been empirically tested for use with PTSD. Such interventions are recommended by Riggs as supplemental approaches to the primary emphasis on treatment of the warrior's PTSD.

After Riggs collected data for his review, Glynn and colleagues (1999) published a study in which behavioral family therapy was provided in addition to exposure therapy for the warrior. Riggs would categorize this study as support treatment, given the emphasis on coping skills training with the primary outcome of interest being reduction in the warrior's PTSD symptoms. The family therapy package included three sessions of orientation and evaluation, two educational sessions about PTSD and mental health treatments, then 11 to 13 sessions focused on skills training (communication and anger control, with a major emphasis on problem solving). The 11 veterans who received both exposure and family therapy achieved approximately double the reduction in PTSD symptoms than the 12 veterans who received exposure therapy alone. However, this difference was not statistically significant. Glynn et al. (1995) provide a more detailed description of interventions evaluated in the 1999 empirical report, including case examples. The 1995 publication also provides recommendations for dealing with avoidance behaviors, physical aggression, substance use, alexithymia, and disclosure of combat experiences during behavioral family therapy.

Monson et al. (2004) recently published data from a pilot study of seven couples. In each case, the husband was a Vietnam veteran with diagnosed PTSD. Three of the seven men had previous divorces. The primary outcome of interest was the man's PTSD severity. However, the cognitive-behavioral couple's therapy that was administered (further detailed in Monson, Guthrie, & Stevens, 2003) also included a strong systemic emphasis. In addition to two psychoeducational sessions about

PTSD and associated relationship problems, this 15-session manualized treatment also included communication skills training and several cognitive intervention sessions targeting the maladaptive patterns associated with the warrior mindset described previously. Results are mixed, with some veterans reporting deterioration in their relationship and increased PTSD, whereas others reported improvement. Ratings by the clinicians and wives were more positive, as were veterans' reports of reduced anxiety and depression. The small sample size and lack of a comparison group or multiple baseline limit interpretation of these findings. However, this pilot project effectively sets the groundwork for larger controlled studies.

Neither systemic or support approaches focus solely on the needs of the family member. Recent studies indicate that loved ones, particularly spouses and other partners, are seeking precisely that piece that has been missing—help for themselves. A survey was conducted at workshops held for families of veterans who were in treatment for PTSD (Lyons & Root, 2001). Nonspouses (children, parents, siblings, friends) reported that their role in helping the veteran deal with the PTSD symptoms was limited and they indicated interest in the types of supportive and systemic services that had traditionally been offered. However, the 30 spouses who completed the survey reported a much different pattern. They described a very active role in helping the veteran manage PTSD symptoms, rating their role as large or "very large...more than the treatment team." They spoke of helping the veteran get to appointments and remember medications, orchestrating the family's lifestyle around the veteran's symptoms to minimize relapses, and taking on roles that the veteran was no longer able to fulfill. Many spouses discussed the difficulty of working outside the home as the primary breadwinner plus inside the home as the primary caretaker for children, aged parents, and/or the veteran. Most had read about PTSD and talked to many providers or organizations about the disorder. While approximately one-third of the spouses expressed some interest in systemic therapy to improve their relationship or reduce shared stress, they reported no desire for more informational sessions about the veteran's illness. What they did request, often with a tone of desperation, were therapies that emphasized the spouse's own needs. They voiced numerous requests for treatments to reduce the spouse's own stress level (not limited to stress that came from the veteran's PTSD symptoms). Many spouses wanted social activities to offset the isolation they felt.

A broader phone survey was conducted to test these findings (Sherman et al., 2005). Eighty-nine women partners of Vietnam men veterans disabled by PTSD were asked to describe services that could help them "better support their loved ones." Even with such other-focused wording of the question, the three interventions most commonly requested involved only the partner, not the veteran. Fifty-four percent requested a partners-only group. Twenty percent asked for more information about trauma and PTSD. Nineteen percent requested individual therapy for themselves. Only 13% sought couple's therapy.

The search for services for themselves reflects the level of personal need these partners feel. The level of caregiver burden they report is extremely high, and is correlated with feeling incapable of controlling the veteran's emotional difficulties or their own coping with his symptoms (Sautter et al., 2006). Their ratings of burden also correlated with their reports that numerous barriers reduced ready access to clinical care (distance, cost, bureaucratic hurdles, etc.). Such barriers may vary from one community to another (Lyons & Root, 2001). In an effort to reduce barriers, home-based interventions are being developed that may be supplemented with phone contacts (Lyons, 2003). Books such as those by Mason (1990, 1998) and Matsakis (1996, 1998) constitute solid self-help options.

THE POSITIVES

We all progress through various stages in our lives, with a series of developmental tasks to accomplish as we mature (learning to develop relationships, becoming self-supporting, etc.). Under normative circumstances, we do not encounter deaths of numerous others and face the possibility of our own death being imminent until late in life. In combat, death is a frequent threat. Facing death so much earlier in life than most people can lead to an acceptance of mortality or numbing of reaction that can appear cold and uncaring. Many families, however, are surprised to see how the warrior shines in times of family or regional tragedies. Combat mode kicks in and the warrior may gear up quickly in time of crisis, efficiently drawing on the skills and coping developed during war to remain calm and task-focused when others are more flustered.

Many who survive the horrors and rigors of war emerge with a new clarity of their own abilities and limitations, a strong sense of values and beliefs, and an ethical maturity that many others do not develop until old age—if then. An entire research field is blossoming focused on the resilience and posttraumatic growth displayed by survivors of war and other traumas (Tedeschi & Calhoun, 2004). Like other survivors, many warriors are successful in imparting some of these survival skills and values to loved ones, helping them develop these characteristics without having to endure the trauma of combat in the process.

The emphasis in this chapter has been on how war changes the warrior. Concurrently, however, partners, children, and others can also change. You may have stretched from old roles and taken on new challenges. You may now long for additional changes or you may be eager to resume familiar patterns, all the more appreciative of tried-and-true traditions within your relationship. Regardless of the direction of change, keep in mind that the relationship will also have to adjust to accommodate to the new you, just as the relationship has to adjust to accommodate the warrior. It is true that the old status quo will be difficult (if not impossible) to reinstate, but doors will be open to many new possibilities.

CONCLUSION

The pressures of war leave a lasting imprint on the warrior. Much of it can be positive, but it can also create hurdles for loved ones. The warrior and loved ones will always be the key players in shaping the future potential for their own relationships. It is hoped that the patterns identified in the first half of this chapter will help your relationship weather the storm. The resources described in the second half of the chapter provide a starting point if more indepth assistance is needed. Although there is no clinical magic that can undo the impact of separation and the experiences of war, clinical resources are available and can help. If you are concerned about the direction your relationship is headed, contact a counselor (http://www.va.gov/rcs/VetCenterDirectory.htm can help you locate local resources). It is much easier to restore a relationship if problems are addressed before they multiply.

REFERENCES

Abrams, P. & Freeman, T. (n.d.). *PTSD: Families Matter* [Video]. North Little Rock, AR: U.S. Department of Veterans Affairs; VA South Central MIRECC. Available from Michael.Kauth@med.va.gov

Ancharoff, M. R., Munroe, J. F., & Fisher, L. M. (1998). The legacy of combat trauma: Clinical implications of intergenerational transmission. In Y. Danieli (Ed.), *International Handbook of Multigenerational Legacies of Trauma* (pp. 257–276). New York: Plenum Press.

Beckham, J. C., Lytle, B. L., & Feldman, M. E. (1996). Caregiver burden in partners of Vietnam War veterans with posttraumatic stress. *Journal of Consulting and Clinical Psychology, 64,* 1068–1072.

Bennett, B. (2005, September 26). Five riveting soldier blogs. *Time, 166,* 82. Retrieved October 3, 2005 from http://www.time.com/time/archive/preview/0,10987,1106331,00.html

Carlson, E. B., & Ruzek, J. (2005). *PTSD and the family: A National Center for PTSD fact sheet.* Retrieved October 3, 2005 from http://www.ncptsd.va.gov/facts/specific/fs_family.html

Caselli, L. T., & Motta, R. W. (1995). The effect of PTSD and combat level on Vietnam veterans' perceptions of child behavior and marital adjustment. *Journal of Clinical Psychology, 51,* 4–12.

Davidson, J., Smith, R., & Kudler, H. (1989). Familial psychiatric illness in chronic posttraumatic stress disorder. *Comprehensive Psychiatry, 30,* 339–345.

Dekel, R., Goldblatt, H., Keidar, M., Solomon, Z., & Polliack, M. (2005). Being a wife of a veteran with posttraumatic stress disorder. *Family Relations, 54,* 24–36.

Dirkzwager, A. J. E., Bramsen, I., Ader, H., & van der Ploeg, H.M. (2005). Secondary traumatization in partners and parents of Dutch peacekeeping soldiers. *Journal of Family Psychology, 19,* 217–226.

Figley, C. R. (1995). *Compassion Fatigue: Coping with Secondary Traumatic Stress Disorder in Those Who Treat the Traumatized.* New York: Brunner/Mazel.

Foa, E. B., Davidson, J. R. T., & Frances, A. (1999). Treatment of posttraumatic stress disorder (Expert consensus guideline series). *Journal of Clinical Psychiatry, 60* (Suppl. 16) 1–76. Retrieved July 3, 2006 from http://www.psychguides.com/ecgs10.php [full guidelines]. Retrieved July 3, 2006 from http://www.psychguides.com/pfg12.php [family guide only]

Foa, E. B., Keane, T. M., & Friedman, M. J. (2000). *Effective Treatments for PTSD: Practice Guidelines from the International Society of Traumatic Stress Studies*. New York: Guilford Press.

Frederikson, L. G., Chamberlain, K., & Long, N. (1996). Unacknowledged casualties of the Vietnam War: Experiences of the partners of New Zealand veterans. *Qualitative Health Research, 6*, 49–70.

Galovski, T., & Lyons, J.A. (2004). Psychological sequelae of combat violence: A review of the impact of PTSD on the veteran's family and possible interventions. *Aggression and Violent Behavior, 9*, 477–501.

Glynn, S. M., Eth, S., Randolph, E. T., Foy, D. W., Leong, G. B., Paz, G. G., et al. (1995). Behavioral family therapy for Vietnam veterans with posttraumatic stress disorder. *Journal of Psychotherapy Practice, 4*, 214–223.

Glynn, S. M., Eth, S., Randolph, E., Foy, D. W., Urbaitis, M., Boxer, L., et al. (1999). A test of behavioral family therapy to augment exposure for combat-related posttraumatic stress disorder. *Journal of Consulting and Clinical Psychology, 67*, 243–251.

Haley, S. A. (1974). When the patient reports atrocities. *Archives of General Psychiatry, 30*, 191–196.

Harkness, L. L. (1991). The effect of combat-related PTSD on children. *National Center for PTSD Clinical Newsletter, 2*(1), 12–13.

Johnson, D. R., Feldman, S., & Lubin, H. (1995). Critical interaction therapy: Couples therapy in combat-related posttraumatic stress disorder. *Family Process, 34*, 401–412.

Jordan, B. K., Marmar, C. R., Fairbank, J. A., Schlenger, W. E., Kulka, R. A., Hough, R. L., et al. (1992). Problems in families of male Vietnam veterans with posttraumatic stress disorder. *Journal of Consulting and Clinical Psychology, 60*, 916–926.

Knox, J., & Price, D. H. (1995). The changing American military family: Opportunities for social work. *Social Service Review, 69*, 479–497.

Lanham, S. L. (2005). *Veterans and Families' Guide to Recovering from PTSD* (3rd ed.) Annandale, VA: Purple Heart Service Foundation. [While supplies last, a free copy should be available through your local Vet Center, http://www.va.gov/rcs/VetCenterDirectory.htm]

Lyons, J. A. (2003). Veterans Health Administration: Reducing barriers to access. In B. H. Stamm (Ed.), *Rural Behavioral Health Care: An Interdisciplinary Guide* (pp. 217–229). Washington, DC: American Psychological Association.

Lyons, J. A., & Root, L. P. (2001). Family members of the PTSD veteran: Treatment needs and barriers. *National Center for PTSD Clinical Quarterly, 10*(3), 48–52.

Maloney, L. J. (1988). Posttraumatic stresses on women partners of Vietnam veterans. *Smith College Studies in Social Work, 58*, 122–143.

Mason, P. H. C. (1990). *Recovering from the War: A Woman's Guide to Helping Your Vietnam Vet, Your Family, and Yourself*. New York: Viking Penguin.

Mason, P. H. C. (1998). *Recovering from the War: A Guide for all Veterans, Family Members, Friends and Therapists* (2nd ed). High Springs, FL: Patience Press.

Matsakis, A. (1996). *Vietnam Wives* (2nd ed). Lutherville, MD: Sidran Press.

Matsakis, A. (1998). *Trust after Trauma: A Guide to Relationships for Survivors and Those Who Love Them*. Oakland, CA: New Harbinger Publications.

Monson, C. M., Guthrie, K. A., & Stevens, S. P. (2003). Cognitive-behavioral couple's treatment for posttraumatic stress disorder. *Behavior Therapist, 26*, 393–402.

Monson, C. M., Schnurr, P. P., Stevens, S. P., & Guthrie, K. A. (2004). Cognitive-behavioral couple's treatment for posttraumatic stress disorder: Initial findings. *Journal of Traumatic Stress, 17*, 341–344.

Peebles-Kleiger, M. J., & Kleiger, J. H. (1994). Re-integration stress for Desert Storm families: Wartime deployments and family trauma. *Journal of Traumatic Stress, 7*, 173–194.

Riggs, D. (2000). Marital and family therapy. In E. B. Foa, T. M. Keane, & M. J. Friedman (Eds.), *Effective Treatments for PTSD: Practice Guidelines from the International Society for Traumatic Stress Studies* (pp. 280–301). New York: Guilford Press.

Riggs, D. S., Byrne, C. A., Weathers, F. W., & Litz, B. T. (1998). The quality of the intimate relationships of male Vietnam veterans: Problems associated with posttraumatic stress disorder. *Journal of Traumatic Stress, 11,* 87–101.

Rosenheck, R. (1985). Father-son relationships in malignant post-Vietnam stress syndrome. *American Journal of Social Psychiatry, 5,* 19–23.

Rosenheck, R. (1986). Impact of posttraumatic stress disorder of World War II on the next generation. *Journal of Nervous and Mental Disease, 174,* 319–327.

Rosenheck, R., & Fontana, A. (1998). Transgenerational effects of abusive violence on the children of Vietnam combat veterans. *Journal of Traumatic Stress, 11,* 731–741.

Rosenheck, R., & Nathan, P. (1985). Secondary traumatization in children of Vietnam veterans. *Hospital and Community Psychiatry, 36,* 538–539.

Ruscio, A. M., Weathers, F. W., King, L. A., & King, D. W. (2002). Male war-zone veterans' perceived relationships with their children: The importance of emotional numbing. *Journal of Traumatic Stress, 15,* 351–357.

Sautter, F., Lyons, J., Manguno-Mire, G., Perry, D., Han, X., Sherman, M., et al. (2006). Predictors of partner engagement in PTSD treatment. *Journal of Psychopathology and Behavioral Assessment, 28*(2), 123–130.

Sherman, M. D., Sautter, F., Lyons, J., Manguno-Mire, G., Han, X., Perry, D., et al. (2005). Mental health treatment needs of cohabiting partners of veterans with combat-related PTSD. *Psychiatric Services, 56,* 1150–1152.

Tarrier, N., Sommerfield, C., & Pilgrim, H. (1999). Relatives' expressed emotion (EE) and PTSD treatment outcome. *Psychological Medicine, 29,* 801–811.

Tedeschi, R. G., & Calhoun, L. (2004). Posttraumatic growth: A new perspective on psychotraumatology. *Psychiatric Times, 21*(4). Retrieved October 7, 2005 from http://www.psychiatrictimes.com/p040458.html

VHA Directive (2001, February 7). *Veterans health care service standards. VHA Directive 2001–006.* Washington, DC: Veterans Health Administration.

Wilson, J. P., & Kurtz, R. R. (1997). Assessing posttraumatic stress disorder in couples and families. In J. P. Wilson & T. M. Keane (Eds.), *Assessing Psychological Trauma and PTSD* (pp. 349–372). New York: Guilford Press.

Index

D

R

T

W